AF371350

Thermoplasmonics
Heating Metal Nanoparticles Using Light

Plasmonics is an important branch of optics concerned with the interaction of metals with light. Under appropriate illumination, metal nanoparticles can exhibit enhanced light absorption, becoming nanosources of heat that can be precisely controlled. This book provides an overview of the exciting new field of thermoplasmonics and a detailed discussion of its theoretical underpinnings in nanophotonics. This topic has developed rapidly in the last decade, and is now a highly active area of research due to countless applications in nanoengineering and nanomedicine. These important applications include photothermal cancer therapy, drug and gene delivery, nanochemistry and photothermal imaging. This timely and self-contained text is suited to all researchers and graduate students working in plasmonics, nano-optics and thermal-induced processes at the nano scale.

Guillaume Baffou is a CNRS Research Scientist at the Institut Fresnel in Marseille. His research is focused on the interface between nano-optics and small-scale thermal effects. Specifically, he has been investigating the interaction between light and plasmonic metal nanoparticles, and the resulting applications in physics, chemistry, and biology. In 2015 he was awarded the bronze medal of the CNRS in recognition of his important contributions to the field.

Thermoplasmonics
Heating Metal Nanoparticles Using Light

GUILLAUME BAFFOU

Institut Fresnel, CNRS, University of Aix-Marseille

CAMBRIDGE
UNIVERSITY PRESS

University Printing House, Cambridge CB2 8BS, United Kingdom

One Liberty Plaza, 20th Floor, New York, NY 10006, USA

477 Williamstown Road, Port Melbourne, VIC 3207, Australia

4843/24, 2nd Floor, Ansari Road, Daryaganj, Delhi – 110002, India

79 Anson Road, #06–04/06, Singapore 079906

Cambridge University Press is part of the University of Cambridge.

It furthers the University's mission by disseminating knowledge in the pursuit of education, learning, and research at the highest international levels of excellence.

www.cambridge.org
Information on this title: www.cambridge.org/9781108418324
DOI: 10.1017/9781108289801

First published 2018

Printed in the United Kingdom by TJ International Ltd. Padstow Cornwall

A catalogue record for this publication is available from the British Library

Library of Congress Cataloging-in-Publication Data
Names: Baffou, Guillaume, 1980– author.
Title: Thermoplasmonics : heating metal nanoparticles using light / Guillaume
Baffou (Institut Fresnel, CNRS, University of Aix-Marseille).
Description: Cambridge, United Kingdom ; New York, NY : Cambridge
University Press, 2017. | Includes bibliographical references and index.
Identifiers: LCCN 2017026228| ISBN 9781108418324 (hardback ; alk. paper) |
ISBN 1108418325 (hardback ; alk. paper)
Subjects: LCSH: Nanoparticles. | Metal clusters. | Heat–Transmission. |
Plasmons (Physics)
Classification: LCC TA1530 .B34 2017 | DDC 620/.5–dc23
LC record available at https://lccn.loc.gov/2017026228

ISBN 978-1-108-41832-4 Hardback

à Camille, Marius et Clovis

Contents

Foreword

In the spring of 1999 I was working at my desk at C. P. M. O. H. (Centre de physique moléculaire optique et hertzienne) in Bordeaux when I received an unexpected phone call from Claude Boccara from E. S. P. C. I. (École supérieure de physique et de chimie industrielles de la ville de Paris) in Paris. Claude had been particularly interested in photoacoustic and photothermal spectroscopies, as he had used them to optimize the mirrors of the VIRGO gravitational wave detector.[1] He asked me about work by a Japanese colleague – it turned out to be Tsuguo Sawada from Tokyo University – who used a thermal-lens microscope to detect very weak concentrations of molecules, potentially even down to single molecules.[2,3] I did not know the work, but as we were thinking of detecting single molecules at room temperature through their absorption instead of their fluorescence, I was immediately thrilled to learn about this new possibility. Some weeks later, I attended a talk by David A. Schultz from San Diego about applications of gold nanoparticles in bioimaging.[4] Putting these two pieces of information together, I proposed that my young colleagues Philippe Tamarat and Abdelhamid Maali should start with the photothermal detection of gold nanoparticles instead of trying the absorption of single molecules at room temperature. Indeed, these gold nanoparticles neither bleach nor blink, and they exist in different sizes. This makes it much easier to optimize a technique on large nanoparticles before attempting to detect smaller ones. Contrary to the work of Tsuguo Sawada and Takehiko Kitamori, who investigated fluid suspensions in which diffusing molecules could enter and leave the detection volume during the detection period, we decided to work on immobilized gold nanoparticles, with which we had previous experience. A further advantage of

[1] Optical and Thermal Characterization of Coatings
J. P. Roger, P. Gleyzes, H. Elrhaleb, D. Fournier, A.C. Boccara,
Thin Solid Films 261 (1995) 132–138
DOI: 10.1016/S0040-6090(95)06533-4

[2] Single- and Countable-Molecule Detection of Non-Fluorescent Molecules in Liquid Phase
M. Tokeshi, M. Uchida, K. Uchiyama, T. Sawada, T. Kitamori,
J. Lumin. 83–84 (1999) 261–264.
DOI: 10.1016/S0022-2313(99)00109-X

[3] Determination of Subyoctomole Amounts of Nonfluorescent Molecules Using a Thermal Lens Microscope: Subsingle Molecule Determination
M. Tokeshi, M. Uchida, A. Hibara, T. Sawada, T. Kitamori,
Anal. Chem. 73 (2001) 2112–2116
DOI: 10.1021/ac001479g

[4] Single-Target Molecule Detection with Nonbleaching Multicolor Optical Immunolabels
S. Schultz, D. R. Smith, J. J. Mock, D. A. Schultz
Proc. Natl. Acad. Sci. 97 (2000) 996–1001
DOI: 10.1073/pnas.97.3.996

photothermal detection was the high-frequency modulation of the heating beam, leading to very efficient rejection of background scattering by the sample and of low-frequency noise, both appreciable features in the complex and heterogeneous environment of biological cells. This project eventually led to our group's first work on photothermal detection, published in 2002.[5] Later work by Brahim Lounis' group[6] and by others has established the thermal-lens microscope as a unique tool for the detection of single absorbing objects down to individual molecules.[7] Thermal lens experiments can be seen as the first application of the coupling of optics to thermal gradients in a material; indeed, of the new field of thermoplasmonics. However, the effects of light on absorbing structures include many effects beyond thermal ones, notably acoustic wave generation, chemical reactions, mechanical transport phenomena through thermophoresis and photophoresis, and many more, which are described and explored in the present volume.

Thanks to improved fabrication techniques such as ion milling and nanolithography from the electronics industry, plasmonics and nanophotonics emerged as unique methods to control light at the nanoscale in the late 1990s. They have steadily developed since then. The collective motion of conduction electrons in metals gives rise to surface plasmon polariton (SPP) excitations, which enable the concentration, manipulation and control of light at subdiffraction scales, sometimes down to less than a nanometer in size. Several new techniques using SPPs were designed in the past 20 years, starting with near-field optics (SNOM), surface-enhanced or, more recently, tip-enhanced Raman scattering (SERS or TERS), and related spectroscopies in the near-infrared or far-infrared, collectively designed as nanophotonics. Nanophotonics find applications in many scientific and applied fields, from lighting and solar energy harvesting to biomedical tracking and sensing. However, unavoidable ohmic losses lead to significant dissipation of optical power in metal structures, and therefore to significant heating of plasmonic assemblies and of their environment. The associated temperature changes and gradients can produce a wide range of physical and chemical phenomena with potential impact in a growing range of research areas, from solid-state physics to chemical physics, catalysis, life sciences and medicine.

The heat released upon optical excitation of plasmonic structures may have disrupting effects, particularly in biological systems. Whereas those effects are sometimes desirable, as in photothermal therapy where apoptosis of sick cells is selectively triggered by heating, temperature rises are mostly damaging to proteins and most other biomolecules. Therefore, unless temperature is carefully measured and controlled, heat release may be a serious limitation to biomedical applications. Metal structures themselves are sensitive to temperature elevations. The temperature-assisted diffusion of surface metal atoms leads to reshaping of

[5] Photothermal Imaging of Nanometer-Sized Metal Particles Among Scatterers
D. Boyer, P. Tamarat, A. Maali, B. Lounis, M. Orrit
Science 297 (2002) 1160–1163
DOI: 10.1126/science.1073765

[6] Photothermal Heterodyne Imaging of Individual Nonfluorescent Nanoclusters and Nanocrystals
S. Berciaud, L. Cognet, G. A. Blab, B. Lounis
Phys. Rev. Lett. 93 (2004) 257402
DOI: 10.1103/PhysRevLett.93.257402

[7] Room-Temperature Detection of a Single Molecule's Absorption by Photothermal Contrast
A. Gaiduk, M. Yorulmaz, P. V. Ruijgrok, M. Orrit
Science 330 (2010) 353–356
DOI: 10.1126/science.1195475

the structures, *e.g.,* to blunting of sharp edges, tips or gaps, and to shape changes leading to irreversible alterations of the desired plasmonic properties. Here again, accurate temperature measurements are required to limit the structural changes and thereby to mitigate long-term loss of desired plasmonic functions.

Thermal effects can also be harnessed for useful goals, as started to be realized in the past ten years. The heat released can give rise to nearly background-free signals to detect the presence of absorbing objects, or to provide the quantum efficiency of heat dissipation (and therefore the luminescence quantum yield) of small nanoparticles or single molecules. Temperature gradients lead to thermophoretic effects, which can be applied to the manipulation and guiding of nanoparticles. An exciting application of those effects is photon nudging,[8] where an asymmetric particle – called a *swimmer* – is directed towards a desired target through absorption of a heating beam at well-chosen times. Other applications involve the assembly of particles, or the laser-printing of nanostructures,[9] through optically controlled heating and chemistry. Besides these potential applications, the field of nanoscale heat transfer, heat generation and thermal fields poses a number of fundamental questions about heat and molecule dynamics in thermal non-equilibrium, the existence of effective temperatures, the applicability of Fourier's law at small enough scales, the characteristics of nanoscale phase transitions, and more, which themselves contain the germs of new perspectives and applications. Guillaume Baffou, who has been a prominent actor in the recent development of thermoplasmonics, gathers his experience and insights in this monography. He follows heat from its production in the metal structure to its transport towards the environment, describes qualitatively and quantitatively the various effects it may generate, and details the applications which have already emerged or can be expected in the near future. The present book establishes a conceptual basis encompassing optical, thermal and material properties and their interdependence, providing the reader with the necessary keys to enter the field of thermoplasmonics. The conceptual basis and the many illustrations of the book will prove equally useful to experimentalists, simulators and theorists working with nano-optics, colloidal systems, thermophoresis and photophoresis, labels and sensing for biomedical applications. By providing an accessible framework for this broad range of systems, phenomena and observations, the book will favor the design and interpretation of new experiments and eventually spawn new applications combining focused laser beams with metal nanostructures. Compiling this rich volume and the variety of complex physics and chemistry it describes made me keenly aware of how much progress has been achieved since the early experiments of the late 1990s.

Michel Orrit

MoNOS, Leiden Institute of Physics, Huygens-Kamerlingh Onnes Laboratory

[8] Harnessing Thermal Fluctuations for Purposeful Activities: The Manipulation of Single Micro-Swimmers by Adaptive Photon Nudging
B. Qian, D. Montiel, A. Bregulla, F. Cichos, H. Yang,
Chem. Science 4 (2013) 1420–1429
DOI: 10.1039/c2sc21263c

[9] Nanolithography by Plasmonic Heating and Optical Manipulation of Gold Nanoparticles
M. Fedoruk, M. Meixner, S. Carretero-Palacios, T. Lohmüller, J. Feldmann,
ACS Nano 7 (2013) 7648–7653.
DOI: 10.1021/nn402124p

Heating metal nanoparticles with light. This is what the field of *thermoplasmonics* deals with, and this is what this book is all about. The chart below displays statistics collected from `webofknowledge.com` in March 2017.[1] It is intended to give an idea of the evolution of the field of thermoplasmonics.

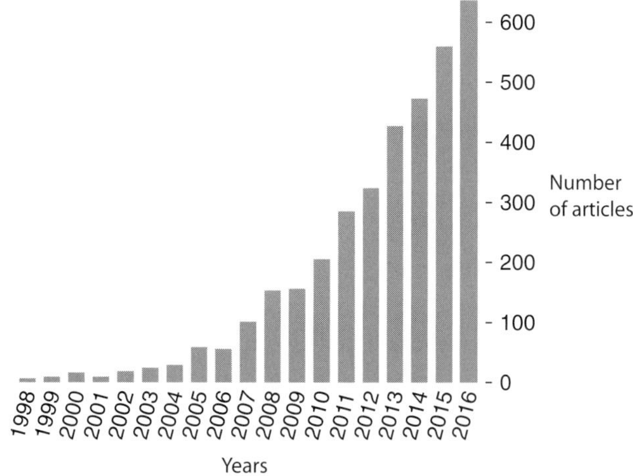

These data show that the field is recent (established around 2002–2003), important and still rapidly growing.

At first glance, it may be surprising that heating metal nanoparticles with light could occupy so many researchers for more than a decade. I can see two main reasons for this. First, gold nanoparticles under illumination are nothing but nanosources of heat, and this is one of the most basic concept in science one can think of. Second, many or all fields of science exhibit some thermal-induced effects (just think about chemistry, fluid dynamics, magnetism, cell biology, phase transitions, polymer science, etc.). Consequently, many applications from a wide variety of backgrounds can be tackled on the nano and micro scales as soon as one uses nanosources of heat, with only one limitation: the imagination. In agreement with this idea, since the early 2000s, heating gold nanoparticles using light has found applications in physics, chemistry, cell biology and biomedicine. Before 2002, the plasmonics community was already active, but it was considering photothermal effects of nanoparticles as side effects that had to be minimized. In 2002–2003, thanks to a couple of

---

[1] The Webofknowledge search criterion was "TOPIC: (plasmon* and nanoparticle* and (thermal or photothermal or heat*))."

pioneering works in photothermal cancer therapy and photothermal microscopy, the plasmonics community realized that the heat generation arising from plasmonic nanoparticles under illumination could also be beneficial, with some imagination.

The most exciting aspect of thermoplasmonics to me is that it enables one to carry out research in many fields of science. Upon reading these pages or just by looking at the table of Contents, you will see that a general understanding of the world of thermoplasmonics involves knowledge in (take a deep breath!) nano-optics, plasmonics, thermodynamics, molecular fluorescence, Raman spectroscopy, X-ray spectroscopy, NV centers and nanodiamonds, cell biology, organic and inorganic chemistry, fluid dynamics, phase transition, magnetism and magnetic recording, biomedicine, cancer therapy, graphene, lanthanides, acoustics, SERS, thermophoresis, fluid superheating, bubble dynamics, black body radiation, ... and I am certainly missing some fields.

This diversity of concepts, along with the fact that thermoplasmonics couples optics and thermodynamics, makes a general understanding of thermoplasmonics difficult. This book was born from this observation. At the time of writing these lines, there are no textbooks devoted to thermoplasmonics. A couple of review articles exist, but they are under 20 pages long, and they only represent introductions to this field of research. Once read, much work remains to be done before one can become experienced in the field of thermoplasmonics, and it is far from easy to acquire all the necessary information. Moreover, even when gathering all that information, one still has to learn a lot of new concepts due to the multidisciplinary nature of thermoplasmonics. Thus, I have done my best to tackle all the facets of the fields in order to draw a comprehensive overview. I have also done my best to explain many concepts from scratch because I expect the potential readers of this book to have very different scientific backgrounds, from physics to chemistry and biology. For instance, I have taken the time to detail the basic principles of Raman spectroscopy, X-ray spectroscopy, NV-center fluorescence, fluid superheating, thermophoresis in liquids, etc. This way, not only first-year Ph. D. students but also experienced researchers from various scientific origins can benefit from this book.

The success of thermoplasmonics not only stems from its multidisciplinarity, it also benefits from the success of the even larger field of *plasmonics*. Of course, researchers working plasmonics are not all working in thermoplasmonics but my experience is that researchers in plasmonics are bound to be concerned at some point with photothermal effects: even when photothermal effects are not desirable but detrimental, they have to be quantified in order to avoid misinterpretation of the results, to discard possible artifacts or to avoid damage to the samples. Heating up is what metal nanoparticles do best under illumination.

So please do not hesitate to contact me if you think you can help improve the quality of this book. Furthermore, thermoplasmonics is a fast-growing field of research with several exciting new applications on the verge of being developed. A second edition of this book will certainly make sense in the future.

Investigating thermoplasmonics for a decade has been a rich endeavor for me. I would have enjoyed owning this book when I started working in this field of research a decade ago. It would have saved me a lot of time. I wrote it with that in mind, and I hope you will enjoy reading this book as much as I have enjoyed writing it.

Acknowledgments

I wish to express my thanks to colleagues and students for their careful reading of these pages and their valuable input. I am grateful to Romain Quidant, who aroused my interest in this whole area of research. I want also to express especial thanks to Michel Orrit for accepting to write the Foreword, for his careful reading of the whole book and for his wise comments. I also received helpful feedback from Xavier Audier, Johann Berthelot, Mauricio Garcia-Vergara, Karl Joulain, Adrien Lalisse, Serge Monneret, Hervé Rigneault, Hadrien Robert, Julien Savatier and Siddharth Sivankutty. This book definitely benefited also from the initiative of Alexandra Elbakyan.

Nanoplasmonics

This introductory chapter deals with basic, general and important notions in nanoplasmonics that will be useful before entering the field of thermoplasmonics. The aim is to provide the reader with simple ideas and mathematical expressions that can be used to explain and understand the plasmonic response of metal nanoparticles.

The first section introduces the physics of the localized plasmon resonance for a dipolar spherical metal nanoparticle, for which closed-form expressions of the optical response exist. The section also describes what happens when enlarging the size of the nanoparticle or breaking its spherical symmetry. The second section explains why *gold* nanoparticles have been preferred in plasmonics compared with nanoparticles made of other materials. This second section also takes the opportunity to discuss a very recent branch of nanoplasmonics aiming to try and find alternative plasmonic materials. Finally, a third section introduces the field of thermoplasmonics by answering common experimental questions.

1.1 Localized Plasmon Resonance

1.1.1 Definitions

A *localized plasmon* (LP) is a normal mode of collective oscillation of the free electrons contained in a metal nanoparticle. A LP *resonance* can be excited using light when the electric field of the incoming light oscillates at a frequency close to the plasmon eigen frequency [49].

What I call a *localized plasmon* (LP) in this book is often coined *localized surface plasmon* (LSP) in the literature. I prefer to remove the word "surface" for the following reason. Apart from LP, there exist *bulk plasmons* (BP) and *surface plasmons* (SP). With a bulk plasmon, the excitation occurs in a metal extending over the three dimensions of space (3D). With a SP, the electronic oscillation occurs at the interface between a metal and a dielectric extending over two dimensions of space (2D). With a LP, the oscillation occurs in a space that is confined in all dimensions of space (0D). While "bulk" means 3D, "surface" means 2D, I find it appropriate to use the adjective "localized" to signify 0D, and not "localized surface."

In this book, I distinguish between the fields of *plasmonics* and *nanoplasmonics*. While plasmonics encompasses the physics of LPs and SPs, nanoplasmonics rather focuses on LPs; hence the title of this chapter.

In a striking coincidence, noble metal nanoparticles feature LP resonances in the UV–visible–NIR range, just like many aromatic compounds, although a metal nanoparticle has nothing to do with an organic molecule. This coincidence of Nature is responsible for the burgeoning activity of nanoplasmonics. Basically, due to this coincidence, metal nanoparticles can be advantageously put in standard optical microscopes with common light sources and detectors.

In the following sections, I will introduce basic notions of LP resonance starting from scratch and complicating the concepts step by step. I will successively discuss:

- the response of a metal sphere to a static electric field;
- the LP response of a dipolar sphere to a time-harmonic electric field;
- the LP response of large spheres and retardation effects;
- the LP response of particles of arbitrary morphology;
- the influence of the surrounding medium.

1.1.2 Dipolar Metal Nanoparticle

As usual in physics, the case of a sphere is simple and very instructive (see Figure 1.1). To introduce the physics of LP resonance, let us consider a metal sphere of radius a standing in a surrounding medium of refractive index $n_s = \sqrt{\varepsilon_s}$, along with a monochromatic incident illumination characterized by the electric field[1]

$$\mathbf{E}_0(\mathbf{r}, t) = \mathrm{Re}\left(\underline{\mathbf{E}}_0(\mathbf{r})\, e^{-i\omega t}\right), \tag{1.1}$$

an angular frequency ω and a wavelength in vacuum $\lambda = 2\pi c/\omega$. For a linearly polarized plane wave illumination, one has

$$\underline{\mathbf{E}}_0(\mathbf{r}) = \hat{\mathbf{u}}\, E_0\, e^{i\,\mathbf{k}\cdot\mathbf{r}} \tag{1.2}$$

where $\hat{\mathbf{u}}$ is the unit vector along the direction of the light polarization, and $k = 2\pi n_s/\lambda$ is the norm of the wave vector $\mathbf{k}$. In the presence of the nanoparticle, the electric field at any location $\mathbf{r}$ reads

$$\mathbf{E}(\mathbf{r}, t) = \mathrm{Re}\left(\underline{\mathbf{E}}(\mathbf{r})\, e^{-i\omega t}\right). \tag{1.3}$$

Static Electric Field

Let us first consider a *static* applied electric field ($\omega \rightarrow 0$). The effect of the electric field is to displace the electrons from equilibrium, creating charge accumulation on the boundaries of the nanoparticle, as represented in Figure 1.2a. The magnitude of this charge displacement can be described by the dipolar moment $\mathbf{p}$ of the nanoparticle. $\mathbf{p}$ is collinear with the applied electric field and reads

$$\mathbf{p} = \varepsilon_0 \alpha \mathbf{E}_0. \tag{1.4}$$

---

[1] In this chapter, underlined letters mean complex number quantities defined in the harmonic regime at the angular frequency ω; boldface letters correspond to vectors.

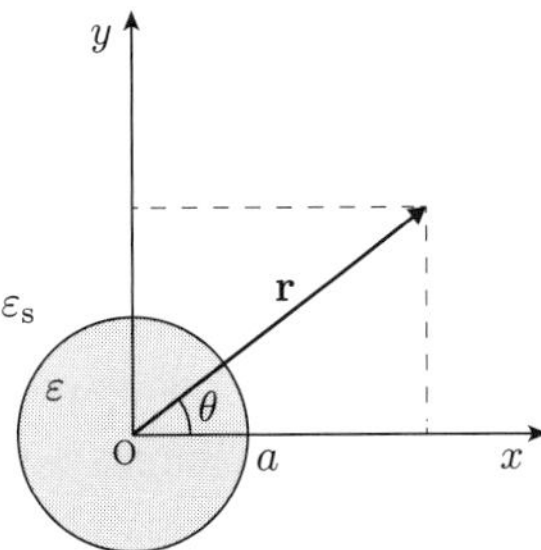

Fig. 1.1 Spatial description of the system under study: a spherical metal nanoparticle of relative permittivity ε in a dielectric medium of relative permittivity $\varepsilon_s = n_s^2$.

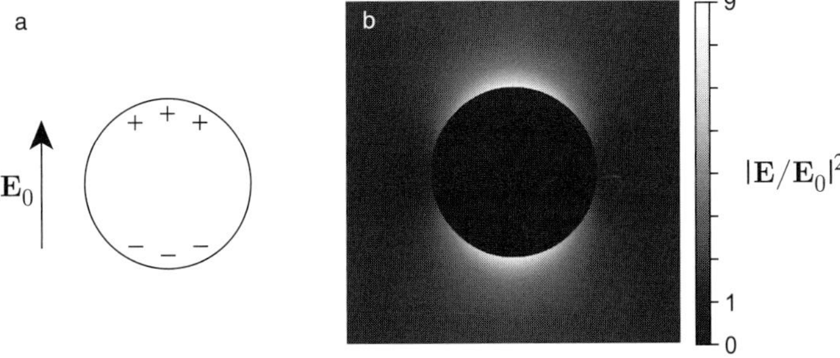

Fig. 1.2 (a) Representation of a spherical metal nanoparticle immersed in a static and uniform electric field. (b) Map of the normalized electric field intensity in the case of a metal sphere in vacuum.

α represents the polarizability of the nanoparticle and scales as a volume. For a metal sphere in a static electric field, one has

$$\alpha = 4\pi \varepsilon_s a^3 \tag{1.5}$$

$$\alpha = 3V\varepsilon_s \tag{1.6}$$

where $V = 4\pi a^3/3$ is the volume of the particle.

In such a simple problem, the electric potential surrounding the sphere reads in spherical coordinates [31]

$$\phi(r,\theta) = E_0 \left(r - \frac{a^3}{r^2} \right) \cos\theta. \tag{1.7}$$

Using this expression of ϕ, the electric field can be calculated anywhere in the surroundings. In particular, one can show that it features a maximum value on the outer boundary of the sphere that is three times as big as the applied electric field intensity

$$E_{\mathrm{max}} = 3\,E_0 \tag{1.8}$$

This figure is independent of n_s and independent of the sphere radius. The total electric near-field *intensity* is thus nine times as big as the applied electric field intensity. Figure 1.2b plots the normalized electric field intensity $|\mathbf{E}/\mathbf{E}_0|^2$ calculated around a metal nanoparticle immerscd in a static electric field, where a maximum value of 9 is evidenced at the vicinity of the nanoparticle, in the direction of the polarization of $\mathbf{E}_0$. The interest of discussing the static case is the following: strong enhancement of the electric near-field is a distinctive feature of metal particles. It occurs even out of the plasmonic resonance, like here in the static case. The effect of a plasmonic resonance will be to drive the near-field enhancement above this threshold value of 9.

In the static case, within the nanoparticle the electric field is rigorously zero, since it is cancelled out by the surface charges accumulated on its boundaries. But this feature no longer holds with non-static electric fields (see next subsection).

The charge separation depicted in Figure 1.2a is responsible for a restoring force within the nanoparticle that tends to move the charges back to their original stable configuration. As in many cases in physics, such a restoring force can be at the origin of a resonance effect (a LP resonance in our case), if the system is excited at the proper frequency. The excitation of a metal sphere with a time-harmonic electric field (*i.e.*, light) and the occurrence of a LP resonance is the purpose of the next subsection.

Time-Harmonic Electric Field (i.e., Light)

The occurrence of a LP resonance of the electronic gas at a given angular frequency ω stems from the restoring force acting between the positive and negative charges facing at each sides of the nanoparticle. For noble metal nanoparticles, such a resonance lies around optical frequencies ($\omega \sim 2\pi \times 10^{14}$ Hz). In this section we will only consider the quasistatic approximation, which assumes that the phase of the electric field oscillation is uniform in the nanoparticle, which amounts to neglecting retardation effects. This assumption is valid for nanoparticles much smaller than the wavelength ($ka < 1$). Within this approximation, one can consider the nanoparticle as a pure dipole (no multipolar term of the charge distribution) and simple closed-form expressions can be derived. In particular, the polarizability of a sphere with an electric permittivity $\varepsilon(\omega)$ reads [31, 55]

$$\underline{\alpha} = 3\,V\,\varepsilon_s\,\frac{\underline{\varepsilon}(\omega) - \varepsilon_s}{\underline{\varepsilon}(\omega) + 2\varepsilon_s}. \tag{1.9}$$

One can also define the enhancement factor $\underline{\xi}$

$$\underline{\xi} = \frac{\underline{\varepsilon}(\omega) - \varepsilon_s}{\underline{\varepsilon}(\omega) + 2\varepsilon_s} \tag{1.10}$$

which plays an important role in the physics of LP resonance. It represents the charge oscillation in the nanosphere in amplitude and phase. This is represented in Figure 1.3. The amplitude $|\underline{\xi}|$ equals 1 at low frequency (large wavelength) and features a resonance when the denominator of $\underline{\xi}$ reaches a minimum for a given angular frequency ω, *i.e.*, when $\mathrm{Re}(\underline{\varepsilon}) \approx -2\varepsilon_s$ (see Figure 1.3a). This situation is possible with metals, since they feature a negative value of the real part of their permittivity. Note that $|\underline{\xi}|$ does not diverge at the resonance due to the remaining imaginary part of $\underline{\varepsilon}$ in the denominator when $\mathrm{Re}(\underline{\varepsilon}) = -2\varepsilon_s$.

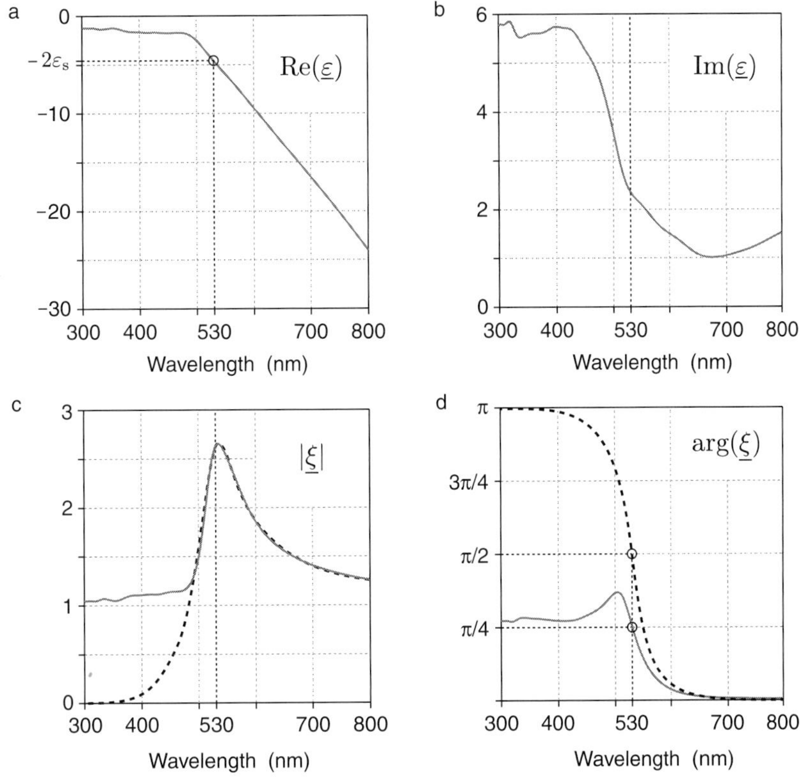

 (a) Real part of the permittivity of gold. The resonance wavelength (530 nm) of a dipolar gold sphere in water is indicated by a vertical dashed line. (b) Imaginary part of the permittivity of gold. (c) Norm of the enhancement factor $\underline{\xi}$. (d) Argument of the enhancement factor $\underline{\xi}$.

The static problem described in the previous section corresponded to $|\varepsilon| \rightarrow \infty$, *i.e.*, $\underline{\xi} = 1$. This is why $\underline{\xi} \rightarrow 1$ at large wavelength.

In general, the response of a resonator (like a spring-mass system) is universal. In particular, the phase delay at resonance is $\pi/2$ and the amplitude of the response vanishes at high excitation frequencies. This is not the case for a plasmon resonance. Figure 1.3c shows the response of a gold nanosphere (solid lines) compared to the response of a regular resonator (dashed lines). The plasmon resonance amplitude goes back to unity at high frequencies (small wavelengths), not to zero. This is due to the presence of another type of electron in the nanoparticles, which can be optically excited at higher photon energy (below $\lambda = 500$ nm) and which are mobile enough to screen the incoming field. These are the d-electrons of the metals as explained in the next section. The response of the nanoparticle when these d-electrons are not present is schematized by a dashed line in Figure 1.3c, which represents the response of a conventional resonator.

A similar unusual resonance behavior is observed in Figure 1.3d. The phase of the oscillation saturates at high frequencies (small wavelengths), whereas it normally approaches π for a normal resonance. The phase of the electronic oscillation *at resonance* is also not conventional. Instead of the usual $\pi/2$ delay in a resonance process, one has a reduced phase shift (see Figure 1.3d). For a gold nanosphere it is close to $\pi/4$, but this is a coincidence. Once again, this is due to the presence of mobile d-electrons that can be excited only with high photon energies. The phase response of the nanoparticle when these d-electrons are not present is schematized by a dashed line in Figure 1.3d.

The maximum amplitude of the electric field on the outer boundary of the sphere, which used to equal $3E_0$ in the static case, now reads

$$\underline{E}_{\max} = (1 + 2\,\underline{\xi})\,E_0. \tag{1.11}$$

Developing this expression highlights two terms: E_0, which is the incoming electric field, and $2\,\underline{\xi}\,E_0$, which is the near-field created by the charges of the nanoparticle, which superimposes constructively with the incoming field to yield a field enhancement. This term further evidences that $\underline{\xi}$ represents the electronic oscillation.

Another interesting quantity is the electric field observed *inside* the nanoparticle. While it was cancelled out by the charge accumulation in the static case, there is a nonzero electric field inside the nanoparticle, especially under plasmonic resonance. For a dipolar sphere, the inner electric field is uniform and equals

$$\underline{E}_{\text{in}} = (1 - \underline{\xi})\,E_0. \tag{1.12}$$

Again, developing this expression highlights two terms: E_0, which is the incoming electric field, and $-\underline{\xi}\,E_0$, which is the screening field created by the charges of the nanoparticle, which superimposes destructively with the incoming field. From Equations (1.11) and (1.12), one can derive two useful parameters that quantify the enhancement of the electric field intensity outside and inside the nanoparticle [39]:

$$\eta_{\text{out}} = |1 + 2\underline{\xi}|^2 = 9 \left| \frac{\varepsilon}{\varepsilon + 2} \right|^2 \tag{1.13}$$

$$\eta_{\text{in}} = |1 - \underline{\xi}|^2 = 9 \left| \frac{1}{\varepsilon + 2} \right|^2. \tag{1.14}$$

The η values are important, first because they report on how a metal nanoparticle acts on the electric field, and then because they stand for universal dimensionless constants for a given metal, if considered at the plasmon resonance frequency ($\eta^{\text{res}} = \eta(\lambda_{\text{res}})$). For gold, one gets $\eta_{\text{out}} = 19$ and $\eta_{\text{in}} = 0.86$ when excited in vacuum at $\lambda = 526$ nm. Figure 1.4 plots the corresponding map of the electric field intensity. A maximum value of 19 is indeed observed on the outer boundary of the gold sphere. This map has to be compared with Figure 1.2b representing the map of the electric near-field in the static case, where a near-field enhancement of only 9 was obtained.

1.1.3 Band Energy Diagram of Metals

The excitation of d electrons is not beneficial in plasmonics as it increases loss (see the increase of $\text{Im}(\varepsilon)$ in Figure 1.3b). This is why metals with fully occupied d-bands are preferred. Fully occupied bands necessarily lie below the Fermi level and can be optically excited only above a given photon energy. Figure 1.5a explains this effect by sketching the

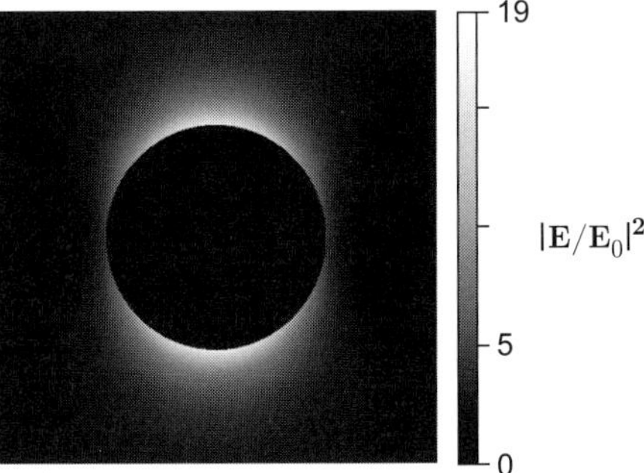

Fig. 1.4 Calculation of the normalized electric field intensity around a gold nanosphere in vacuum illuminated at its plasmon resonance frequency $\lambda = 526$ nm.

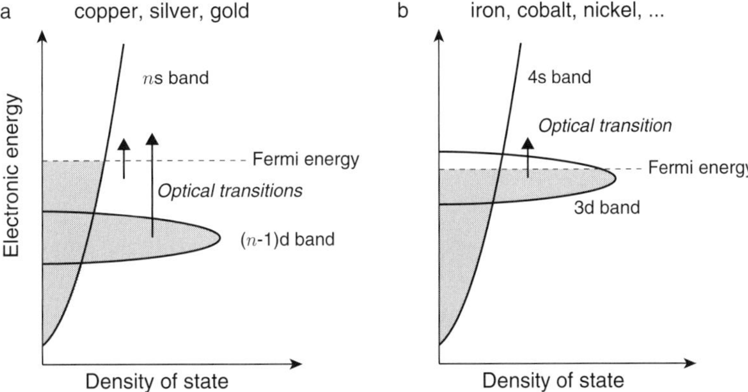

Fig. 1.5 (a) Schematic band energy diagram of metals of the 11th column of the Mendeleev's table, characterized by a fully occupied *d*-band. *d* electrons can be optically excited via an interband transition only above a certain photon energy threshold. (b) Schematic band energy diagram of metals featuring a partially occupied *d*-band, such as iron, copper, and nickel. *d* electrons can be excited by any photon energy.

band energy diagram of the family of elements corresponding to the 11th column of the Mendeleev's table, that is copper, silver and gold. *d*-bands of these materials are fully occupied, their electronic configurations being $4s^1 3d^{10}$, $5s^1 4d^{10}$ and $6s^1 5d^{10}$ respectively.

For metals with a partially occupied *d*-band, *d*-electrons can be excited for any photon energy, which creates additional non-radiative interaction and energy loss, detrimental in plasmonics (see Figure 1.5b).

1.1.4 Optical Cross Sections

The polarizability $\underline{\alpha}$ and the enhancement factor $\underline{\xi}$ of a nanoparticle are valuable parameters, since they render the nanoparticle response. However, they are usually not the parameters of interest in plasmonics. One usually prefers to deal with *optical cross sections*. The optical cross sections of a nanoparticle are directly related to the polarizability, but they are more useful since they can be used to estimate scattered or absorbed energies just from the knowledge of the light irradiance I (power per unit area). The absorption and scattering cross sections are defined such that the powers absorbed and scattered by the nanoparticle under plane wave illumination read:

$$P_{\text{abs}} = \sigma_{\text{abs}}\, I \tag{1.15}$$

$$P_{\text{sca}} = \sigma_{\text{sca}}\, I. \tag{1.16}$$

We also define the extinction cross section as the sum of the two above-mentioned cross sections:

$$\sigma_{\text{ext}} = \sigma_{\text{abs}} + \sigma_{\text{sca}}. \tag{1.17}$$

Absorption cross sections can be defined for any particle interacting with light. For a dipolar sphere, they can be simply expressed as functions of the polarizability $\underline{\alpha}$ [53]:

$$\sigma_{\text{sca}} = \frac{k^4}{6\pi}\, |\underline{\alpha}|^2 \tag{1.18}$$

$$\sigma_{\text{ext}} = k\,\text{Im}(\underline{\alpha}) \tag{1.19}$$

$$\sigma_{\text{abs}} = \sigma_{\text{ext}} - \sigma_{\text{sca}}. \tag{1.20}$$

Figure 1.6 plots the extinction, absorption and scattering cross sections of a gold nanoparticle, 50 nm in diameter, in water. The line shapes are very similar to those of $\underline{\xi}$ (see Figure 1.3c), featuring a resonance around the same wavelength. However, the cross sections vanish at large wavelengths due to the factor k in their expressions.

For small spheres (typically smaller than 40 nm in diameter), $\sigma_{\text{abs}} \gg \sigma_{\text{sca}}$ and one can assume that $\sigma_{\text{abs}} \approx \sigma_{\text{ext}} = k\,\text{Im}(\underline{\alpha})$. Still in this approximation, and using Equation (1.10), one can write the absorption cross sections as functions of the enhancement factor $\underline{\xi}$.

$$\sigma_{\text{sca}} \approx \frac{8\pi}{3} k^4 \varepsilon_s^2 |\underline{\xi}|^2 a^6 \tag{1.21}$$

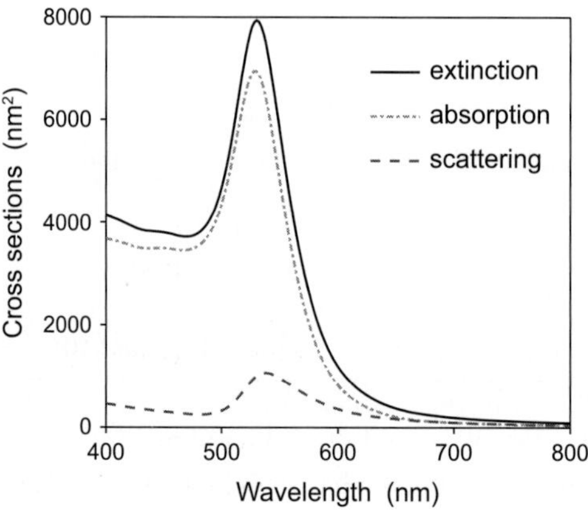

Fig. 1.6   Extinction, scattering and absorption cross sections of a 50 nm gold nanosphere in water.

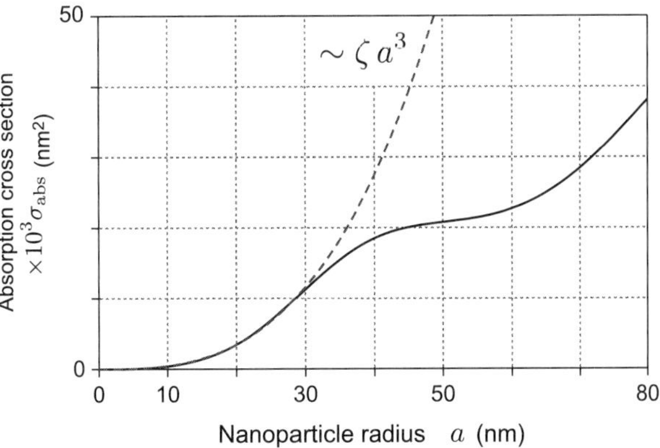

Fig. 1.7 Absorption cross section of a spherical gold nanoparticle in water as a function of its diameter (solid line) for an illumination wavelength of 532 nm. Fitting function for small diameter (dashed line). Reproduced with permission from Reference [52]. Copyright 2015, American Chemical Society.

$$\sigma_{\mathrm{abs}} \approx 4\pi\varepsilon_{\mathrm{s}}\mathrm{Im}(\underline{\xi})a^3. \tag{1.22}$$

As a consequence, the absorption cross section of small spherical nanoparticles scales with the nanoparticle volume (a^3). This rule is valid for nanoparticles smaller than typically 60 nm, as observed in Figure 1.7. In this range of size, one can thus define a constant factor ζ such that the absorption cross section of a gold nanosphere reads

$$\sigma_{\mathrm{abs}} = \zeta a^3, \tag{1.23}$$

The use of this number avoids the need to conduct numerical simulations. For instance, for a gold nanosphere in water illuminated at 532 nm [52]:

$$\zeta = 0.430\,\mathrm{nm}^{-1}. \tag{1.24}$$

1.1.5 Influence of the Particle Size: Retardation Effects

When increasing the size of a particle, the first trend will naturally be an increase of both the absorption and scattering cross sections. Interestingly, while absorption is dominant for small particles, scattering becomes more significant upon increasing the size of a particle since it scales as a^6, while absorption scales as a^3. For spherical gold particles in water, this transition occurs for a diameter $2a = 88$ nm, as represented in Figure 1.8a. For $2a = 88$ nm, the scattering and absorption cross section maxima are equal, as represented in Figure 1.8b. Note the small shift between the resonances in absorption and scattering. As a rule of thumb in plasmonics, the resonance wavelength in scattering does not necessarily match the one in absorption.

Another effect observed in larger particles is the red shift of the plasmonic resonance (see Figure 1.8c). This effect comes from the fact that the particle can no longer be considered as a point-like oscillating dipole. In the previous section, we considered that all the charges in the nanoparticle along with the inner electric field were uniformly oscillating in phase. This

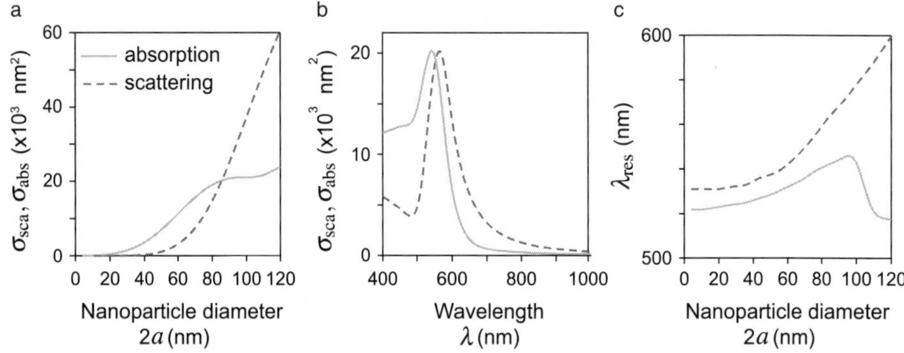

Fig. 1.8 Scattering and absorption cross sections of a gold sphere in vacuum calculated using Mie theory. (a) Plots of the cross sections as functions of the nanoparticle diameter, calculated at the resonance wavelength $\lambda_{res}(a)$. (b) Plots as a function of the wavelength for a nanoparticle diameter $2a = 88$ nm. (c) Resonance wavelength $\lambda_{res}(2a)$.

is the so-called quasistatic approximation. If the metal sphere is enlarged, such an ideal case no longer holds. First, if the particle size becomes of the order of the incoming wavelength, the exciting electric field $\mathbf{E}_0$ may be different in phase from one location to another, which will create some nonuniformities of the charge oscillation within the particle. But a finite wavelength λ is not what primarily invalidates the quasistatic approximation. Indeed, there is another typical length in this problem, the so-called plasmon wavelength, which is shorter than the incoming light wavelength. For gold, it is around 50 nm. If the particle is larger than this dimension, some retardation effects occur due to the electromagnetic interaction between distant charges within the particle. For this reason, for gold, the quasistatic approximation no longer applies for particles larger than typically 40–60 nm. Figure 1.8c plots the resonance wavelengths in absorption and scattering of spherical gold nanoparticles as functions of their radius, evidencing a rule of thumb in plasmonics: enlarging a nanoparticle red-shifts the plasmon resonance.

There is no simple analytical expression of the response of a metal sphere of arbitrary size. However, simple numerical simulations can be carried out using Mie theory, as described in Section 3.1, on page 81.

1.1.6 Influence of the Particle Morphology

We have seen that *enlarging* a particle could shift the plasmon resonance to the red, but it is not the most efficient strategy for playing with the resonance wavelength. With gold one can hardly shift the resonance above 600 nm, and the size of the nanoparticle cannot be arbitrarily increased for some applications, in particular for biomedical applications. The other strategy consists in deviating from a spherical shape. At least three categories of shape modification can be used to substantially red-shift the resonance:

- The most common approach to shift the resonance is to modify the *aspect ratio* of the nanoparticles [32], using rods or discs for instance, illuminated with a polarization along their longer dimension. Figure 1.9 plots absorption cross sections of a dipolar gold sphere

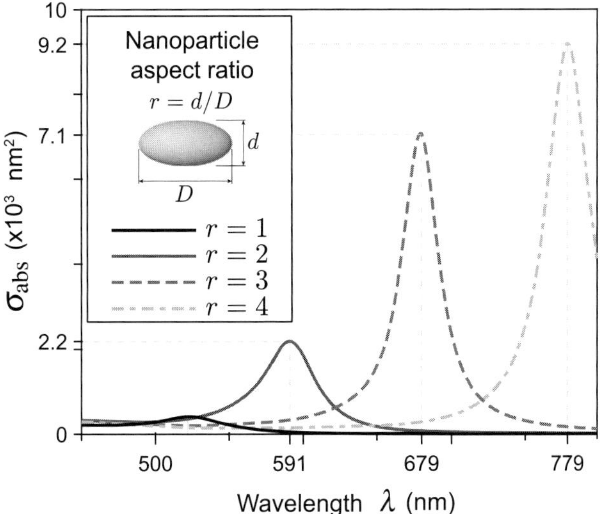

Fig. 1.9 Absorption spectra of a sphere, 20 nm in diameter, immersed in water ($n_s = 1.33$), progressively elongated into a spheroid at constant volume. The aspect ratio ranges from $r = 1$ to $r = 4$. The incident linear polarization is along the long axis of the ellipsoid.

that is progressively elongated, at constant volume. This asymmetry creates a strong red shift of the resonance, along with a strong enhancement of the resonance amplitude. Figures 1.16 and 1.17 further on plot the shift of the resonance as a function of the aspect ratio of a spheroid nanoparticle.

- Core–shell dielectric-gold nanoparticles, called gold nanoshells [22], also exhibit a strong red-shift of the plasmonic resonance. This category of nanoparticles was introduced by Naomi Halas in the late '90s [3].
- Dimer nanoparticles, which consist of two metal structures separated by a nanometric gap. The gap distance can be maintained if the nanoparticles are deposited or fabricated on a solid substrate, or it can results from the presence of a nanometric linker between couple of nanoparticles in solution. The presence of a nanometric gap results in the very close proximity of accumulated charges separated by a nanometric distance (see Figure 1.10b(ii)), hence the presence of a huge electromagnetic field within the gap (see Figure 1.10b(iv) where a field enhancement of three orders of magnitude results from the presence of a gap).

Basically, anything that deviates from a sphere exhibits a red-shifted resonance and a response enhancement. The effects of elongation and gaps are illustrated in Figure 1.10.

1.1.7 Influence of the Surrounding Medium

The position and amplitude of the plasmonic resonance of a nanoparticle strongly depends on the refractive index n_s of its surrounding medium. As the resonance occurs when the real part of the denominator of $\underline{\xi}$ equals zero (see Equation (1.10)), *i.e.*, when $-\varepsilon' = 2n_s^2$, $-\varepsilon'$

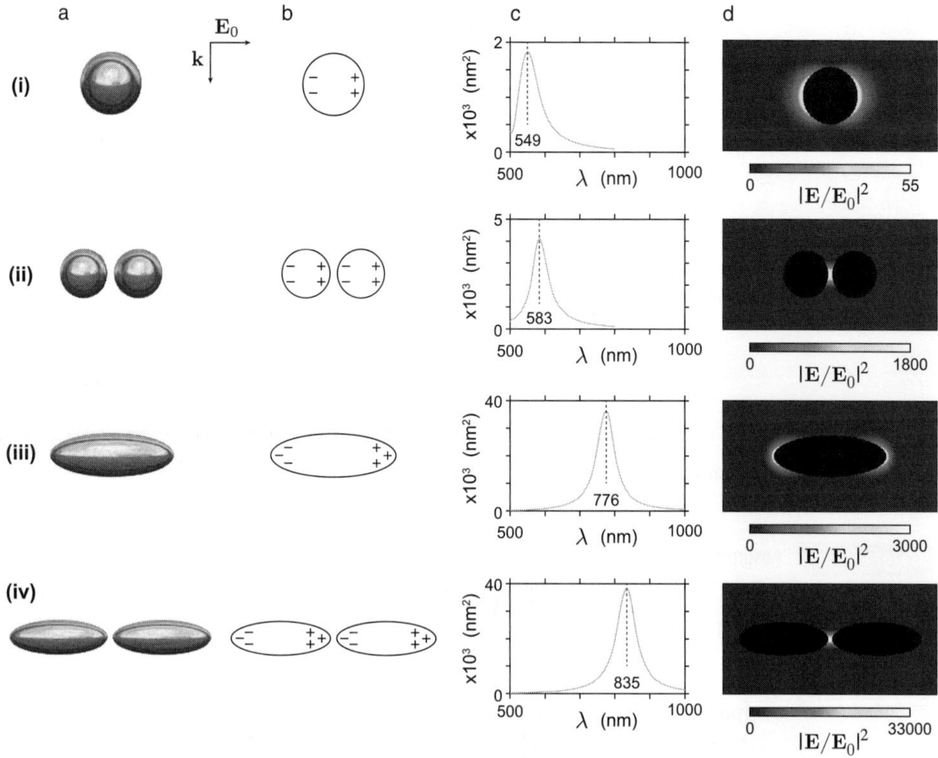

Fig. 1.10　Numerical simulations of gold nanoparticles in water, reproduced from Reference [5]. (a) Gold nanoparticle geometries. (b) Representation of the corresponding charge accumulation on the nanoparticle boundaries. (c) Corresponding scattering spectra. (d) Corresponding maps of the normalized electric field intensity. (i) Case of a sphere. (ii) Case of a sphere dimer. (iii) Case of a prolate spheroid. (iv) Case of a spheroid dimer. Reproduced from Reference [5] with permission from the Centre National de la Recherche Scientifique (CNRS) and The Royal Society of Chemistry.

has to be larger to meet a resonance upon increasing n_s. As $\varepsilon'(\lambda)$ is a *decreasing* function (see Figure 1.3a or Figure 1.13c), it yields an increase of the resonance wavelength.

In Figure 1.11, absorption spectra of a sphere and a prolate spheroid of identical volumes are plotted for three common refractive indices: 1, 1.33 and 1.5. The effect of the refractive index on the enhancement is the same for a sphere and a spheroid. However, the red-shift is much stronger with a spheroid.

When discussing the influence of n_s on the plasmonic resonance, caution has to be used. One has to clarify whether n_s is varied at constant incident intensity I (power per unit area) or at constant incident electric field E_0. When plotting the absorption cross section, one implicitly considers a fixed incoming intensity I. But the electric field depends on the refractive index

$$I = \frac{n_s c}{2}\varepsilon_0 \mathbf{E}_0^2 \tag{1.25}$$

where c is the speed of light. Consequently, at fixed incident electric field E_0, the relative efficiencies of the plasmonic resonances would not match what can be observed in

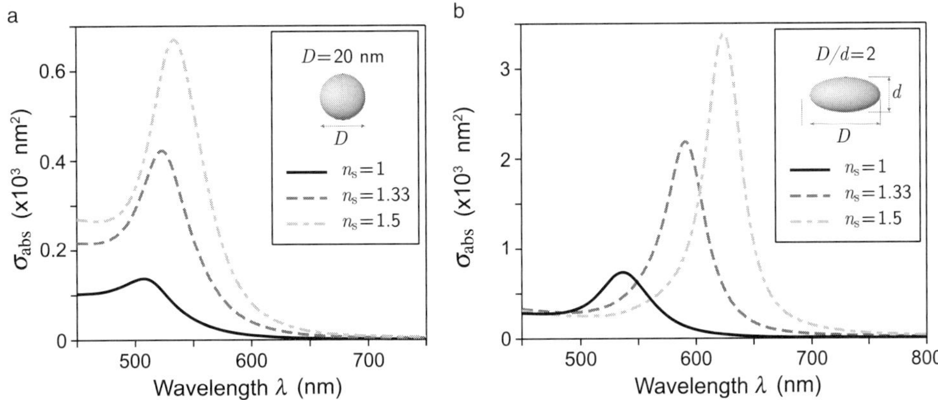

Fig. 1.11 Absorption cross section spectra of a spheroid, 20 nm in effective diameter for various surrounding refractive indices, (a) with an aspect ratio of 1 (sphere) and (b) with a prolate aspect ratio of 2.

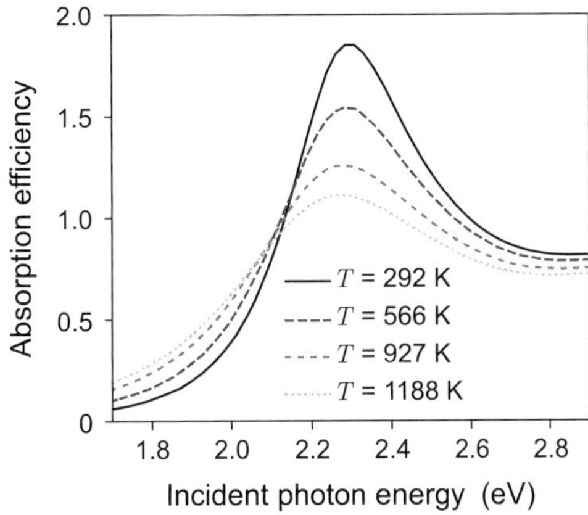

Fig. 1.12 Absorption efficiency of a 20 nm gold nanosphere embedded in silica at different temperatures [1].

Figure 1.11. Reasoning at constant electric field does not really make sense when considering photothermal effects. It is, however, more appropriate when considering near-field enhancement [39].

1.1.8 Influence of the Temperature

The LP resonance of a nanoparticle is dictated by the permittivity of the material, which is strongly dependent not only on the wavelength but also on the temperature. As a rule of thumb, increasing the temperature increases the electron–photon interaction, which results in larger loss and poorer plasmonic performances. This problem has not been much explored. A. Alabastri published a couple of articles on this subject [1, 2]. Figure 1.12 plots the dependence of the absorption efficiency of a 20 nm gold nanosphere embedded in silica from room temperature to 1188 K ($\sim$ 900°C). These data have been calculated using a model involving several damping processes: electron–electron, electron–phonon and electron–surface scattering. On these plots, one can observe how the increase of temperature lowers the value of the absorption while broadening the resonance spectrum. This behavior mainly results from an increase of ε''. ε' remains weakly temperature-dependent. The conclusion is that the effect of the temperature remains weak. If the nanoparticles are not embedded in any matrix, gold nanoparticles are bound to undergo reshaping around 500 K before their permittivity substantially varies.

In 2000, Dalacu and Martinu were already investigating experimentally the temperature dependence of plasmonic resonance of gold nanoparticles [17]. To avoid reshaping, they embedded the nanoparticles in a SiO_2 matrix. This way, the authors could heat up to 1100°C, close to the melting temperature of gold.

The temperature dependence of plasmonic resonance was also investigated in 2013 by Yeshchenko *et al.* [75].

In 2016, the group of Shalaev contributed to this topic by measuring the complex refractive index of gold as a function of the temperature using gold films and ellipsometry [59]. The temperature range remained below 500°C to prevent thermal damage to the film, which was not stabilized by any coating. The main conclusion was that the imaginary part of the permittivity changes significantly with increasing temperature while the real part remains almost intact. This confirms that a temperature increase is associated with weaker plasmonic performances.

1.2 Gold and Other Materials in Plasmonics

This section is devoted to discussing the nature of the materials used in plasmonics. In particular, it focuses on gold and explains why most of the applications in plasmonics are based on the use of gold nanoparticles. But this section also highlights a recent research activity aimed at seeking new plasmonic materials that could replace gold for some applications in the near future.

1.2.1 Why Gold?

Noble metals such as silver, gold And copper all yield plasmonic resonances in the UV–visible–NIR range. Yet gold remains the metal of choice for most applications in plasmonics. Here is why.

- Gold yields LP resonances not only in the visible range, but also in the NIR where human tissues are weakly absorbing, opening the path for biomedical applications.
- Gold is biocompatible and a priori weakly cytotoxic.
- The chemistry of gold is friendly. In particular, the sulfur–gold (S–Au) bond is covalent and particularly strong, enabling easy molecular functionalization of gold nanoparticles provided the molecules of interest are endowed with thiol groups (–SH).
- Gold does not oxidize in the presence of oxygen.

- Optical simulations involving gold nanoparticles simply based on the use of the gold bulk permittivity are usually in very good agreement with experimental results, provided the data set of Johnson and Christy [34] is used (not Palik's [56]). This is not the case, for instance, for silver, where the nature of the particles are not well defined due to the systematic presence of a metal oxide or sulfide layer on the nanoparticle surface. This is not the case either for aluminum. Its optical response strongly depends on the degree of crystallinity of the nanoparticle [63]. It seems that gold permittivity is not that dependent on the degree of crystallinity.

Figure 1.13 gathers the optical constants of gold as functions of the wavelength. All the possible quantities of interest have been considered: the enhancement factor $\underline{\xi}$ of a gold sphere (see Equation (1.10)), the optical index $\underline{n}$ and the permittivity $\underline{\varepsilon}$:

$$\underline{n} = n + i\kappa \tag{1.26}$$

$$\underline{\varepsilon} = \varepsilon' + i\varepsilon''. \tag{1.27}$$

These quantities are linked by the relations:

$$\underline{n}^2 = \underline{\varepsilon} \tag{1.28}$$

$$\varepsilon' = n^2 - \kappa^2 \tag{1.29}$$

$$\varepsilon'' = 2n\kappa. \tag{1.30}$$

Note that the imaginary part ε'' is positive. This sign is dictated by the complex number convention $e^{-i\omega t}$, as it is preferred in optics. The opposite convention $e^{+i\omega t}$, also possible but rather used in quantum physics or in engineering, would have yielded a negative value of the imaginary part of the permittivity.

Until recently gold has been the best candidate for investigation in plasmonics for the above-mentioned reasons. However, following recent advances in nanoplasmonics (related to high-temperature applications, hot-electron processes, nanochemistry, sensing and active plasmonics), new materials have been introduced, reducing the supremacy of gold and silver in plasmonics. The variety of possible materials in nanoplasmonics is now so wide that selecting the best material for a specific application at a specific wavelength may become a difficult task.

1.2.2 Other Materials in Plasmonics (Heteroplasmonics)

Despite all its benefits, gold has some drawbacks that can turn into real limitations for some applications. In particular it is expensive, it has a weak catalytic activity and it features a weak melting temperature (1064°C). Finding alternative materials is not easy [9], mostly because most metals exhibit localized plasmon resonances in the UV range.

As stated by Blaber and coworkers in [10]: "as nanofabrication techniques become increasingly fast and accurate, the performance of plasmonic systems relies less and less on structure fabrication and more on the fundamental limitations of the underlying materials" themselves.

It was proposed to call this emerging field of research heteroplasmonics based on the use of materials that differ from the most common materials Au and Ag [39]. This neologism

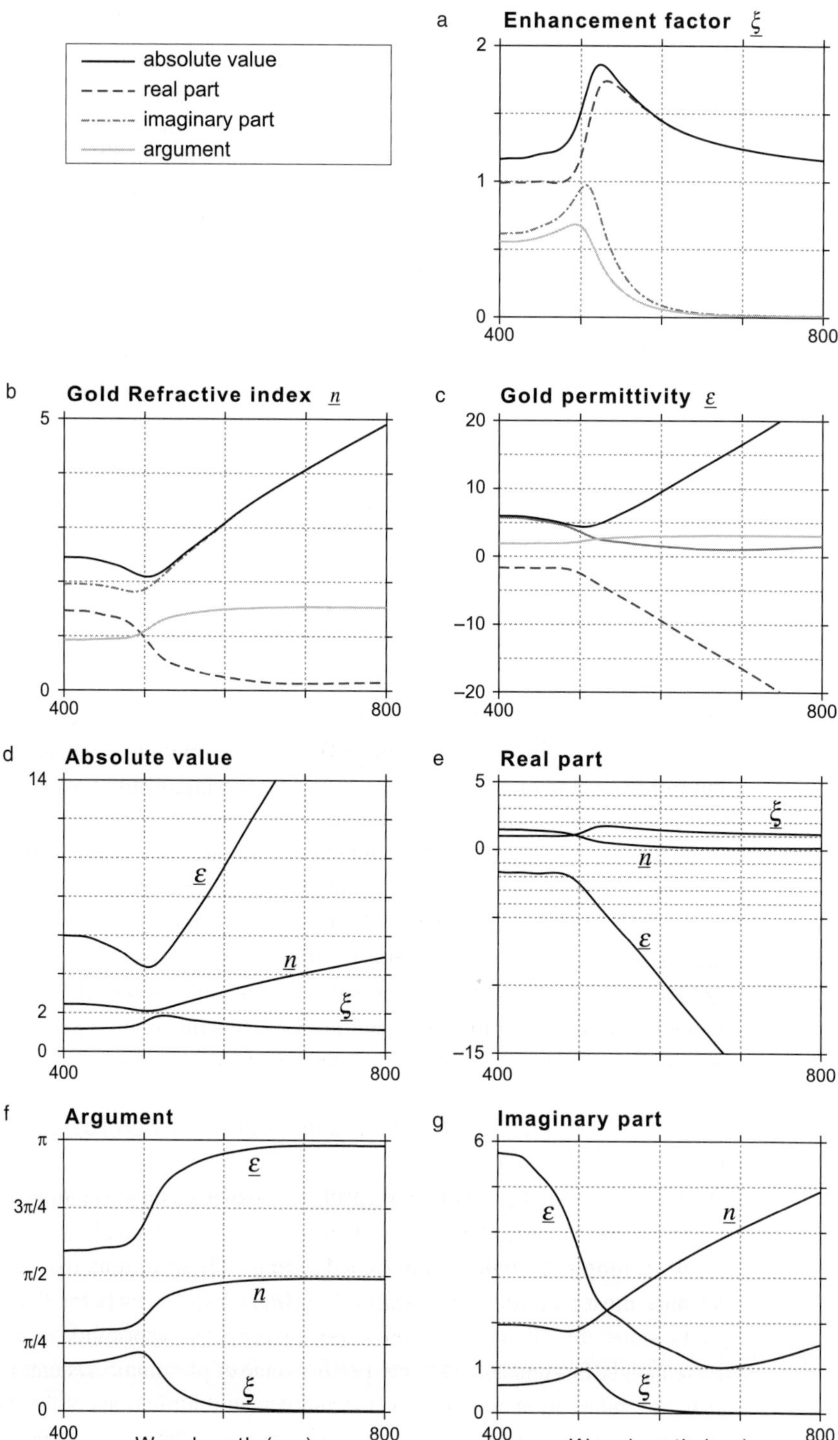

Fig. 1.13 Optical constants of gold as functions of the wavelength, namely the absolute value, argument (phase), real part and imaginary part of the enhancement factor, the refractive index and the permittivity. Calculation from the data sets provided by Johnson and Christy [34].

	Mg											Al					
		Ti TiN		Cr				Ni	Cu		Ga						
		Y	ZrN		Mo		Ru	Rh	Pd	Ag		In					
			HfN	Ta TaN	W					Pt	Au						

Fig. 1.14   Mendeleev's table representing the elements (and metal nitrides) that have been the subject of dedicated articles considering them as plasmonic materials.

is inspired by the denomination *heteroatom* in chemistry, which relates to any atom that differs from the most common atoms (H and C) constituting organic molecules.

So far, the LP resonance of many transition metals, and even other elements, have been investigated (see Figure 1.14), namely magnesium (Mg) [64, 61], yttrium (Y) [66], titanium (Ti) [61], titanium nitride (TiN) [26, 57], zirconium nitride (ZrN) [28, 29, 54], hafnium nitride (HfN) [54], tantalum (Ta) [42], tantalum nitride (TaN) [54], chromium (Cr) [61], molybdenum (Mo) [27], tungsten (W) [61, 27], ruthenium (Ru) [61], rhodium (Rh) [61, 69], nickel (Ni) [14], palladium (Pd) [61, 45, 30], platinum (Pt) [61], copper (Cu) [61, 67], silver (Ag), gold (Au), aluminum (Al) [36, 37], gallium (Ga) [71, 61, 74, 73, 38] and indium (In) [61, 60].

Let us comment on some of the main alternative materials in plasmonics.

Silver. Theoretically, silver should lead to gigantic plasmonic resonance [39]. However, numerical estimations of its LP resonance properties usually overestimate what is observed experimentally. This discrepancy stems for metal oxidation by sulfur atoms (sulfidation) [51, 24], which is difficult to control and which strongly damps the plasmonic resonance. For this reason, it is hard to benefit from the outstanding plasmonic properties of silver in practice [7]. Studies of the plasmonic damping due to sulfidation of silver aroused interest over the last decade. The shift of the resonance peak wavelength has been measured to be 65 nm after 36 h in contact with air, a shift attributed to the contamination from sulfur and the formation of AgS_2 on the nanoparticle surface [51]. In some circumstances, Ag nanoparticles can even feature a lower scattering yield than Au [24]. The formation of an oxide layer does not passivate the particles. Oxidation does not stop and may lead to total destruction of the particles after some time.

Aluminum. Aluminum is envisioned to be the ideal candidate for numerous applications in UV plasmonics [36, 37, 41, 47, 48, 60, 50]. Indeed, aluminum exhibits a strong plasmon resonance in the ultraviolet (around 140 nm), similar to silver. Among all the plasmonic materials that have been considered so far, aluminum demonstrates the strongest near-field enhancement and photothermal efficiency [39]. Al suffers from the same problem as silver and oxidizes even more rapidly [61]. However, the native oxide Al_2O_3 results in the formation of a passivation layer of typically 3 nm that prevents complete denaturation of the material [12]. The permittivity of Al can be corrected in order to take into account oxidation using the Bruggeman effective medium approximation, as recently proposed by Knight *et al.* [37]. The passivation does not strongly modify the plasmonic resonance. In some circumstances, experimental observations can even reveal sharper plasmonic resonance than predicted numerically, despite the presence of the oxidation layer [63]. The origin of this observation is the highest degree of crystallinity in Al nanoparticles compared to the metal layers used to measure the optical constants of metals. Unlike gold, the degree of crystallinity of aluminum seems to strongly affect its optical properties. The only problem with aluminum is the wavelength range of the plasmonic resonance that lies in the UV.

Refractory metals. The weak melting point of gold becomes a problem for applications where a laser excitation of the plasmonic resonance may induce a detrimental heating of the plasmonic nanoparticles. The famous example is the application of plasmonics in heat-assisted magnetic recording (see Chapter 6.4, on page 236). In this application the metal nanoparticle located on the writing head of a hard disk drive is submitted to a strong temperature increase of a few hundred degrees, which compromises the lifetime of the plasmonic structure if made of gold. To circumvent this problem, the plasmonics community started investigating new materials called refractory metals, leading to the research thematics called *refractory plasmonics* [27]. The adjective "refractory" refers to a material that can sustain significant temperature increase, such as tungsten (W), molybdenum (Mo), tantalum (Ta) or metal nitrides (TiN, ZrN, HfN). A more pronounced interest is observed for metal nitrides [28, 29, 54] because they undergo plasmonic resonances in the visible range, just like gold.

Metal nitrides. Metal nitrides constitute a promising family, namely TiN, ZrN and to a lesser extent HfN and TaN [28, 29, 54]. They have drawn attention for applications in plasmonics mainly promoted by the group of Boltasseva and Shalaev. Metal nitrides have been proposed to replace gold for applications in high-temperature plasmonics due to their high melting point ($2930°C$ for TiN and $2952°C$ for ZrN compared to $1064°C$ for Au). In particular, this group investigated the near-field [25] and photothermal [26] properties of TiN in two recent publications. Metal nitrides feature LP resonances in the visible range, like gold. ZrN nanospheres even exhibits a dramatic resonance around 460 nm that widely exceeds that of gold nanospheres, and TiN demonstrates a broadband heat generation in the visible region. These two observations partly explain the recent interest of metal nitrides in plasmonics. Another advantage of such composite materials is their permittivities that can be varied depending on the relative contents of metal and nitrogen [57]. This degree of freedom enables adjustable plasmonic properties. Metal nitrides are also CMOS compatible. It was recently demonstrated that metal nitrides are very poor near-field enhancers compared to gold [40].

UV metals. There exists a family of metals that feature almost identical plasmonic properties [39], lying in the UV range, namely chromium, cobalt, iron, manganese, nickel, palladium, platinum, titanium. These metals are not really of interest in plasmonics for this reason. Only palladium and platinum are relevant thanks to their catalytic properties for applications in nanochemistry, active plasmonics or sensing.

Alkali and Alkaline earth metals. Alkali metals (first column of the Mendeleev's table) and Alkaline earth metals (second column) are known to exhibit gigantic plasmonic resonances. Unfortunately, they react vigorously or even explosively with water even at ambient

Table 1.1	
Alloys	**Intermetallics**
A_xB_{1-x}	A_mB_n

temperature, and their resonance lies in the UV range. This is why they are only sparsely investigated, even theoretically [8]. Only recently have the plasmonic properties of hybrid magnesium/palladium nanoparticles been proved to be valuable for applications in active plasmonics, by the group of Harald Giessen [64].

Alloys and intermetallics. To tune the plasmonic properties of nanoparticles, in addition to play with the nanoparticle morphology, one can also play with its composition if made of alloys or intermetallics [10]. Many studies have investigated the properties of Ag–Au alloys, where the LP resonance varies progressively from the elemental Ag value of 380 nm to the elemental Au value of 520 nm upon varying the Ag/Au stoichiometry [43, 46]. Importantly, this does not imply that the property of a Ag–Au alloy nanoparticle can be simply determined by a linear combination of optical constants. A refined description in terms of movement of the optical gap and Fermi level is required.

Other metal–metal alloys have been marginally investigated, such as Ni–Au, Cu–Zn, Al–Ga, Au–Ga. Alloys of alkali and noble metals are foreseen to be interesting [9], especially K–Au. Fernando *et al.* have published a comprehensive review of alloy nanoparticles and their optical and catalytic properties [20].

Intermetallics are metallic materials made of various elements, but unlike normal alloys, where the atoms are randomly distributed, intermetallics are composed of a periodic alternation of atoms leading to a unique crystalline structure. For instance, $AuAl_2$ [16], Co_2FeGa [6] or Au_3Zn [11] are intermetallics. While alloys usually enable an arbitrary modification of the stoichiometry, intermetallics do not, by definition (see Table 1.1).

Graphene. Graphene is a two-dimensional arrangement of carbon atoms covalently bonded according to a honeycomb lattice. As itself, graphene does not exhibit plasmonic properties. However, the optical properties of graphene can be strongly modified by applying a gate voltage, by doping or by chemical functionalization. All these approaches make the optical properties of graphene much richer.

In particular, doping graphene (chemically or by applying an electrical bias) has a strong effect on its optical properties [13] and contributes to make appear a longitudinal surface plasmon resonance that can be shifted from radio wave frequencies to the infrared region, depending on the doping level. Further shifting the plasmonic properties of graphene in the visible region is still out of reach, although several possible strategies to extend them towards the visible and near infrared have been proposed, such as a reduction in the size of the graphene structures [21] and a further increase in the level of doping. Shrinking in size of piece of graphene is bound to yield electronic resonances in the visible range just because one ends up at some point with aromatic molecules (benzene, naphatelene, anthracene, tetracene, etc.). But this phenomenon is no surprise and should not be considered as an application of graphene or plasmonics.

Numerical simulations have shown that graphene seems to be a good plasmonic material for THz applications. However, at near-infrared frequencies, losses in graphene may still be comparable to noble metals. This makes graphene less attractive as an alternative plasmonic material at the telecommunications and visible wavelengths [70].

Doped semiconductor nanocrystals. Recently, doped semiconductor nanocrystals have been shown to exhibit LP resonances [15]. As their free electron density remains smaller than the one of metals, the resonance remains in the infrared. Many materials have been proposed, such as copper selenide, tungsten oxide, copper telluride, germanium telluride, zinc oxide [15, 19, 44]. These materials also offer the possibility to engineer core–shell semiconductor-gold nanoparticles with tunable plasmonic properties. For instance, Ding *et al.* fabricated hybrid Au–Cu_9S_5 nanoparticles which are supposed to enhanced optical absorption for photothermal cancer therapy applications [18].

1.2.3 Quality Factors and Figures of Merit of Plasmonic Materials

To compare the efficiencies of plasmonic materials with each other, quality factors (also coined "figures of merit" or "metrics") have been defined. The most famous quality factor in plasmonics reads $-\varepsilon'/\varepsilon''$, but depending on the context different quality factors have been proposed, which leads to some ambiguities in the literature. For instance, Wang and Shen claim that "for a given resonant frequency, the quality factor of the resonance is determined only by the complex dielectric function of the metal, independent of the nanostructure shape or the dielectric environment" [68]. This statement is in contradiction with the definition of the quality factor given by West *et al.* in Reference [70], where the authors claim that "quality factor for localized surface plasmon resonances depends significantly on the shape of the metal nanoparticles." It is sometimes unclear whether the authors consider the quality factor of a plasmonic *material*, or the quality factor of a plasmonic *nanoparticle*. It is also unclear whether some quality factors are only valid for spheres, or only valid at resonance. This section is intended to clarify and critically discuss the domain of research dealing with quality factors in plasmonics.

The Quality Factor $-\varepsilon'/\varepsilon''$

The most employed figure of merit in plasmonics is defined by the ratio of the real part of the permittivity divided by the imaginary part [10].

$$Q_{\mathrm{LP}} = -\frac{\varepsilon'}{\varepsilon''} \tag{1.31}$$

The minus sign is intended to compensate for the negative value of ε' and obtain a positive number. Although its origin is not clear, this expression highlights two general rules in plasmonics: $|\varepsilon'|$ has to be high while ε'' has to be small. Figure 1.15 lists the maximum Q_{LP} values and the corresponding frequencies at which they occur for all non-group-f metals.

The Quality Factor Q: Jackson's Definition

Looking at the reference textbook of Jackson, *Classical electrodynamics* [31], is usually a good starting point when addressing a new problem in electromagnetism. In Chapter 8,

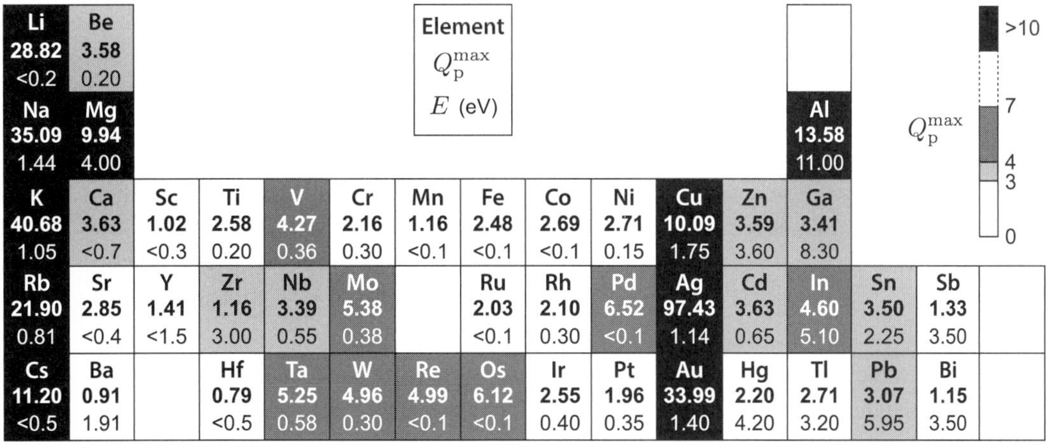

Fig. 1.15 Quality factors $Q_p = -\varepsilon'/\varepsilon''$ of metals displayed on the Mendeleev's table calculated in Reference [10].

dealing with resonant cavities, Jackson introduces a simple definition of what a quality factor should be:

$$Q = 2\pi \frac{\langle \text{Stored energy} \rangle_t}{\text{Dissipated energy over one cycle}} \tag{1.32}$$

$$Q = \omega \left\langle \frac{\text{Stored energy}}{\text{Dissipated power}} \right\rangle_t \tag{1.33}$$

Starting from this definition, Wang and Shen derive an expression of Q in nanoplasmonics valid only at resonance, in the quasistatic approximation, for low-loss metals ($\varepsilon'' \ll |\varepsilon'|$):

$$Q = -\frac{\omega \frac{d\varepsilon'}{d\omega}}{2\varepsilon''}. \tag{1.34}$$

The authors stress the fact that this Q factor of the resonance is determined only by the complex dielectric function of the metal material, and is independent of the nanostructure shape or the dielectric environment.

The way it is defined, this quality factor quantifies the sharpness of the resonance spectra. As noticed by Jackson, the definition (1.32) based on energy considerations can be also recast into the famous expression

$$Q = \omega_0/\Gamma \tag{1.35}$$

where ω_0 is the resonance frequency and Γ is the full-width half-maximum of the resonance spectra. This definition also determines how fast the plasmon oscillation naturally damps due to loss. Q can be understood as the average number of self oscillations during decay.

In his review article on plasmonics in 2011, Stockman explained that this quality factor (Equation (1.34)) also quantifies the near-field enhancement: "It also shows how many times the local optical field at the surface of a plasmonic nanoparticle exceeds the external field" [65]. This statement is enigmatic and surprising because the near-field enhancement strongly depends on the shape of the nanoparticle. Since this statement is not supported by a reference, it is difficult to understand its origin.

In my opinion, the way it is defined, the quality factor (1.34) is not supposed to quantify the near-field enhancement. One can easily understand that the width of a resonance depends of a derivative of a quantity with respect to ω. However, a near-field enhancement at a given frequency should only depend on ε, not on how quantities vary with respect to ω around the resonance.

The Figure of Merit E_{max}/E_0

Another quantity is often used to quantify the efficiency of a material in plasmonics, the figure of merit defining the near-field enhancement, defined by the ratio of the complex amplitudes:

$$\mathrm{FOM} = \frac{\text{Maximum electric near-field over space}}{\text{Incoming electric field}}. \tag{1.36}$$

I prefer to call this quantity a *figure of merit* to avoid any confusion and to differentiate it from the *quality factor* Q defined in the previous paragraph.

Boltasseva and Shalaev stated that this definition leads to an expression of FOM that simply reads in the case of a sphere [70]:

$$\mathrm{FOM} = -\frac{\varepsilon'}{\varepsilon''} \tag{1.37}$$

and in the case of a prolate spheroid:

$$\mathrm{FOM} = \frac{\varepsilon'^2}{\varepsilon''}. \tag{1.38}$$

The authors support this formalism with an obscure proceedings reference that I could not find [62], which makes it difficult to understand the origin of this result. Two observations can be made regarding these equations. First, Equation (1.38) is puzzling because one cannot continuously go from Equation (1.38) to Equation (1.37), whereas one can go continuously from a sphere to a spheroid. Second, Equation (1.38) is the most famous figure of merit in plasmonics. However, the origin of this quality factor is enigmatic. I couldn't find any reference deriving its expression. Apparently, it cannot be derived from Equation (1.36), even for a sphere. As demonstrated in Reference [39] and as discussed hereinafter, the near-field enhancement of a sphere, at resonance, in the quasistatic regime rather reads FOM $=$ $3|\underline{\varepsilon}|/\varepsilon''$. So the quantity $-\varepsilon'/\varepsilon''$ does not quantitatively render the near-field enhancement. I believe $-\varepsilon'/\varepsilon''$ *qualitatively* indicates that $|\varepsilon'|$ has to be maximized while ε'' has to be minimized in order to achieve efficient plasmonic resonance. But it cannot be considered as a *quantitative* figure of merit.

Reference [39] also derives an expression of the figure of merit for spheroids, which is different from Equation (1.38). At resonance, it reads

$$\mathrm{FOM} = \frac{1}{L}\frac{|\underline{\varepsilon}|^2}{\varepsilon''} \tag{1.39}$$

where L is a form factor called the depolarization factor (see the definition in Chapter 3, Equation (3.7) on page 84). It depends only on the aspect ratio of the spheroid. It equals $1/3$ for a sphere.

The Faraday Number

The Faraday number of a plasmonic nanoparticle is defined by a ratio of two electric field amplitudes [39]:

$$\mathsf{Fa} = \left| \frac{\text{Maximum electric near-field over space}}{\text{Incoming electric field}} \right|^2. \tag{1.40}$$

It differs from the definition (1.36) only by a power of two: this definition involves the electric field intensity rather than simply the electric field amplitudes. For a sphere surrounded by a dielectric medium of permittivity ε_s, in the quasistatic regime, an exact expression can be derived [39]:

$$\mathsf{Fa}^0 = 9 \left| \frac{\varepsilon}{\underline{\varepsilon} + 2\varepsilon_\mathrm{s}} \right|^2. \tag{1.41}$$

Superscript 0 means "for a sphere." Note that Fa^0 does not depend on the size of the sphere, as long as the quasistatic approximation remains valid, that is as long as the sphere is small enough ($<$ 60 nm in diameter [52]). Consequently, Fa^0 can be considered as a figure of merit of the material itself, and can be used to compare different materials with each other, at any wavelength, not necessarily at the resonance wavelength.

At resonance, $\varepsilon' = -2\varepsilon_\mathrm{s}$, and Equation (1.41) takes a simpler form:

$$\mathsf{Fa}^0(\lambda_\mathrm{res}) \approx 9 \left| \frac{\varepsilon_\mathrm{res}}{\varepsilon''_\mathrm{res}} \right|^2 = 9 \left(1 + \left| \frac{\varepsilon'_\mathrm{res}}{\varepsilon''_\mathrm{res}} \right|^2 \right). \tag{1.42}$$

As stated previously, the usual figure of merit in nanoplasmonics reads $-\varepsilon'/\varepsilon''$, which differs from the figures of merit (1.42). Although one ends up with the same general trend (that is, ε' has to be large and ε'' has to be low), Fa appears as refined figures of merit, under plasmonic resonance. Out of plasmonic resonance, the simplified Equation (1.42) is no longer valid, and one has to consider the more general definitions of Fa (Equation (1.41)) which is valid at any wavelength.

The interest of the Faraday number compared to $-\varepsilon'/\varepsilon''$ is that it gives quantitative estimations of the near-field enhancement (at least for spheres in the quasistatic regime), it enables quantitative comparisons between different materials, it is valid at any wavelength (not only the close to resonance) and it takes into account the effect of the surrounding medium, as it depends not only on ε' and ε'' but also on ε_s.

Table 1.2 draws up the values of the Faraday numbers Fa^0 at the plasmonic resonance for most materials that have been used in plasmonics so far, along with the plasmon resonance wavelengths. Interestingly, according to this table, gold does not seem to be the best candidate for applications in plasmonics, as already discussed in the previous section.

Silver and aluminum feature outstanding plasmonic capabilities, at least according to this figure of merit that only focuses on near-field enhancement. However, in practice, such

Table 1.2 Dimensionless numbers characterizing the ability of a material to enhance the near-field (Fa^0), generate heat (Jo^0), along with the corresponding resonance wavelengths λ_{res}^{in} and λ_{res}^{out}. Optical constants were taken from references indicated in the first column. T_s (°C) is the melting temperature at atmospheric pressure.

			λ_{res}^{out} (nm)	Fa^0	λ_{res}^{in} (nm)	Jo^0	T_s (°C)
[34]	gold	Au	528	19.6	507	6.32	1064
[56]	silver (Palik)	Ag	357	118	354	52.0	961
[34]	silver (J&C)	Ag	355	476	354	111	"
[58]	aluminum (Rakic)	Al	140	1290	140	477	660
[34]	copper	Cu	585	15.2	538	2.65	1085
[35]	cobalt	Co	366	10.9	< 200	>12	1495
[35]	chromium	Cr	289	10.6	< 200	>11	1907
[35]	iron	Fe	337	9.38	< 200	>11	1538
[56]	molybdenum	Mo	154	18.3	140	41.3	2623
[35]	manganese	Mn	380	9.82	< 200	>9.2	1246
[35]	nickel	Ni	218	11.6	141	21.4	1455
[35]	palladium	Pd	223	13.3	< 200	>21	1555
[56]	platinum	Pt	323	10.5	210	12.0	1768
[56]	rhodium	Rh	199	18.4	173	31.5	1964
[56]	tantalum	Ta	735	11.8	640	2.45	3017
[56]	tantalum	Ta	177	11.6	77	52.3	"
[35]	titanium	Ti	274	11.2	< 200	>13	1668
[56]	titanium nitride	TiN	566	12.0	488	5.55	2930
[56]	tungsten	W	159	14.2	131	25.6	3422
[56]	zirconium nitride	ZrN	467	28.6	437	16.6	2952

giant enhancements are never observed, mainly due to the presence of an oxide or sulfide layer that rapidly (timescale of hours) grows on the nanoparticle surface. For aluminum, this oxidation leads to a stable passivation layer. But in the case of silver, sulfidation does not stop once a sulfide layer is formed, which is more problematic.

The figures of merit derived in this subsection all stem from data sets taken from the literature. An important question is how reliable these data sets are. This is the subject of the following Subsection 1.2.4.

For arbitrary geometries, no simple expression of Fa exists. Numerical simulations have to be conducted to compute Equation (1.40), except for spheroids where the expression of the Faraday number reads

$$Fa = \left| \frac{\varepsilon}{\varepsilon_s + L_j(\varepsilon - \varepsilon_s)} \right|^2 \tag{1.43}$$

where $j \in \{x, y, z\}$ represents the polarization of the incident light and where L_j is the depolarization factor defined in Chapter 3, in Equation (3.7) on page 84. Figure 1.16 displays the values of Faraday numbers at LP resonance for prolate spheroids (cigars) of different

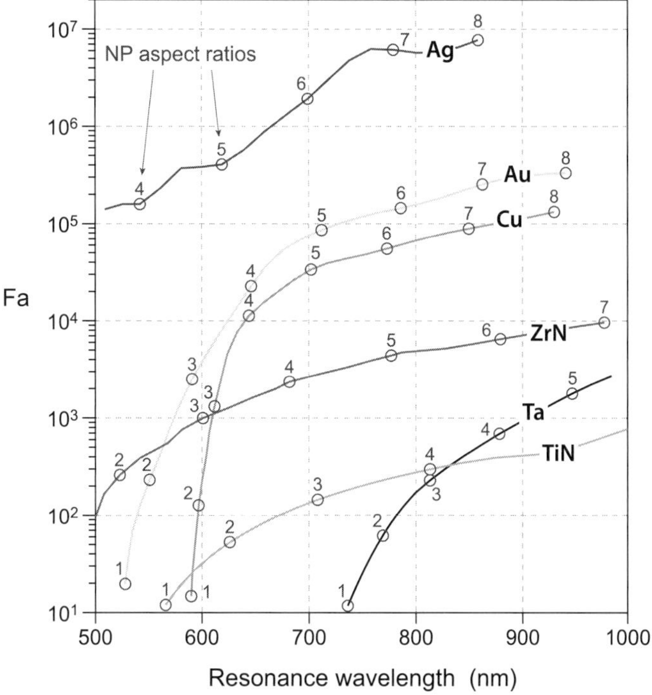

Fig. 1.16   Faraday numbers Fa at plasmonic resonance for spheroid nanoparticles of various aspect ratios and for various materials. Reproduced with permission from Reference [39]. Copyright 2015, American Chemical Society.

aspects ratios (from 1 to 8) and different materials, to quantify the effect of an elongation on the near-enhancement as represented. These results highlight the ability of gold nanoparticles to feature gigantic near-field enhancement when elongated.

The Joule Number

Fundamentally, trying to define a *single* figure of merit to quantify the ability of a material to behave as a good plasmonic material does not make much sense, because plasmonic nanoparticles can be used for different reasons. In particular, they can be used to enhance the near field, like in SERS, but also to generate heat, like in photothermal cancer therapy. And a good near-field enhancer won't necessarily be an efficient heat generator.

Following this remark, in parallel with the Faraday number, the Joule number Jo has been defined to quantify the ability of a plasmonic material to generate heat and is defined by [39]

$$\text{Jo} = \frac{e\,\varepsilon''}{n_\text{s}}\,|E_\text{in}/E_0|^2. \tag{1.44}$$

This expression contains the imaginary part of the permittivity of the material ε'' and the photon energy e in eV, *i.e.*, $e = \hbar\omega/\hbar\omega_0$ with $\hbar\omega_0 = 1$ eV. One can also conveniently use the relation $e = \lambda_{\text{ref}}/\lambda$ where $\lambda_{\text{ref}} \approx 1240$ nm. The arbitrary factor e was used to obtain a dimensionless number.

For a sphere in the quasistatic approximation, there exist equivalent simple expressions of Jo:

$$\text{Jo}^0 = 9\,\frac{e\,\varepsilon''}{n_s}\,\left|\frac{\varepsilon_s}{\varepsilon + 2\varepsilon_s}\right|^2 \tag{1.45}$$

$$= \frac{\lambda_{\text{ref}}\,\sigma_{\text{abs}}}{2\pi\,V} \tag{1.46}$$

where σ_{abs} is the absorption cross section of the nanoparticle, V denotes the nanoparticle volume and $\lambda_{\text{ref}} \approx 1240$ nm. Superscript 0 means "for a sphere." The latter expression, derived by considering the relation $qV = \sigma_{\text{abs}}\,I$, highlights the relation between Jo and σ_{abs}, two quantities related to heat generation in plasmonics. The Joule number turns out to scale as the absorption cross section divided by the nanoparticle volume. This is interesting because the common approach rather consists in dividing σ_{abs} by the projected surface of the nanoparticle to obtain a dimensionless number quantifying the absorption efficiency. This formalism suggests that a division by the nanoparticle *volume* makes more sense. Equation (1.46) also enables the extension of the definition of Jo to any nanoparticle geometry, not only for spheres.

At resonance, $\varepsilon' = -2\varepsilon_s$, and Equation (1.45) takes a simpler form:

$$\text{Jo}^0(\lambda_{\text{res}}) \approx 9\,e\,\frac{n_s^3}{\varepsilon''_{\text{res}}} = 9\,e\,\frac{|\varepsilon'_{\text{res}}|^{3/2}}{\varepsilon''_{\text{res}}}. \tag{1.47}$$

Noteworthily, this expression tells us that ε'' has to be minimized even when considering heat generation. ε'' is yet related to loss and according to Equation (1.44), one could have expected that a large value of ε'' would be beneficial for the photothermal properties of nanoparticles, but that is not the case. While heat generation seems to be proportional to ε'' in Equation (1.44), the inner electric field $\mathbf{E}_{\text{in}}$ is actually damped by a factor $|1/\varepsilon''|^2$, hence the $1/\varepsilon''$ general dependence of Jo, and in general of heat generation in plasmonics.

Table 1.2 draws up the values of the Joule numbers Jo^0 at the plasmonic resonance for most materials that have been used in plasmonics so far, along with the plasmon resonance wavelengths. Interestingly, according to this table, gold does not seem to be the best candidate for applications in plasmonics, as already discussed in the previous section.

According to Definition (1.46), Joule numbers can also be calculated for any geometry. In particular, they can be calculated for prolate spheroids (cigars) to quantify the effect of an elongation on the heat generation. This is represented in Figure 1.17. The Jo numbers have been calculated for a series of aspect ratios (from 1 to 8) and for various materials. These results highlight the ability of gold nanoparticles to feature strong heat generation when elongated.

1.2.4　How Reliable Are Optical Constant Data Sets?

Tables of optical constants are necessary to conduct numerical simulations in plasmonics. Such information can be found either in articles or in handbooks. Noteworthily, variations of the values can be evidenced from one source to another. For instance, Johnson and

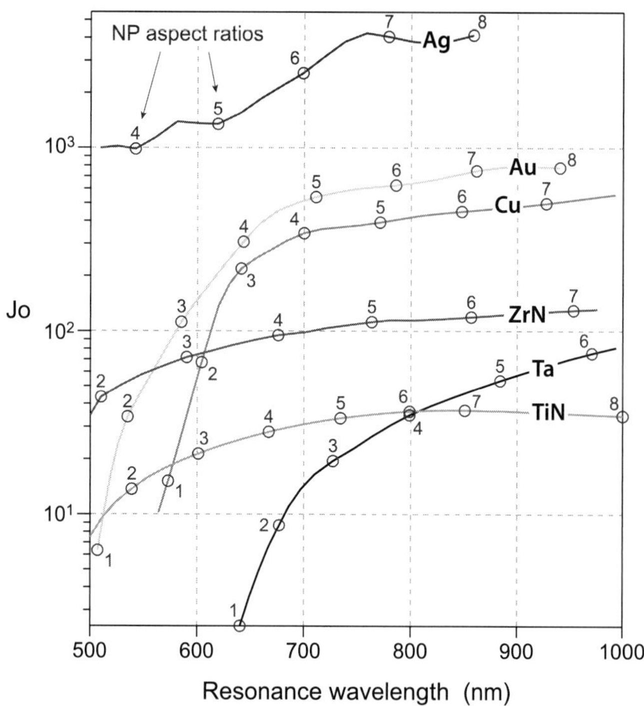

Fig. 1.17 Joule numbers Jo at plasmonic resonance for spheroid nanoparticles of various aspect ratios and for various materials. Reproduced with permission from Reference [39]. Copyright 2015, American Chemical Society.

Christy (henceforth referred to as J&C) on the one hand, and Palik's handbook on the other, both provide a large set of permittivity values for gold over a wide range of wavelengths, but the values slightly differ for some wavelengths. Johnson and Christy's values are known to much better fit experimental observations for gold, for instance, compared to Palik's values.

Johnson and Christy published two seminal articles reporting on the measurements of the optical constants of many metals, namely Cu, Ag and Au in 1972 [34], and Ti, V, Cr, Mn, Fe, Co, Ni and Pd in 1974 [35].

Even stronger discrepancies between J&C and Palik data sets are encountered with silver. Figure 1.18 plots the Joule spectra calculated using J&C and Palik data sets. A substantial difference is observed. As stated in Reference [33], small variations in the optical constants can be magnified several fold under plasmonic resonance, leading to inaccuracies in the modeling and interpretation of results. It is probable that the degree of sulfidation of the metal layer was not the same during the different series of experiments at the basis of these data sets. It is usually admitted that J&C values tend to overestimate the plasmonic response while Palik's values tend to underestimate it [33]. Moreover, Palik's data sets actually combine the work of four research groups using different sample preparation methods, which yields some inconsistencies. Jiang *et al.* [33] recently published a comprehensive review of the problem and proposed to re-evaluate the optical constants of silver measured so far. Note that new measurements of the optical constants of silver have just been reported by Yang *et al.* [72].

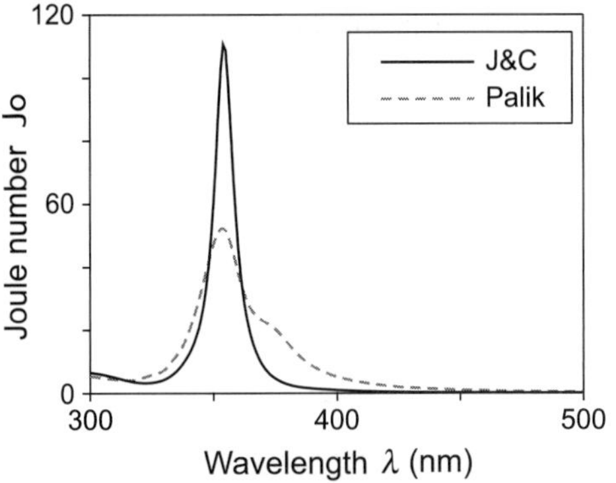

Fig. 1.18 Joule spectra for silver calculated from two different data sets of permittivity: Johnson & Christy [34], and Palik's Handbook [56]. Reproduced with permission from Reference [39]. Copyright 2015, American Chemical Society.

Some problems may also arise when using alloys, because the permittivity naturally depends on the stoichiometry. For this reason, in Reference [26], the group of Boltasseva preferred to measure the permittivity of their own metal nitride (TiN) sample by ellipsometry, prior to conducting their numerical simulations. Interestingly, the measured values of the permittivities of TiN were rather different from the permittivities of Palik's handbook. Moreover, they have shown that the permittivity of TiN depends on sample annealing during material deposition, as annealing induces a change of the Ti/N stoichiometry. For instance, higher substrate annealing temperature (800°C compared to 400°C) led to much better photothermal efficiencies.

1.3 Getting Started in Thermoplasmonics

1.3.1 Introduction

As explained in this chapter, the interaction between a nanoparticle and an incoming light can be described by two parameters: the scattering and absorption cross sections. While the scattering cross section tells how light will be re-radiated by the nanoparticle, the absorption cross section is directly related to heat generation. Thus, absorption is what primarily matters in *thermoplasmonics*, the domain of nanoplasmonics that uses plasmonic nanoparticles as nanosources of heat [23, 4], which is the main subject of this book.

Before entering the details of thermoplasmonics, here are some basic considerations regarding what an experimental set up in thermoplasmonics looks like.

1.3.2 Which Sample?

Two main sample geometries are used:

- A two-dimensional geometry, where nanoparticles are lying on a substrate (*e.g.*, a glass coverslip). These nanoparticles can be colloidal nanoparticles that have been deposited, or spin-coated or chemically attached to the sample. Or they can be metal structures that have been lithographied using for instance e-beam lithography or focused ion beam etching.

- a three-dimensional geometry where nanoparticles are dispersed in volume. This is the case for colloidal nanoparticles dispersed in a solution, in a polymer matrix, or in a living organism (within the tumor in a mouse, for instance) for biomedical applications.

1.3.3 Which Microscope Configuration?

For some applications that involve the heating over *millimetric or centimetric* areas, no microscope is required. The source of light (laser or sun light) is directly sent onto the sample without passing through an objective lens. But for most applications, the use of a microscope and an objective lens is mandatory to investigate the thermal induced effects on the micro- and nanoscales and possibly at the single nanoparticle level.

When using an optical microscope and a laser beam to heat, two configurations can be advantageously employed and implemented on a single microscope (as explained in Figure 1.19).:

- heating by focusing the laser beam. In this case, the laser is collimated and sent at the entrance of the objective lens.
- heating over an extended area. In this configuration, the laser is focused at the entrance of the objective lens and the diameter of the illuminated area (typically from a few microns

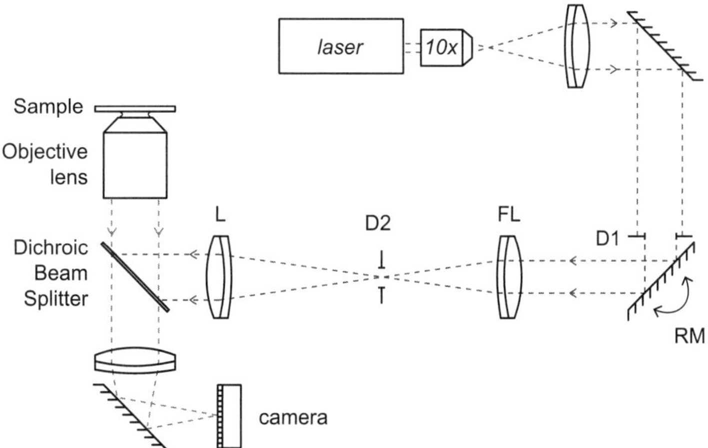

Fig. 1.19 Microscope configuration enabling the heating of a light-absorbing sample using two configurations: in a first configuration, two lenses (L and FL) are placed according to a 4-*f* configuration to send a collimated laser beam through the microscope objective. With this configuration, the laser is focused at the sample location, and the focus can be moved in *x* and *y* by playing with the angle of the mirror RM. In a second configuration, the flip lens FL is removed so that the laser is focused by the lens L at the entrance of the objective lens. This way, the laser beam illuminates a wide area at the sample location, the diameter of which can be adjusted using the diaphragm D2.

to a few 100s of microns) depends on the magnification and can be adjusted using a diaphragm.

1.3.4 Which Laser?

Heating on the microscale requires large power. This is an important rule of thumb in thermoplasmonics. For instance, if one considers a 20-nm nanosphere immersed in water, in order to increase its temperature by 1 K, it has to generate a power of 75 nW (see Equations (2.48)). It may seem ridiculously weak, but it is not! – for at least three reasons:

- First, compared to the volume of the nanoparticle it corresponds to a power density of 10^{16} W·m^{-3}, which is around seven orders of magnitude larger than the power density experienced by the resistance of a common toaster.
- Second, to generate 75 nW in a 20-nm gold nanosphere, one has to use a light irradiance (power per unit area) of 10^8 W·m^{-2}, five orders of magnitude larger than the Sun light irradiance.
- Third, to achieve such an irradiance, one needs to focus a laser beam with a power of 0.1 mW (over an area of 1 μm in diameter), which can be fine for many lasers. But if the laser is not focused (second configuration mentioned in the previous subsection) in order to heat several nanoparticles at once, let's say if the diameter of the illuminated area is 10 μm, then the required power of the laser beam is now 10 mW. And if you wish now to heat your nanoparticle not by 1 K but 10 K, you will need 100 mW.

Given that some power is generally lost between the laser output and the sample plane, the rule of thumb is that you need lasers with a power of a few 100s of mW to conduct experiments in thermoplasmonics.

The most versatile candidate to me is a Ti–Sapphire laser. It offers many advantages: the wavelength is tunable (from 700 to 900 nm and even more), and matches the resonance of many gold nanoparticles (nanorods, nanoshells, etc.). With some laser models, one can even deflect the pump beam at 532 nm and use it as the heating laser. By chance, this wavelength matches the resonance of gold nanospheres. The power can be huge, typically around 1 W in the infrared and around 10 W at 532 nm. You can switch between a continuous mode (cw) and a pulsed mode (depending on the mode, from 100 fs to a few ps). This is the kind of laser I recommend, although it is expensive.

References

[1] Alabastri, A., Tuccio, S., Giugni, A., Toma, A., Liberale, C., Das, G., De Angelis, F., Di Fabrizio, E., and Zaccaria, P. 2013. Molding of Plasmonic Resonances in Metallic Nanostructures: Dependence of the Non-Linear Electric Permittivity on System Size and Temperature. *Materials*, **6**, 4879–4910.

[2] Alabastri, A., Toma, A., Malerba, M., De Angelis, F., and Proietti Zaccaria, R. 2015. High Temperature Nanoplasmonics: The Key Role of Nonlinear Effects. *ACS Photonics*, **2**, 115–120.

[3] Averitt, R. D., Westcott, S. L., and Halas, N. J. 1999. Linear Optical Properties of Gold Nanoshells. *J. Opt. Soc. Am. B*, **16**(10), 1824–1832.

[4] Baffou, G., and Quidant, R. 2013. Thermo-Plasmonics: Using Metallic Nanostructures as Nano-Sources of Heat. *Laser & Photon. Rev.*, **7**(2), 171–187.

[5] Baffou, G., and Quidant, R. 2014. Nanoplasmonics for Chemistry. *Chem. Soc. Rev.*, **43**, 3898–3907.

[6] Basit, L., Wang, C., Jenkins, C. A., Balke, B., Ksenofontov, V., Fecher, G. H., Felser, C., Mugnaioli, E., Kolb, U., Nepijko, S. A., Schönhense, G., and Klimenkov, M. 2009. Heusler Compounds as Ternary Intermetallic Nanoparticles: Co_2FeGa. *Journal of Physics D: Applied Physics*, **42**(8), 084018.

[7] Bell, A. P., Fairfield, J. A., McCarthy, E. K., Mills, S., Boland, J. J., Baffou, G., and McCloskey, D. 2015. Quantitative Study of the Photothermal Properties of Metallic Nanowire Networks. *ACS Nano*, **9**(5), 5551–5558.

[8] Blaber, M. G., Arnold, M. D., Harris, N., Ford, M. J., and Cortie, M. B. 2007. Plasmon Absorption in Nanospheres: A Comparison of Sodium, Potassium, Aluminium, Silver and Gold. *Physica B*, **394**, 184–187.

[9] Blaber, M. G., Arnold, M. D., and Ford, M. J. 2009. Optical Properties of Intermetallic Compounds from First Principles Calculations: A Search for the Ideal Plasmonic Material. *J. Phys.: Condens. Matter*, **21**, 144211.

[10] Blaber, M. G., Arnold, M. D., and Ford, M. J. 2010. A Review of the Optical Properties of Alloys and Intermetallics for Plasmonics. *J. Phys.: Condens. Matter*, **22**, 143201.

[11] Cable, R. E., and Schaak, R. E. 2007. Solution Synthesis of Nanocrystalling M–Zn (M=Pd, Au, Cu) Intermetallic Compounds via Chemical Conversion of Metal Nanoparticle Precursors. *Chem. Mater.*, **19**, 4098–4104.

[12] Chan, G. H., Zhao, J., Schatz, G. C., and Van Duyne, P. 2008. Localized Surface Plasmon Resonance Spectroscopy of Triangular Aluminum Nanoparticles. *J. Phys. Chem. C*, **112**, 13958–13963.

[13] Chen, C. F., Park, C. H., Boudouris, B. W., Horng, J., Geng, B., Girit, C., Zettl, A., Crommie, M. F., Segalman, R. A., Louie, S. G., and Wang, F. 2011a. Controlling Inelastic Light Scattering Quantum Pathways in Graphene. *Nature*, **471**, 617–620.

[14] Chen, J., Albella, P., Pirzadeh, Z., Alonso-González, P., Huth, F., Bonetti, S., Bonanni, V., Å kerman, J., Nogués, J., Vavassori, P., Dmitriev, A., Aizpurua, J., and Hillenbrand, R. 2011b. Plasmonic Nickel Nanoantennas. *Small*, **7**(16), 2341–2347.

[15] Comin, A., and Manna, L. 2014. New Materials for Tunable Plasmonic Colloidal Nanocrystals. *Chem. Soc. Rev.*, **43**, 3957–3975.

[16] Cortie, M. B., Maaroof, A., Smith, G. B., and Ngoepe, P. 2006. Nanoscale Coatings of $AuAl_x$ and $PtAl_x$ and their Mesoporous Elemental Derivatives. *Curr. Appl. Phys.*, **6**(3), 440–443.

[17] Dalacu, D., and Martinu, L. 2000. Temperature Dependence of the Surface Plasmon Resonance of Au/SiO_2 Nanocomposite Films. *Appl. Phys. Lett.*, **77**(26), 4283–4285.

[18] Ding, X., Hao Liow, C., Zhang, M., Huang, R., Li, C., Shen, H., Liu, M., Zou, Y., Gao, N., Zhang, Z., Li, Y., Wang, Q., Li, S., and Jiang, J. 2014. Surface Plasmon Resonance

Enhanced Light Absorption and Photothermal Therapy in the Second Near-Infrared Window. *J. Am. Chem. Soc.*, **136**(44), 15684.

[19] Faucheaux, J. A., Stanton, A. L. D., and Jain, P. K. 2014. Plasmon Resonances of Semiconductor Nanocrystals: Physical Principles and New Opportunities. *J. Phys. Chem. Lett.*, **5**(6), 976–985.

[20] Ferrando, R., Jellinek, J., and Johnston, R. L. 2008. Nanoalloys: From Theory to Applications of Alloy Clusters and Nanoparticles. *Chem. Rev.*, **108**, 845–910.

[21] García de Abajo, F. J. 2014. Graphene Plasmonics: Challenges and Opportunities. *ACS Photonics*, **1**, 135–152.

[22] Gobin, A. M., Lee, M. H., Halas, N. J., James, W. D., Drezek, R. A., and West, J. L. 2007. Near-Infrared Resonant Nanoshells for Combined Optical Imaging and Photothermal Cancer Therapy. *Nano Lett.*, **7**(7), 1929–1934.

[23] Govorov, A. O., and Richardson, H. H. 2007. Generating Heat with Metal Nanoparticles. *Nano Today*, **2**(1), 30.

[24] Guler, U., and Turan, R. 2010. Effect of Particle Properties and Light Polarization on the Plasmonic Resonances in Metallic Nanoparticles. *Opt. Express*, **18**(16), 17322–17338.

[25] Guler, U., Naik, G. V., Boltasseva, A., Shalaev, V. M., and Kildishev, A. V. 2012. Performance Analysis of Nitride Alternative Plasmonic Materials for Localized Surface Plasmon Applications. *Appl. Phys. B*, **107**, 285–291.

[26] Guler, U., Ndukaife, C., Naik, G. V., Nnanna, A. G. A., Kildishev, A. V., Shalaev, V. M., and Boltasseva, A. 2013. Local Heating with Lithographically Fabricated Plasmonic Titanium Nitride Nanoparticles. *Nano Lett.*, **13**, 6078–6083.

[27] Guler, U., Boltasseva, A., and Shalaev, V. M. 2014. Refractory Plasmonics. *Science*, **344**, 263–264.

[28] Guler, U., Shalaev, V. M., and Boltasseva, A. 2015a. Nanoparticle Plasmonics: Going Practical with Transition Metal Nitrides. *Mater. Today*, **18**(4), 227–237.

[29] Guler, U., Kildishev, A. V., Boltasseva, A., and Shalaev, V. M. 2015b. Plasmonics on the Slope of Enlightenment: The Role of Transition Metal Nitrides. *Faraday Discuss.*, **178**, 71.

[30] Huang, X., Tang, S., Mu, X., Dai, Y., Chen, G., Zhou, Z., Ruan, F., Yang, Z., and Zheng, N. 2011. Freestanding Palladium Nanosheets with Plasmonic and Catalytic Properties. *Nature Nanotech.*, **6**, 28–32.

[31] Jackson, J. D. 1999. *Classical Electrodynamics*. Wiley.

[32] Jain, P. K., Lee, K. S., El-Sayed, I. H., and El-Sayed, M. A. 2006. Calculated Absorption and Scattering Properties of Gold Nanoparticles of Different Size, Shape, and Composition: Applications in Biological Imaging and Biomedicine. *J. Phys. Chem. B*, **110**(14), 7238–7248.

[33] Jiang, Y., Pillai, S., and Green, M. A. 2015. Re-Evaluation of Literature Values of Silver Optical Constants. *Opt. Express*, **23**(3), 2133–2144.

[34] Johnson, P. B., and Christy, R. W. 1972. Optical Constants of the Noble Metals. *Phys. Rev. B*, **6**, 4370.

[35] Johnson, P. B., and Christy, R. W. 1974. Optical Constants of Transition Metals: Ti, V, Cr, Mn, Fe, Co, Ni and Pd. *Phys. Rev. B*, **9**(12), 5056–5070.

[36] Knight, M. K., Liu, L., Wang, Y., Brown, L., Mukherjee, S., King, N. S., Everitt, H. O, Nordlander, P., and Halas, N. J. 2012. Aluminum Plasmonic Nanoantennas. *Nano Lett.*, **12**, 6000–6004.

[37] Knight, M. W., King, N. S., Liu, L., Everitt, H. O., Nordlander, P., and Halas, N. J. 2014. Aluminum for Plasmonics. *ACS Nano*, **8**(1), 834–840.

[38] Knight, M. W., Coenen, T., Yang, Y., Brenny, B. J. M., Losurdo, M., Brown, A. S., Everitt, H. O., and Polman, A. 2015. Gallium Plasmonics: Deep Subwavelength Spectroscopic Imaging of Single and Interacting Gallium Nanoparticles. *ACS Nano*, **9**(2), 2049–2060.

[39] Lalisse, A., Tessier, G., Plain, J., and Baffou, G. 2015. Quantifying the Efficiency of Plasmonic Materials for Near-Field Enhancement and Photothermal Conversion. *J. Phys. Chem. C*, **119**, 25518–25528.

[40] Lalisse, A., Tessier, G., Plain, J., and Baffou, G. 2016. Plasmonic Efficiencies of Nanoparticles Made of Metal Nitrides (TiN, ZrN) Compared with Gold. *Sci. Rep.*, **6**, 38647.

[41] Lecarme, O., Sun, Q., Ueno, K., and Misawa, H. 2014. Robust and Versatile Light Absorption at Near-Infrared Wavelengths by Plasmonic Aluminum Nanorods. *ACS Photonics*, **1**, 538–546.

[42] Lindquist, N. C., Nagpal, P., McPeak, K. M., Norris, D. J., and Oh, S. H. 2012. Engineering Metallic Nanostructures for Plasmonics and Nanophotonics. *Rep. Prog. Phys.*, **75**, 036501.

[43] Link, S., Wang, Z. L., and El-Sayed, M. A. 1999. Alloy Formation of Gold-Silver Nanoparticles and the Dependence of the Plasmon Absorption on their Composition. *J. Phys. Chem. B*, **103**, 3529–3533.

[44] Liu, X., and Swihart, M. 2014. Heavily-Doped Colloidal Semiconductor and Metal Oxide Nanocrystals: An Emerging New Class of Plasmonic Nanomaterials. *Chem. Soc. Rev.*, **43**, 3908–3920.

[45] Long, R., Rao, Z., Mao, K., Li, Y., Zhang, C., Liu, Q., Wang, C., Li, Z. Y., Wu, X., and Xiong, Y. 2015. Efficient Coupling of Solar Energy to Catalytic Hydrogenation by Using Well-Designed Palladium Nanostructures. *Angew. Chem. Int. Ed.*, **127**, 2455–2460.

[46] Mallin, M. P., and Murphy, C. J. 2002. Solution-Phase Synthesis of Sub-10 nm Au–Ag Alloy Nanoparticles. *Nano Lett.*, **2**(11), 1235–1237.

[47] Martin, J., and Plain, J. 2015. Fabrication of Aluminium Nanostructures for plasmonics. *J. Phys. D: Appl. Phys.*, **48**, 184002.

[48] Martin, J., Kociak, M., Mahfoud, Z., Proust, J., D., Gérard, and Plain, J. 2014. High-Resolution Imaging and Spectroscopy of Multipolar Plasmonic Resonances in Aluminum Nanoantennas. *Nano Lett.*, **14**(10), 5517–5523.

[49] Mayer, S. A. 2007. *Plasmonics*. Springer.

[50] McClain, M. J., Schlather, A. E., Ringe, E., King, N. S., Liu, L., Manjavacas, A., Knight, M. W., Kumar, I., Whitmire, K. H., Everitt, H. O., Nordlander, P., and Halas, N. J. 2015. Aluminum Nanocrystals. *Nano Lett.*, **15**(4), 2751–2755.

[51] McMahon, M. D., Lopez, R., Meyer, H. M., Feldman, L. C., and Haglund, R. F. 2005. Rapid Tarnishing of Silver Nanoparticles in Ambient Laboratory Air. *Appl. Phys. B*, **80**, 915–921.

[52] Metwally, K., Mensah, S., and Baffou, G. 2015. Fluence Threshold for Photothermal Bubble Generation Using Plasmonic Nanoparticles. *J. Phys. Chem. C*, **119**, 28586–28596.

[53] Myroshnychenko, V., Rodríguez-Fernández, J., Pastoriza-Santos, I., Funston, A. M., Novo, C., Mulvaney, P., Liz-Marzán, L. M., and García de Abajo, F. J. 2008. Modeling the Optical Response of Gold Nanoparticles. *Chem. Soc. Rev.*, **37**, 1792–1805.

[54] Naik, G. V., Kim, J., and Boltasseva, A. 2011. Oxides and Nitrides as Alternative Plasmonic Materials in the Optical Range. *Opt. Mater. Express*, **1**(6), 1090–1099.

[55] Novotny, L., and Hecht, B. 2006. *Principles of Nano-Optics*. Cambridge University Press.

[56] Palik, E. D. (ed). 1998. *Handbook of Optical Constants of Solids*. Academic Press, Elsevier.

[57] Patsalas, P., Kalfagiannis, N., and Kassavetis, S. 2015. Optical Properties and Plasmonic Performances of Titanium Nitride. *Materials*, **8**, 3128–3154.

[58] Rakic, A. D., Djurisic, A. B., Elazar, J. M., and Majewski, M. 1998. Optical Properties of Metallic Films for Vertical-Cavity Optoelectronic Devices. *Appl. Opt.*, **37**(22), 5271–5283.

[59] Reddy, H., Guler, U., Kildishev, A. V., Boltasseva, A., and Shalaev, V. M. 2016. Temperature-Dependent Optical Properties of Gold Thin Films. *Opt. Mater. Express*, **6**(9), 2776.

[60] Ross, M. B., and Schatz, G. C. 2014. Aluminum and Indium Plasmonic Nanoantennas in the Ultraviolet. *J. Phys. Chem. C*, **118**, 12506–12514.

[61] Sanz, J. M., Ortiz, D., Alcaraz de la Osa, R., Saiz, J. M., González, F., Brown, A. S., Losurdo, M., Everitt, H. O., and Moreno, F. 2013. UV Plasmonic Behavior of Various Metal Nanoparticles in the Near- and Far-Field Regimes: Geometry and Substrate Effects. *J. Phys. Chem. C*, **117**, 19606–19615.

[62] Shalaev, V. M. 2000. Unknown title. *Proceedings of the International School on Quantum Electronics*, 239–243.

[63] Sobhani, A., Manjavacas, A., Cao, Y., McClain, M. J., García de Abajo, F. J., Nordlander, P., and Halas, N. J. 2015. Pronounced Linewidth Narrowing of an Aluminum Nanoparticle Plasmon Resonance by Interaction with an Aluminum Metallic Film. *Nano Lett.*, **15**(10), 6946–6951.

[64] Sterl, F., Strohfeldt, N., Walter, R., Griessen, R., Tittl, A., and Giessen, H. 2015. Magnesium as Novel Material for Active Plasmonics in the Visible Wavelength Range. *Nano Lett.*, **15**, 7949–7955.

[65] Stockman, M. I. 2011. Nanoplasmonics: Past, Present, and Glimpse into Future. *Opt. Express*, **19**(22), 22029–22106.

[66] Strohfeldt, N., Tittl, A., Schäferling, M., Neubrech, F., Kreibig, U., Griessen, R., and Giessen, H. 2014. Yttrium Hydride Nanoantennas for Active Plasmonics. *Nano Lett.*, **14**(3), 1140–1147.

[67] Sun, Q. C., Ding, Y., Goodman, S. M., Funke, H. H., and Nagpal, P. 2014. Copper Plasmonics and Catalysis: Role of Electron–Phonon Interactions in Dephasing Localized Surface Plasmons. *Nanoscale*, **6**, 12450.

[68] Wang, F., and Shen, Y. R. 2006. General Properties of Local Plasmons in Metal Nanostructures. *Phys. Rev. Lett.*, **97**, 206806.

[69] Watson, A. M., Zhang, X., Alcaraz de la Osa, R., Sanz, J. M., González, F., Moreno, F., Finkelstein, G., Liu, J., and Everitt, H. O. 2015. Rhodium Nanoparticles for Ultraviolet Plasmonics. *Nano Lett.*, **15**, 1095–1100.

[70] West, P. R., Ishii, S., Naik, G. V., Emani, N. K., Shalaev, V. M., and Boltasseva, A. 2010. Searching for Better Plasmonic Materials. *Laser & Photon. Rev.*, **6**, 795–808.

[71] Wu, P. C., Khoury, C. G., Kim, T. H., Yang, Y., Losurdo, M., Bianco, G. V., Vo-Dinh, T., Brown, A. S., and Everitt, H. O. 2009. Demonstration of Surface-Enhanced Raman Scattering by Tunable, Plasmonic Gallium Nanoparticles. *J. Am. Chem. Soc.*, **131**, 12032–12033.

[72] Yang, H. U., D'Archangel, J., Sundheimer, M. L., Tucker, E., Boreman, G. D., and Raschke, M. B. 2015. Optical Dielectric Function of Silver. *Phys. Rev. B*, **91**, 235137.

[73] Yang, Y., Callahan, J. M., Kim, T. H., Brown, A. S., and Everitt, H. O. 2013. Ultraviolet Nanoplasmonics: A Demonstration of Surface-Enhanced Raman Spectroscopy, Fluorescence, and Photodegradation Using Gallium Nanoparticles. *Nano Lett.*, **13**, 2837–2841.

[74] Yarema, M., Wörle, M., Rossell, M. D., Erni, R., Caputo, R., Protesescu, L., Kravchyk, K. V., Dirin, D. N., Lienau, K., von Rohr, F., Schilling, A., Nachtegaal, M., and Kovalenko, M. V. 2014. Monodisperse Colloidal Gallium Nanoparticles: Synthesis, Low Temperature Crystallization, Surface Plasmon Resonance and Li-Ion Storage. *J. Am. Chem. Soc.*, **126**, 12422–12430.

[75] Yeshchenko, O. A., Bondarchuk, I. S., Gurin, V. S., Dmitruk, I. M., and Kotko, A. V. 2013. Temperature Dependence of the Surface Plasmon Resonance in Gold Nanoparticles. *Surf. Sci.*, **608**, 275–281.

2

Thermodynamics of Metal Nanoparticles

This chapter is intended to settle the theoretical background of thermoplasmonics and to derive the main physical underlying rules that govern a temperature field generated by nanoparticles. Although a lot of equations will be introduced, the aim of this part is not really to give the tools to compute a temperature field; that is the aim of the next chapter. This chapter is rather intended to provide the reader with the proper physical intuitions. This is why most of this chapter will be based on considerations of ideal cases, such as point-like or spherical heat sources.

After a first section introducing the general theoretical framework and the governing differential equations of thermodynamics, the remaining sections are devoted to covering all the usual situations encountered in thermoplasmonics, namely a continuous wave illumination and the associated steady state, a pulsed illumination, a time-harmonic illumination at the angular frequency ω, a single nanoparticle, multiple nanoparticles, nanoparticles in a uniform medium and nanoparticles in a more complex medium composed of an interface or of multiple layers.

2.1 Mechanisms and Governing Equations

In general, when heat is locally generated, one way or another, the corresponding energy can escape in the environment via three processes: (i) diffusion, (ii) convection and (iii) radiation. This also applies to nanoparticle heating in the context of plasmonics. This section introduces the physics of heat generation in plasmonic nanoparticles, and describes the three possible subsequent processes of diffusion, convection and radiation, although the main kind of energy transfer in plasmonics remains diffusion.

2.1.1 Heat Generation

Upon illumination, the free charges of a metal nanoparticle oscillate at the frequency of the electric field of the incident light. This electronic oscillation, which is nothing but an electronic current in a metal, generates energy dissipation via the Joule effect. Hence, the heat power density in the metal nanoparticle reads at any time t and any location $\mathbf{r}$

$$q(\mathbf{r}, t) = \mathbf{J}(\mathbf{r}, t) \cdot \mathbf{E}(\mathbf{r}, t) \tag{2.1}$$

where $\mathbf{J}$ is the electronic current density (charge per unit time and area) and $\mathbf{E}$ the electric field inside the nanoparticle. Let us name ω the angular frequency of illumination. All the physical quantities of the problem oscillate at this frequency. From now on, we shall use the complex-number notation defined for any physical quantity A as:

$$A(\mathbf{r}, t) = \mathrm{Re}(\underline{A}(\mathbf{r})\, e^{-i\omega t}) \tag{2.2}$$

$$A(\mathbf{r}, t) = \frac{1}{2}(\underline{A}(\mathbf{r})\, e^{-i\omega t} + \underline{A}^{\star}(\mathbf{r})\, e^{i\omega t}). \tag{2.3}$$

Using this formalism, (2.1) reads

$$q = \frac{1}{4}(\underline{\mathbf{J}}\, e^{-i\omega t} + \underline{\mathbf{J}}^{\star}\, e^{i\omega t}) \cdot (\underline{\mathbf{E}}\, e^{-i\omega t} + \underline{\mathbf{E}}^{\star}\, e^{i\omega t}) \tag{2.4}$$

$$q = \frac{1}{4}(\underline{\mathbf{J}} \cdot \underline{\mathbf{E}}\, e^{-2i\omega t} + \underline{\mathbf{J}}^{\star} \cdot \underline{\mathbf{E}}^{\star}\, e^{2i\omega t} + \underline{\mathbf{J}}^{\star} \cdot \underline{\mathbf{E}} + \underline{\mathbf{J}} \cdot \underline{\mathbf{E}}^{\star}). \tag{2.5}$$

The parameter of practical interest in not the time-dependent quantity $q(\mathbf{r}, t)$, but rather its time average $\bar{q}(\mathbf{r}) = \langle q(\mathbf{r}, t)\rangle_t$. Indeed, the oscillations of q at the angular frequency ω are always much faster than any subsequent effect observed in thermal plasmonics. From (2.5), one gets

$$\bar{q} = \frac{1}{4}(\underline{\mathbf{J}}^{\star} \cdot \underline{\mathbf{E}} + \underline{\mathbf{J}} \cdot \underline{\mathbf{E}}^{\star}) = \frac{1}{2}\mathrm{Re}(\underline{\mathbf{J}}^{\star} \cdot \underline{\mathbf{E}}). \tag{2.6}$$

From now on, we will use the notation q to signify $\bar{q}$ for the sake of simplicity. The heat source density can be expressed as a function of the polarization vector $\mathbf{P}$ using the relation $\mathbf{J} = \partial_t \mathbf{P}$, which reads in complex-number notations $\underline{\mathbf{J}} = -i\omega\underline{\mathbf{P}}$:

$$q = \frac{1}{2}\mathrm{Re}(i\omega\underline{\mathbf{P}}^{\star} \cdot \underline{\mathbf{E}}) \tag{2.7}$$

$$q = -\frac{\omega}{2}\mathrm{Im}(\underline{\mathbf{P}}^{\star} \cdot \underline{\mathbf{E}}). \tag{2.8}$$

Finally, by noticing that $\underline{\mathbf{P}} = \varepsilon_0(\underline{\varepsilon} - 1)\underline{\mathbf{E}}$, one obtains

$$\boxed{q(\mathbf{r}) = \frac{\omega}{2}\varepsilon_0 \mathrm{Im}(\underline{\varepsilon})|\underline{\mathbf{E}}(\mathbf{r})|^2.} \tag{2.9}$$

The heat power density within a nanoparticle is thus proportional to the square of the amplitude of the electric field. This feature was already mentioned in Chapter 1, Equation (1.44).

Note that we chose the sign convention $e^{-i\omega t}$ while using the complex-number notation (see Equation (2.2)), as it is preferred in optics. The opposite convention $e^{+i\omega t}$ is used in quantum physics and in engineering. It is important to notice that, in optics, the choice of one convention or another dictates the sign of the imaginary part of the permittivity. For the regular convention $e^{-i\omega t}$, the sign of the imaginary part of ε is positive, which yields, as expected, a positive value of the heat power density in (2.9). With the other convection, Equation (2.9) would feature a negative sign because $\mathbf{J}_\omega$ would equal $+i\omega\mathbf{P}_\omega$.

Figure 2.1 plots distributions of normalized electric field intensity inside the nanoparticles, a quantity directly related to the heat source densities $q(\mathbf{r})$ via Equation (2.9). Electric field intensities outside the nanoparticles are also plotted as a comparison. Various morphologies

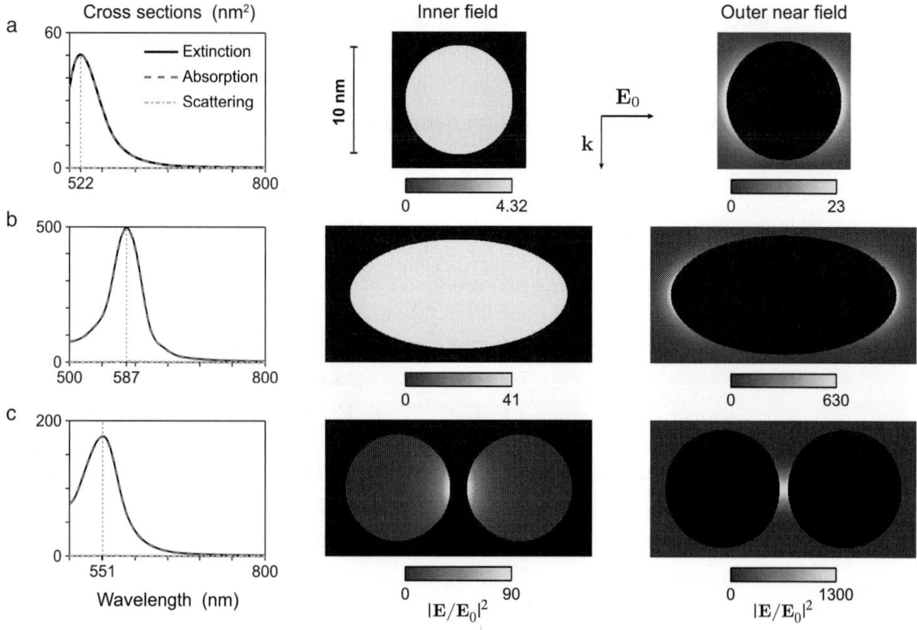

Fig. 2.1 Extinction, scattering and absorption spectra related to different gold nanoparticles morphologies and the associated inner and outer maps of the normalized electric field intensities. All the nanoparticles are in water (surrounding refractive index $n_s = 1.33$). Maps have been plotted for the plasmon resonance wavelength in extinction, represented in the spectra. (a) Sphere, 10 nm in diameter. (b) Ellipsoid, 10 nm wide, 20 nm long. (c) sphere dimer composed of two 10 nm spheres separated by a gap of 1 nm.

of nanoparticles are considered. The nanoparticles are here particularly small (see next paragraph for the case of larger nanoparticles), typically smaller than 50 nm and, in this case, the electric field intensity within the nanoparticle is fully uniform. This rule no longer holds if gaps are present (Figure 2.1c), even for small nanoparticles. Plasmonic gaps make the electric field highly confined at the vicinity of the gaps both inside and outside the nanoparticle.

Figure 2.2 plots distributions of normalized electric field intensity for 10 times larger nanoparticles. In this case, retardation effects occurs and the electric field intensity is no longer uniform within the nanoparticles. It is rather located at the boundaries of the nanoparticles. The distributions of electric field intensities can be also radically different inside and outside the nanoparticles [4]. This is blatant in the case of the ellipsoid (Figure 2.1b). While the inner electric field is oriented North-South, the outer electric field is enhanced East–West, along the polarization of the incident light. This effects comes from the fact that heat is generated where the electric current can freely flow along the direction of light polarization. The spectra associated to the cases of the ellipsoid and the dimer show that the absorption resonance can even disappear at the plasmonic resonance. Large nanoparticles are often inefficient photothermal convertors.

Finally, the total heat power Q delivered by a nanoparticle is nothing but the integral of q over the spatial domain $\mathcal{V}$ delimited by the volume of the nanoparticle:

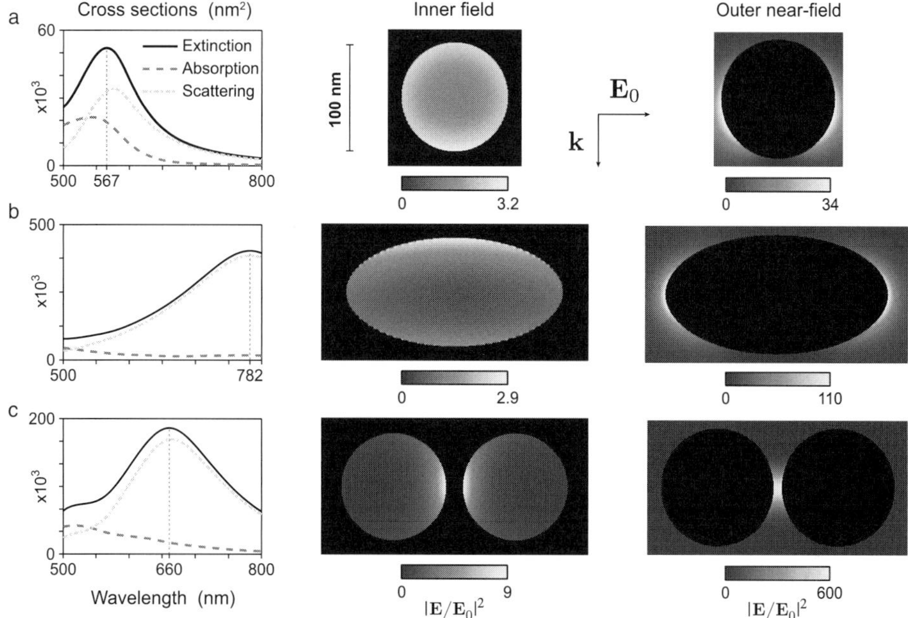

Fig. 2.2 Extinction, scattering and absorption spectra related to different gold nanoparticles morphologies and the associated inner and outer maps of the normalized electric field intensities. All the nanoparticles are in water (surrounding refractive index $n_s = 1.33$). Maps have been plotted for the plasmon resonance wavelength in extinction, represented in the spectra. (a) Sphere, 100 nm in diameter. (b) Ellipsoid, 100 nm wide, 200 nm long. (c) sphere dimer composed of two 100 nm spheres separated by a gap of 1 nm.

$$Q = \frac{\omega}{2}\varepsilon_0 \mathrm{Im}(\varepsilon_\omega) \int_{\mathcal{V}} |\mathbf{E}_\omega|^2 d\mathbf{r}. \tag{2.10}$$

The quantity is most of the time *the* important parameter to determine in order to compute the temperature field.

2.1.2 Diffusive Heat Transfer

The heat source density introduced in the previous section is responsible for a temperature increase of the nanoparticle and of the surrounding medium via heat diffusion. This subsection derives the constitutive equations that govern the temperature field.

Heat Diffusion Equation

Let us consider a medium that can be described by a thermal conductivity κ, a mass density ρ and a specific[1] heat capacity at constant pressure c_p. These parameters may be space-dependent throughout the system. They are, however, considered as independent

---

[1] The adjective "specific" means "per unit mass."

of the temperature in this section (for the temperature dependence of thermodynamical constants of water, see Appendix B).

In such a medium, a local temperature variation $\delta T(\mathbf{r}, t)$ results in a local variation of the thermal energy density $\delta u_{\text{th}}(\mathbf{r}, t)$ (energy per unit volume). These two quantities are proportional.

$$\delta u_{\text{th}} = \rho c_{\text{p}} \delta T. \tag{2.11}$$

The proportionality factor ρc_{p} (an energy per unit volume and kelvin) quantifies the ability of a medium to *store* thermal energy.

Moreover, Fourier's law states that the heat flux density vector $\mathbf{J}_{\text{th}}(\mathbf{r}, t)$ (power per unit area) is proportional to the temperature gradient:

$$\mathbf{J}_{\text{th}} = -\kappa \nabla T. \tag{2.12}$$

This expression defines the thermal conductivity κ of a medium. κ quantifies the ability of a medium to *conduct* thermal energy.

The energy density and the energy flux introduced above are linked by a statement of energy conservation,

$$\partial_t u_{\text{th}} + \nabla \cdot \mathbf{J}_{\text{th}} = q \tag{2.13}$$

where $q(\mathbf{r}, t)$ is a source term that represents the heat power density (power per unit volume). Using (2.11) and (2.12) leads to the famous heat diffusion equation:

$$\rho c_{\text{p}} \, \partial_t T - \nabla \cdot \kappa \nabla T = q. \tag{2.14}$$

In the case of a uniform medium, κ is constant and the Laplacian operator ∇^2 can be used in (2.14):

$$\boxed{\rho c_{\text{p}} \, \partial_t T - \kappa \nabla^2 T = q.} \tag{2.15}$$

In plasmonics, the heat power density $q(\mathbf{r}, t)$ is supposed to be nonzero only within the metal nanoparticles and is given by the expression (2.9) derived in the previous section. Outside the metal particles, in the surrounding medium, one has therefore to solve the homogeneous heat diffusion equation:

$$\boxed{\partial_t T - D \nabla^2 T = 0.} \tag{2.16}$$

In this equation, the thermal diffusivity $D = \kappa / \rho c_{\text{p}}$ is used.

Validity of the Heat Diffusion Equation

The question of the validity of the heat diffusion equation for nanoscale systems is a natural concern. The problem with small systems is to know to what extent physical local variables such as temperature may be considered functions of position, *i.e.*, to what extent a local equilibrium can be assumed anywhere in the system, despite the discrete nature of matter at the molecular scale. The problem is not really related to the definition of local variables such as the temperature. A temperature value can always be defined, no matter the system and its size. The problem rather stems from the possible large fluctuations of local variables, which would make them unusable in the constitutive equations of thermodynamics. If one considers a small volume element δV composed of N particles in average, then the rms value of the fluctuation is $\sigma_N = N^{1/2}$. As an example, let us consider a volume as small as $(10\ \text{nm})^3$. For a gas phase in ambient conditions, the relative fluctuations $\sigma_N/N = 1/N^{1/2}$ are on the order of 40%, but for a dense medium (liquid of solid), for the same volume this value is on the order of 0.5%. Consequently, when considering the medium surrounding nanoparticles, if this medium is dense, the heat diffusion equation is not compromised by fluctuation in principle. Note that the heat diffusion equation can be invalidated, even with a dense surrounding medium, when it results from a ballistic regime. This occurs when the energy carriers are phonons, electrons or plasmons. This mode of transportation becomes dominant when the smallest characteristic length of the system is smaller than the mean free path of the energy carriers. Such non-diffusive processes are mainly involved when describing the physics of heat transport inside the plasmonic materials themselves, more rarely when considering heat transport in the surrounding medium.

2.1.3 Convective Heat Transfer

Convection is one of the three mechanisms of heat transfer, besides conduction (previous subsection) and thermal radiation (next subsection). Convection refers to the heat transfer induced by the motion of a fluid. In the context of plasmonics, nanoparticles are often in contact with fluids, such as air or water. The presence of air is not supposed to affect the heat transfer in the surroundings as it mainly occurs via conductive heat transfer via the solid substrate. In liquids, heating gold nanoparticles may produce a thermal-induced fluid convection, as represented below (see the upward fluid velocity lines). However, due to the weak Rayleigh number of water, thermal-induced fluid convection is not supposed to distort the temperature distribution within the liquid, regardless of the temperature increase or the size of the heat source [12]. Consequently, numerical simulations of temperature distribution in fluids can be performed without any consideration on thermal-induced fluid convection.

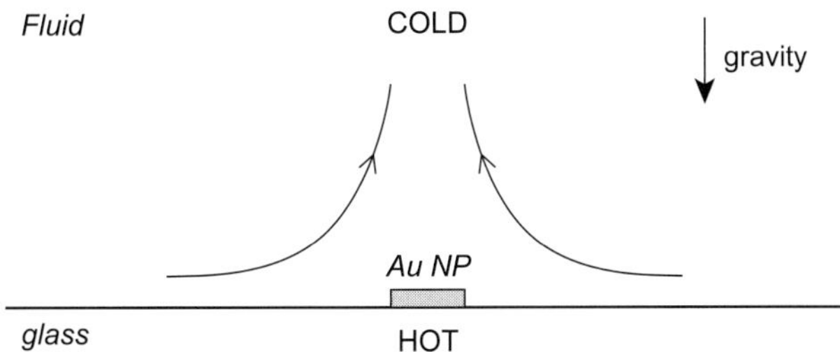

2.1.4 Thermal Radiation

Planck's Law

Any object with a temperature different from absolute zero radiates light. Planck's law gives the spectral radiance of the light radiated by a body in thermal equilibrium at the temperature T. It represents an electromagnetic power emitted per unit area of the surface of the body, per unit solid angle and per unit frequency:

$$\mathcal{L}_\nu^0(\nu, T) = \frac{2h\nu^3}{c^2} \frac{1}{\exp\left(\dfrac{h\nu}{k_B T}\right) - 1} \tag{2.17}$$

so that the power radiated by a surface area dS over a solid angle $d\Omega$ and over a light frequency range $d\nu$ reads $dP = \mathcal{L}_\nu^0 dS d\Omega d\nu$.

The spectral radiance can also be expressed per unit wavelength,[2] $\mathcal{L}_\lambda^0$, so that $dP = \mathcal{L}_\lambda^0 dS d\Omega d\lambda$. But be careful: to obtain $\mathcal{L}_\lambda^0$ it is not sufficient to replace $h\nu$ by hc/λ in Equation (2.17). Otherwise, one just obtains $\mathcal{L}_\nu^0(\lambda, T)$. The relations $\mathcal{L}_\nu^0 d\nu = \mathcal{L}_\lambda^0 d\lambda$ and $d\nu/d\lambda = -c/\lambda^2$ imply:

$$\mathcal{L}_\lambda^0(\lambda, T) = \frac{2hc^2}{\lambda^5} \frac{1}{\exp\left(\dfrac{hc}{\lambda k_B T}\right) - 1}. \tag{2.18}$$

Figure 2.3 plots spectral radiances $\mathcal{L}_\lambda^0(\lambda, T)$ as a function of the wavelength at different temperatures. By calculating the derivative of $\mathcal{L}_\lambda^0(\lambda, T)$ as a function of λ to find the maximum of the function, one obtains the Wien's displacement law giving the central wavelength $\lambda_{\max}$ of the emitted spectrum as a function of the temperature:

$$\lambda_{\max}(T) = \frac{2.898 \times 10^{-3} \ [\text{m} \cdot \text{K}]}{T} \tag{2.19}$$

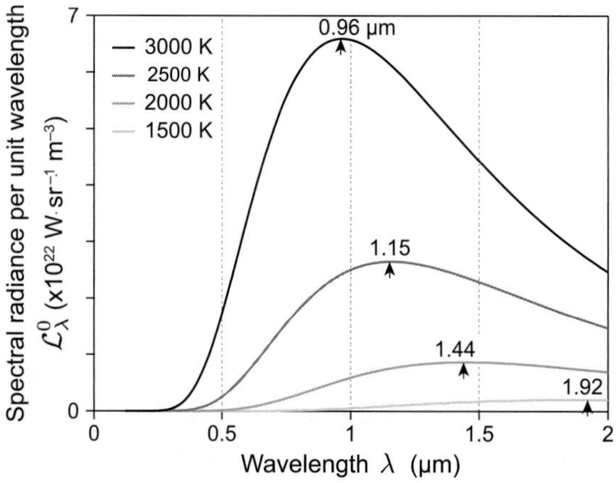

Fig. 2.3   Spectral radiance in wavelength $\mathcal{L}_\lambda^0$ of black body radiation, as a function of the wavelength.

[2] The superscript 0 means "black body." The spectral radiance for an object that is not a black body is different, as explained in the next paragraph.

where T is the absolute temperature. One can also show the dependence of the maximum spectral radiance:

$$\mathcal{L}^0_\lambda(\lambda_{\max}(T), T) \propto T^5. \tag{2.20}$$

By integrating the spectral radiance over λ, one can also show that the total emitted power per unit area reads:

$$I = \sigma T^4 \tag{2.21}$$

where $\sigma = 5.67 \times 10^{-8}$ W·m^{-2}·K^{-4} is the Stefan–Boltzmann constant.

Emissivity

In the previous paragraph, only black bodies were considered, but most objects do not behave as black bodies in the sense that their actual spectral radiance $\mathcal{L}_\lambda$ (resp. $\mathcal{L}_\nu$) is damped compared to the Planck's law (2.17) (resp. (2.18)). The damping factor is called the spectral emissivity ϵ_λ (resp. ϵ_ν):

$$\mathcal{L}_\lambda(\lambda, T) = \epsilon_\lambda(\lambda)\mathcal{L}^0_\lambda(\lambda, T) \tag{2.22}$$

with $\epsilon_\lambda(\lambda) \leq 1$. Interestingly, Kirchoff's law states that the spectral emissivity equals the spectral absorptivity of the surface:

$$\boxed{\epsilon_\lambda = \alpha_\lambda.} \tag{2.23}$$

This means that the more an object absorbs light, the more it radiates light upon heating.

2.2 Steady State

Let us consider one or several metal nanoparticles embedded in a uniform medium. From now on, κ without subscript stands for the thermal conductivity of the metal forming plasmonic nanoparticles and κ_s for the thermal conductivity of the surrounding medium. In the steady state, the heat diffusion equations inside (2.15) and outside (2.16) the nanoparticle(s) derived in the previous section can be recast as

$$-\kappa\nabla^2 T = q \tag{2.24}$$
$$\nabla^2 T = 0. \tag{2.25}$$

A steady state occurs when considering a continuous wave illumination. Equation (2.24) is called the Poisson equation while Equation (2.25) is called the Laplace equation. ∇^2 is named the Laplacian operator. It involves second derivatives of space. Depending on the selected set of space variables, different expressions of ∇^2 can be conveniently used:

Cartesian coordinates:

$$\nabla^2 T = \frac{\partial^2 T}{\partial x^2} + \frac{\partial^2 T}{\partial y^2} + \frac{\partial^2 T}{\partial z^2} \tag{2.26}$$

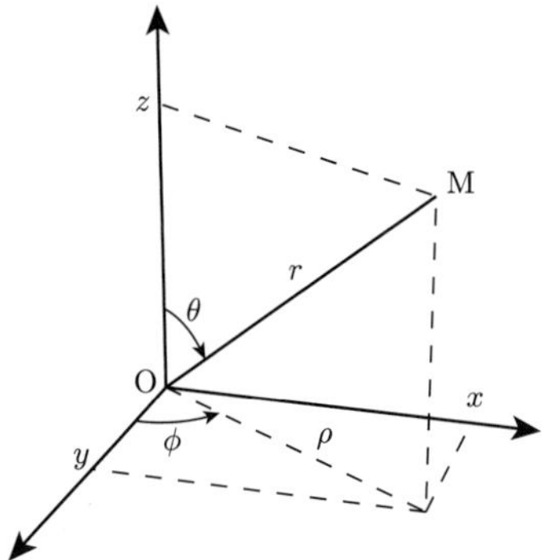

Fig. 2.4   Various coordinates defining the vector $\mathbf{r} = \overrightarrow{OM}$.

Cylindrical coordinates:

$$\nabla^2 T = \frac{1}{\rho}\frac{\partial}{\partial \rho}\left(\rho\frac{\partial T}{\partial \rho}\right) + \frac{1}{\rho^2}\frac{\partial^2 T}{\partial \varphi^2} + \frac{\partial^2 T}{\partial z^2} \tag{2.27}$$

Spherical coordinates:

$$\nabla^2 T = \frac{1}{r^2}\frac{\partial}{\partial r}\left(r^2\frac{\partial T}{\partial r}\right) + \frac{1}{r^2 \sin\theta}\frac{\partial}{\partial \theta}\left(\sin\theta\frac{\partial T}{\partial \theta}\right) + \frac{1}{r^2 \sin^2\theta}\frac{\partial^2 T}{\partial \varphi^2} \tag{2.28}$$

where (x, y, z), (ρ, θ, z) and (r, θ, φ) are the Cartesian, radial and spherical coordinates as defined in Figure 2.4.

2.2.1 Thermal Point Source

Let us first consider a thermal point source

$$q(\mathbf{r}) = Q\,\delta(\mathbf{r}) \tag{2.29}$$

where Q scales as a power. As the problem possesses a central symmetry, all scalar quantities only depend on the radial coordinate r. Anywhere in the system, except at $r = 0$, the temperature profile is governed by the Laplace equation (Equation (2.25)) that reads according to Equation (2.28)

$$\frac{1}{r^2}\frac{\partial}{\partial r}\left(r^2\frac{\partial T}{\partial r}\right) = 0. \tag{2.30}$$

This equation naturally yields a general solution in the form

$$T(r) = \frac{c_1}{r} + c_2, \quad (c_1, c_2) \in \mathbb{R}^2. \tag{2.31}$$

The integration constant c_2 is simply the temperature infinitely far from the source (the ambient temperature), named T_∞. The integration constant c_1 can be determined using the divergence theorem, which states that for a continuously differentiable vector field $\mathbf{A}$

$$\oiint_S \mathbf{A} \cdot \mathbf{n}\,\mathrm{d}^2 r = \iiint_V \nabla \cdot \mathbf{A}\,\mathrm{d}^3 r. \tag{2.32}$$

Stating $\mathbf{A} = \nabla T$ and using Equations (2.24) and (2.29) leads to $c_1 = Q/4\pi\kappa$, hence the general solution of the temperature profile

$$T(r) = \frac{Q}{4\pi\kappa r} + T_\infty. \tag{2.33}$$

Note that the function

$$G(r) = \frac{1}{4\pi\kappa r} \tag{2.34}$$

is nothing but the Green's function of the problem that satisfies the boundary condition $\lim_{r\to\infty} G(r) = 0$. This $1/r$ dependence evidenced here is fundamental in thermo-plasmonics, and more generally in thermodynamics. It implies that any temperature distribution far from its heat sources slowly vanishes as $1/r$. This profile is plotted in Figure 2.5a for $\frac{Q}{4\pi\kappa r} = 1$.

2.2.2 Spherical Metal Particle

The previous subsection was devoted to a point-like source of heat. Let us now slightly increase the degree of complexity by considering the instructive case of a *spherical* nanoparticle of radius a. We have seen in Figure 2.1 that for nanoparticles smaller than the plasmon wavelength (on the order of 50 nm for gold), the electric field can be considered as uniform within the nanoparticle, yielding a uniform heat source density q, since q is proportional to the square amplitude of the electric field; see Equation (2.9). Hereinafter, we shall consider this case, using the notation $q(r) = q_0$ in the metal nanoparticle, for $r < a$. The equations governing $T(r)$ become thus (in spherical coordinates)

$$-\kappa \frac{1}{r^2}\partial_r(r^2\partial_r T) = q_0 \quad \text{for} \quad r < a \tag{2.35}$$

$$\partial_r(r^2\partial_r T) = 0 \quad \text{for} \quad r > a. \tag{2.36}$$

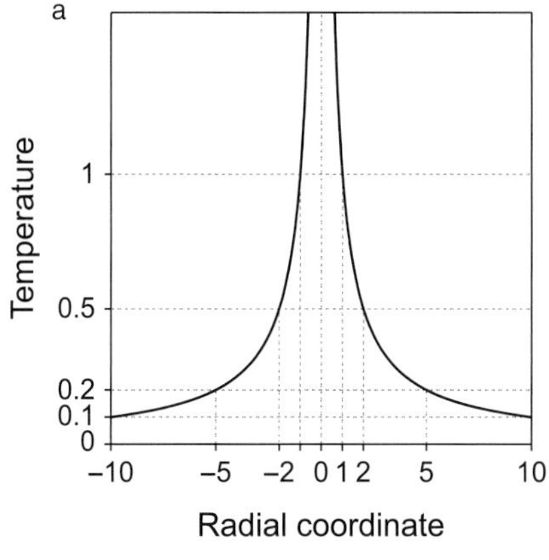
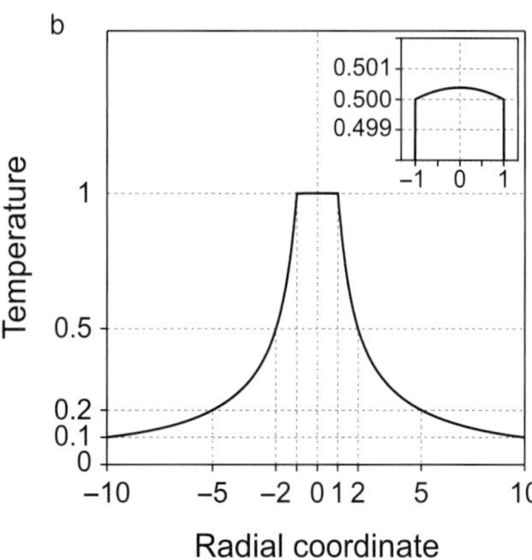

Fig. 2.5 (a) Temperature profile around a point-like source of heat, according to equation (2.33) with $Q/4\pi\kappa_s = 1$. (b) Temperature profile with a spherical source of heat made of gold, according to Equations (2.46) and (2.47) with $a = 1$ and $Q/4\pi\kappa_s = 1$. The inset is a zoom of the top of the line shape.

The general solutions are

$$T(r) = \frac{c_1}{r} + c_2 \quad \text{for} \quad r > a \tag{2.37}$$

$$T(r) = -\frac{q_0 \, r^2}{6\kappa} + \frac{c_3}{r} + c_4 \quad \text{for} \quad r < a \tag{2.38}$$

where c_1, c_2, c_3, c_4 are constants of integration. c_2 stands for the ambient temperature, that is named T_∞. c_3 is necessarily zero since no divergence is expected at $r = 0$ in this problem. Then, c_1 can be determined by a statement of energy conservation. Energy conservation imposes that the total heat power crossing the interface equals the heat power Q delivered by the nanoparticle:

$$4\pi a^2 |\mathbf{j}_{\text{th}}(a)| = Q \tag{2.39}$$

$$4\pi a^2 |\kappa_s \nabla T(a^+)| = Q \tag{2.40}$$

$$4\pi a^2 \kappa_s \frac{c_1}{a^2} = Q \tag{2.41}$$

$$c_1 = \frac{Q}{4\pi \kappa_s} \tag{2.42}$$

Finally, in order to determine the constant c_4, a boundary condition at the metal/surroundings interface has to be stated. If one assumes that no thermal surface resistivity exists on the nanoparticle surface, the temperature has to be continuous across the metal/surrounding interface:

$$T(a^+) = T(a^-) \tag{2.43}$$

$$\frac{Q}{4\pi \kappa_s a} = -\frac{q_0 a^2}{6\kappa} + c_4. \tag{2.44}$$

As $Q = q_0 4\pi a^3 / 3$ (see Equation (2.10)), one gets

$$c_4 = T_\infty + \frac{Q}{4\pi \kappa_s a} \left(1 + \frac{\kappa_s}{2\kappa} \right). \tag{2.45}$$

The temperature profile in the system reads thus:

$$T(r) = \frac{Q}{4\pi \kappa_s r} + T_\infty \quad \text{for} \quad r > a \tag{2.46}$$

$$T(r) = \frac{Q}{4\pi \kappa_s a} \left[1 + \frac{\kappa_s}{2\kappa} \left(1 - \frac{r^2}{a^2} \right) \right] + T_\infty \quad \text{for} \quad r < a. \tag{2.47}$$

The temperature profile outside the nanoparticle scales thus as $1/r$ while it features a parabolic profile on the inside. This profile is plotted in Figure 2.5(b) for $\frac{Q}{4\pi \kappa_s} = 1$ and $a = 1$. Importantly, one can see that the temperature profile within the nanoparticle is quasi-uniform. This comes from the fact that the ratio κ_s/κ is on the order of 1/500. This effect of temperature uniformization is not limited to the case of a small spherical nanoparticle where q_0 is uniform. The fact that the temperature is uniform within a nanoparticle is a general trend in nanoplasmonics that is rather independent of the nanoparticle size or morphology. Therefore, for a spherical plasmonic nanoparticle, one can assume

$$
\begin{aligned}
T(r) &= \frac{Q}{4\pi \kappa_{\mathrm{s}} r} + T_\infty \quad \text{for} \quad r > a \\
T(r) &= \frac{Q}{4\pi \kappa_{\mathrm{s}} a} + T_\infty \quad \text{for} \quad r < a.
\end{aligned}
\tag{2.48}
$$

2.2.3 Spherical Nanoparticle with a Kapitza Resistance

To end up with the temperature profile given by expressions (2.46) and (2.47), the assumption of no surface resistivity was made. However, a surface resistivity may occur depending on the nature of the nanoparticle/surroundings interface, and in particular on the nature of a molecular coating. The surface resistivity can reach appreciable values when the liquid does not wet the solid. For instance, in aqueous medium, a molecular coating of long hydrophobic chains leads to a poor surface thermal conductivity g (or a high resistivity $1/g$). The direct consequence is the occurrence of a temperature discontinuity δT at the interface between the nanoparticle and the surrounding medium. g [W m^{-2} K^{-1}] is defined such that the heat flux density across the metal outer boundary reads

$$
|\mathbf{J}_{\mathrm{th}}| = g\delta T.
\tag{2.49}
$$

In the present case of a spherical nanoparticle, this new relation leads thus to a boundary condition that is different from Equation (2.43):

$$
\Delta T = T(a^-) - T(a^+) = \frac{Q}{4\pi a^2 g}.
\tag{2.50}
$$

This is the expression of the temperature discontinuity that occurs at the nanoparticle interface, which yields another value of the integration constant c_4 and a modification of the temperature profile inside the nanoparticle that simply consists of a uniform offset compared to Equation (2.47):

$$
T(r) = \frac{Q}{4\pi \kappa_{\mathrm{s}} a} \left[1 + \frac{\kappa_{\mathrm{s}}}{2\kappa} \left(1 - \frac{r^2}{a^2} \right) + \frac{\kappa_{\mathrm{s}}}{ga} \right] + T_\infty \quad \text{for} \quad r < a.
\tag{2.51}
$$

The term at the origin of the temperature offset in the nanoparticle is κ_{s}/ga. One usually defines the Kapitza length as

$$
\ell_{\mathrm{th}} = \kappa_{\mathrm{s}}/g.
\tag{2.52}
$$

The occurrence of a significant temperature discontinuity can be predicted by comparing ℓ_{th} and the nanoparticle radius a. If $\ell_{\mathrm{th}} \ll a$, one has $\kappa_{\mathrm{s}}/ga \ll 1$ and no temperature discontinuity is expected at the nanoparticle interface.

If one neglects the term $\kappa_{\mathrm{s}}/2\kappa$ in Equation (2.51) (which is in principle always valid in nanoplasmonics, as explained earlier), one ends up with this simpler expression:

$$
T(r) = T_{\mathrm{NP}} = \frac{Q}{4\pi \kappa_{\mathrm{s}} a} \left[1 + \frac{\ell_{\mathrm{th}}}{a} \right] + T_\infty \quad \text{for} \quad r < a.
\tag{2.53}
$$

Figure 2.6 compares the temperature profiles of a gold sphere without and with surface resistivity. Note that *a finite thermal conductivity only affects the inner temperature*

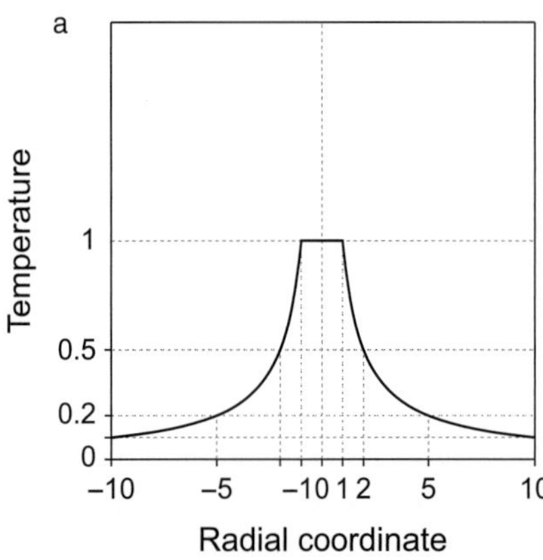
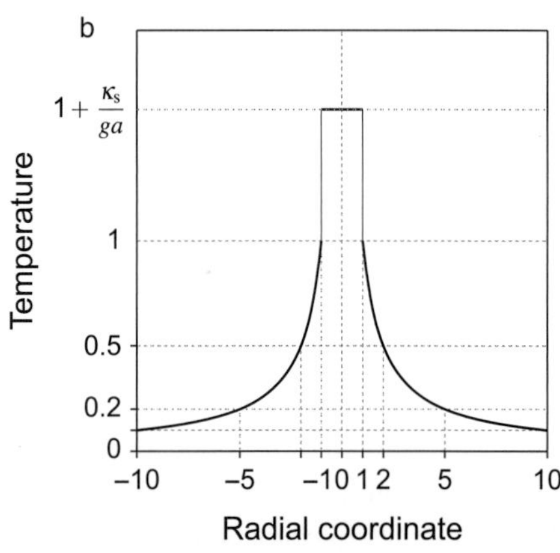

Fig. 2.6   (a) Temperature profile around a spherical source of heat, according to Equation (2.48) and Equations (2.48), with $a = 1$ and $Q/4\pi\kappa_s = 1$. (b) Same system, including a finite value of the surface thermal conductivity g, according to Equation (2.53).

distribution, not the temperature outside the nanoparticle. This is very important since, in most applications, only the temperature outside the nanoparticle matters. In other words, any finite surface conductivity usually has no effect on the heat release and any thermal induced process of interest that occurs in the surrounding medium remains unaffected (at least under cw illumination).

For non-spherical nanoparticles, no general closed-form expression of the temperature profile exists. However, the statement of temperature uniformity within the nanoparticle still applies in general. The temperature inside a nanoparticle under illumination may be nonuniform only for specific cases. In particular, when a nanoparticle with a large aspect ratio is nonuniformly illuminated: typically a metal nanowire or metal flake illuminated at one end [5].

In order to estimate the temperature increase of a nanoparticle of arbitrary morphology, and the temperature profile in its surroundings, numerical simulations have to be carried out. This is described in Chapter 3.

2.2.4 Presence of a Substrate

What has been described so far was related to heat diffusion in a uniform, infinite medium of thermal conductivity (κ_s). Experimentally, this typically corresponds to metal colloids dispersed in a solvent.

Another class of experiments is rather based on the use of metal colloids or lithographic metal structures lying on a planar substrate. In this case, the thermodynamic problem is more complex since the surrounding medium consists of two media featuring different thermal conductivities.

This section explains how the presence of a nonuniform medium can affect heat diffusion in plasmonics, and how such a problem can be simply taken into account without carrying out heavy numerical simulations.

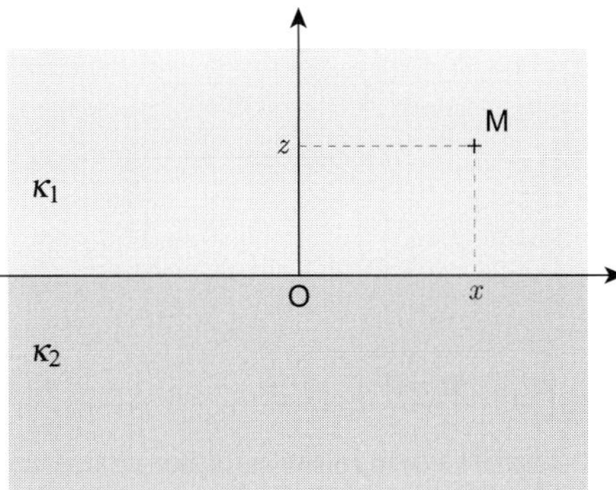

Fig. 2.7 Schematic of the system composed of two infinite media of thermal conductivities κ_1 and κ_2 separated by a planar interface $(0; x, y)$ and where $\mathbf{r} = \overrightarrow{OM} = (x, y, z)$.

In a typical experiment, metal nanoparticles are lying on a planar substrate, usually glass, and surrounded by air or immersed in a liquid medium such as water. We will therefore consider the surrounding medium as two semi-infinite domains separated by a planar interface at $z = 0$. The upper medium $(z > 0)$ features a thermal conductivity κ_1, while the lower medium $(z < 0)$ features a thermal conductivity κ_2. In order to derive the first important physical rules, we shall consider a thermal point source at the location $\mathbf{r}_0$ and move it throughout the system. Figure 2.7 sketches the overall system. The temperature distribution will be the solution of the Laplace equation.

This problem is equivalent to solving the electrostatic problem of a point charge facing a planar interface, since it is also governed by the Laplace equation [22]. In our problem, the temperature T stands for the electric potential, the thermal conductivity stands for the electric permittivity and the heat power stands for the electric charge. The most common approach to solve this electrostatic problem is the famous *image method*. This method consists in replacing the effect of the interface by a fictitious charge image located on the other side of the interface. For a heat source located at $\mathbf{r}_0 = (0, 0, z_0)$ and delivering a heat power Q, the temperature distribution throughout the system reads [6, 22]

$$T(\mathbf{r}) = Q\, G(\mathbf{r}; \mathbf{r}_0) \tag{2.54}$$

where

$$G(\mathbf{r}; \mathbf{r}_0) = \frac{1}{4\pi\,\kappa_1} \left(\frac{1}{R} - \left(\frac{\kappa_2 - \kappa_1}{\kappa_2 + \kappa_1} \right) \frac{1}{R'} \right) \quad \text{for} \quad z \geq 0 \tag{2.55}$$

$$G(\mathbf{r}; \mathbf{r}_0) = \frac{1}{4\pi\,\kappa_1} \frac{1}{R} \left(\frac{2\kappa_1}{\kappa_2 + \kappa_1} \right) \quad \text{for} \quad z \leq 0 \tag{2.56}$$

where

$$R = \sqrt{x^2 + y^2 + (z - z_0)^2}$$

$$R' = \sqrt{x^2 + y^2 + (z + z_0)^2}.$$

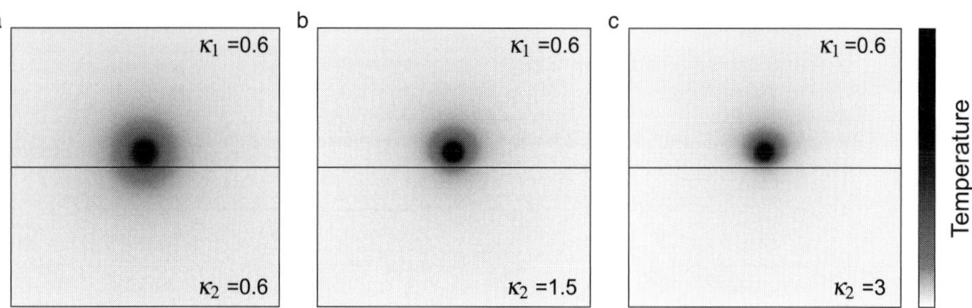

Fig. 2.8   Temperature distributions around a point source near an interface between two media of various thermal conductivities.

G is the Green's function of the problem as it is the solution of the problem for a thermal point source that fulfills the boundary conditions. Maps of temperature distributions according to these expressions are plotted in Figure 2.8. The proximity of a medium of higher thermal conductivity tends to reduce the overall temperature increase, especially in this medium [6, 31].

Interestingly, if one considers the limit $z_0 \rightarrow 0$ in Equations (2.55) and (2.56), *i.e.*, if the heat source is located right at the interface, the temperature profile recovers a point symmetric and reads in any of the two media:

$$T(\mathbf{r}) = \frac{Q}{4\pi\,\bar{\kappa}\,|\mathbf{r}|} \tag{2.57}$$

where $\bar{\kappa}$ simply is the average thermal conductivity

$$\bar{\kappa} = \frac{\kappa_1 + \kappa_2}{2}. \tag{2.58}$$

This important conclusion is at the basis of a current assumption in thermoplasmonics: when conducting numerical simulations involving plasmonic structures lying on a substrate, one can consider the structures as embedded in a uniform medium of average thermal conductivity, which simplifies a lot the numerical simulations. This is an assumption in the sense that the nanoparticles are not exactly at the interface, as they are standing in the upper medium and not within the substrate. The assumption is justified for lithographic nanostructures since the thickness of the particle is usually much smaller than their lateral size.

The case of an extended nanoparticle located away from the interface does not lead to simple closed-form expressions and numerical simulations are required [31] (see Chapter 3).

Finally, let us note that if the surrounding medium is composed of more than two media separated by planar interfaces (a three-layer system for instance), the image method no longer applies. However, analytical solutions exist involving integral formulations. The expression of the Green function of a three-layer medium is given in Appendix C.

2.2.5 Multiple Nanoparticles, Collective Effects, and Thermal Homogenization

In the following, we shall consider nanoparticles in a uniform environment characterized by a thermal conductivity κ_s. As explained in the previous section, if the nanoparticles are all located at the interface between two different media, κ_s can be assumed to be the averaged thermal conductivity of the two media.

Consider N identical spherical nanoparticles of radius a labelled $i \in [\![1, N]\!]$, delivering powers Q_i and located at positions $\mathbf{r}_i$. With a good approximation, the temperature increase experienced by each nanoparticle reads

$$\delta T_j = \frac{Q_j}{4\pi \kappa_s a} + \sum_{\substack{k=1 \\ k\neq j}}^{N} \frac{Q_k}{4\pi \kappa_s |\mathbf{r}_j - \mathbf{r}_k|} \tag{2.59}$$

and the temperature increase anywhere in the surrounding medium reads

$$\delta T(\mathbf{r}) = \sum_{k=1}^{N} \frac{Q_k}{4\pi \kappa_s |\mathbf{r} - \mathbf{r}_k|}. \tag{2.60}$$

In the following, we shall focus on the nanoparticle temperatures δT_j, *i.e.*, on Equation (2.59). According to this equation, δT_j originates from two contributions, a self contribution and a contribution stemming for the $N - 1$ neighboring nanoparticles. Let us name these two contributions respectively $\delta T_j^{\mathrm{self}}$ and $\delta T_j^{\mathrm{ext}}$:

$$\delta T_j = \delta T_j^{\mathrm{self}} + \delta T_j^{\mathrm{ext}}. \tag{2.61}$$

Depending on the relative significance of these two terms, two regimes may occur. First, if $\delta T_j^{\mathrm{self}}$ is dominant for any nanoparticle j, the system is said to be in the temperature confinement regime: temperature increases occur only at the vicinity of the nanoparticles. On the contrary, if $\delta T_j^{\mathrm{ext}}$ is dominant, thermal collective effects occur, characterized by an overall temperature offset which tends to homogenize the temperature distribution on the macrometric scale, despite the nanometric nature of the heat sources. This effect has been the main subject of four articles so far in 2006, 2009 and 2013 [24, 14, 30, 7]. These two regimes are illustrated in Figure 2.9 with regular two-dimensional arrays of nanoparticles.

In order to easily predict the occurrence of one regime or another, a simple dimensionless parameter has to be estimated, which reads

$$\zeta_1 = \frac{p}{a \ln N} \quad \text{if } m = 1 \tag{2.62}$$

$$\zeta_m = \frac{p}{a\, N^{(m-1)/m}} \quad \text{if } m \geq 2 \tag{2.63}$$

where p is the average neighboring nanoparticle interdistance, a the nanoparticle radius, or the typical size of the nanoparticle in the case of a non-spherical nanoparticle, N the number of nanoparticles in the system and m the dimensionality ($m = 1$ for a linear chain of nanoparticles, $m = 2$ for a two-dimensional array, $m = 3$ for nanoparticles dispersed in solution). In particular, for a two-dimensional ($m = 2$) distribution of nanoparticles of typical size L, since $N \sim (L/p)^2$, ζ_2 reads

$$\zeta_2 = \frac{p^2}{3 L a}. \tag{2.64}$$

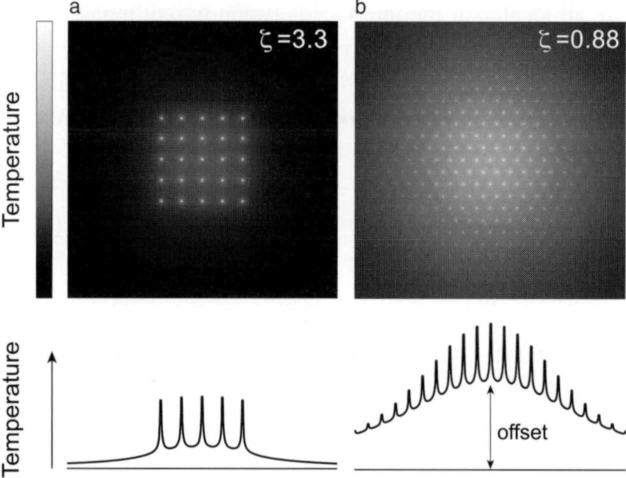

Fig. 2.9　Temperature distribution around arrays of gold nanoparticles. (a) Low sparsity characterized by the localization regime (the temperature rise is located around each source of heat. (b) Homogenization regime where collective effects occurs and where an overall temperature offset appears.

If $\zeta_m \gg 1$, the system is in the localization regime and the temperature distribution of the system resembles hot spots confined around each nanoparticle location (Figure 2.9a). On the contrary, if $\zeta_m \leq 1$, the temperature distribution features an overall offset (Figure 2.9b). Thermal collective effects can even lead to a fully smooth temperature distribution if $\zeta_m \ll 1$. In this case, one speaks about *temperature homogenization*.

Expression (2.63) is important in nanoplasmonics since many experimental works are based on multiple nanoparticle illumination, from sensing to nanochemistry and thermophotovoltaics.

The occurrence of a collective/homogenization regime is a counter intuitive effect that is often discarded in the literature, while it can play an important and dominant role. Such an effect can yield artifactual measurements and misleading conclusions [1, 11, 28, 32] because it can yield much higher temperature rises than what can be predicted if considering only one nanoparticle. In any experiment where numerous nanoparticles are illuminated and where a temperature increase may matter, ζ_m must be estimated and, ideally, the temperature of the illuminated area should be monitored, which can be done in most cases with a simple infrared camera.

Note that collective effects can be dominant even if *e.g.*, 10 nm nanoparticles are separated by $p = 10$ μm in average [28]. It suffices to illuminate an area of 1 cm^2, *i.e.*, the whole coverslip on which the nanoparticles are deposited, to get $\zeta_2 < 1$.

In the next chapter, dedicated to computation methods in thermal plasmonics, a full section will be devoted to estimating the term ΔT_j^{ext} in different illumination conditions, and especially under thermal homogenization. In this case, although the heat source density physically consists of nanometric particles, it can be considered as a continuous function (see Section 3.3.3).

2.2.6 Two-Dimensional Heat Sources

From the previous Subsection 2.2.5, one can deduce that a collection of densely packed nanoparticles can be considered as a continuous heat source density. For this reason, when running numerical simulations related to such systems, there is no need to consider an assembly of nanoparticles, for instance using COMSOL where 2D or 3D heat source densities will be specified.

Numerical modelling is made easier when the nanoparticles are *uniformly* distributed on a *planar* substrate. This is a common situation in plasmonics. This is for instance the case for nanostructures fabricated by e-beam lithography according to a regular pattern, or for colloidal nanoparticles that have been drop-casted or spin-coated on the substrate or nanoparticles made using block copolymer micellar lithography (BCML) [29]. A typical example of uniform nanoparticle distribution made by BCML is shown in Figure 2.10a. For instance, when illuminating a uniform distribution of gold nanoparticles with a uniform and cw[3] laser beam, one can consider the heat source as a single, circular hot plate (see Figure 2.10b,c):

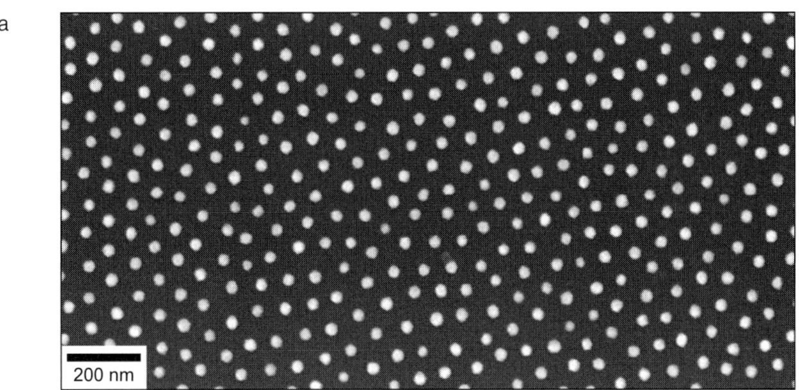

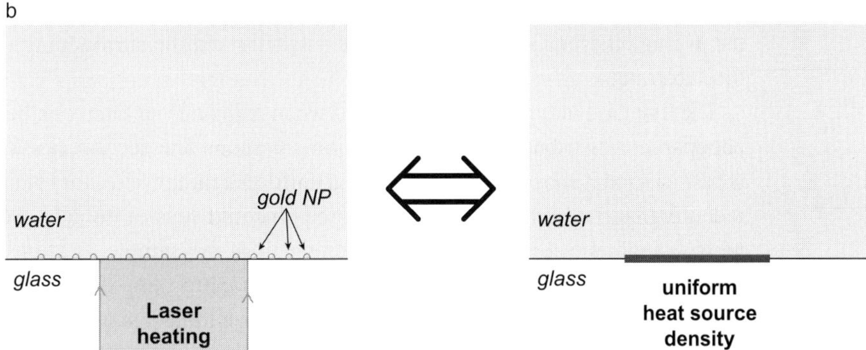

Fig. 2.10 (a) Example of a uniform nanoparticle distribution obtained by block copolymer micellar lithography (courtesy of Julien Polleux). (b) Equivalence between a system composed of nanoparticles and a uniform heat source density, valid if $\zeta_2 \ll 1$.

[3] This rule does not hold under pulsed illumination.

Most of the time with this kind of sample, thermal collective effects occurs ($\zeta_2 \ll 1$) and the heat source density can be considered as smooth, despite the discrete nature of the heat source that consists of many nanoparticles. This simplifies the approach and the calculation of the temperature increase, especially if the temperature increase is to be determined only at the centre of the illuminated area, and if the light beam is uniform or Gaussian.

Reference [7] gives expressions to estimate the temperature increase at the centres of uniform distributions of nanoparticles, when the nanoparticles are uniformly distributed on a planar substrate. As a function of the nature of the nanoparticle distribution and of the illumination profile, this central temperature increase reads:

Case 1: Gaussian illumination of an infinite array

$$\delta T \approx \frac{\sigma_{\mathrm{abs}}\, P}{\bar{\kappa}} \sqrt{\frac{\ln 2}{4\pi}\, \frac{1}{HA}} \left(1 - \frac{4\,\sqrt{\ln(2)\,A}}{\pi\,H}\right)$$

Case 2: Uniform and circular illumination of an infinite array

$$\delta T \approx \frac{\sigma_{\mathrm{abs}}\, P}{\bar{\kappa}}\, \frac{1}{\pi}\, \frac{1}{DA} \left(1 - \frac{2\sqrt{A}}{\sqrt{\pi}\, D}\right)$$

$$\delta T \approx \frac{\sigma_{\mathrm{abs}}\, I}{\bar{\kappa}}\, \frac{1}{4}\, \frac{D}{A} \left(1 - \frac{2\sqrt{A}}{\sqrt{\pi}\, D}\right)$$

Case 3: Uniform illumination of a finite-size square array

$$\delta T \approx \frac{\sigma_{\mathrm{abs}}\, I}{\bar{\kappa}}\, \frac{\ln(1 + \sqrt{2})}{\pi}\, \frac{S}{A}.$$

In these expressions, A is the surface of a unit area of the nanoparticle lattice. In other words, A^{-1} is the number of nanoparticles per unit area. H is the FWHM of the Gaussian beam. P is the power of the laser. I is the irradiance of the laser (power per unit area). $\bar{\kappa}$ is the average thermal conductivity of the substrate and the surroundings medium on top of the substrate.

The first case is typically encountered when using a laser Gaussian beam with a uniform nanoparticle distribution, a very common situation. The second case concerns situations where caution has been used to achieve a uniform beam profile, for instance by focusing an incoming laser beam at the entrance of the objective lens of the microscope and adjusting an iris diaphragm to adjust the beam diameter at the sample location (see Figure 1.19 on page 29). The third case rather concerns lithographic samples where nanoparticles have been fabricated according to a square lattice over a finite square area.

Even when heating a 2D distribution of nanoparticles, a general concern is what the 3D temperature field looks like, and in particular how it decays in the z direction. For a uniform, circular heat source density in the (Ox, y) plane centred in O, the temperature decay along the Oz direction reads [8]

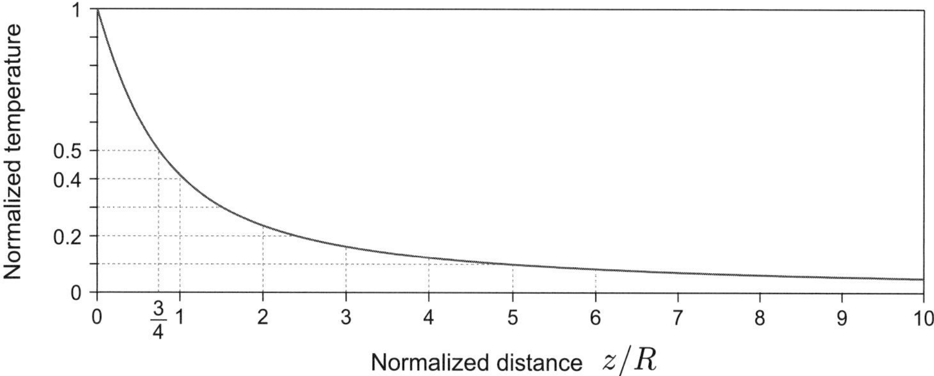

Fig. 2.11 (a) Temperature profile above a planar uniform and circular heat source density, in the z direction according to Equation (2.65).

$$T(z) = \frac{q_0}{2\kappa}\left(\sqrt{R^2 + z^2} - z\right) \tag{2.65}$$

where q_0 is the 2D heat source density (power per unit area), R is the radius of the circular 2D heat source density and κ is the average thermal conductivity of the two media separated by the plane of the heat source density (glass and water for instance). Here is a plot of this profile:

2.3 Transient Evolution

Let us consider a point source in a surrounding medium of thermal diffusivity D_s, and delivering a constant heat power from $t = 0$:

$$Q(t) = Q_0 \ \text{ for } t \geq 0 \tag{2.66}$$

$$Q(t) = 0 \quad \text{ for } t < 0 \tag{2.67}$$

as represented in Figure 2.12a. The temperature evolution in the environment can be expressed using a Green's function formalism and Equation (2.131):

$$T(r, t) = \frac{Q_0}{c_p\, \rho} \int_0^t \frac{1}{(4\pi D_s\, t')^{3/2}} \exp\left(-\frac{r^2}{4\, D_s\, t'}\right) dt' \tag{2.68}$$

Such an integral formulation can be recast into a closed-form expression using the error function defined as

$$\text{erf}(x) = \frac{2}{\sqrt{\pi}} \int_0^x e^{-u^2}\, du. \tag{2.69}$$

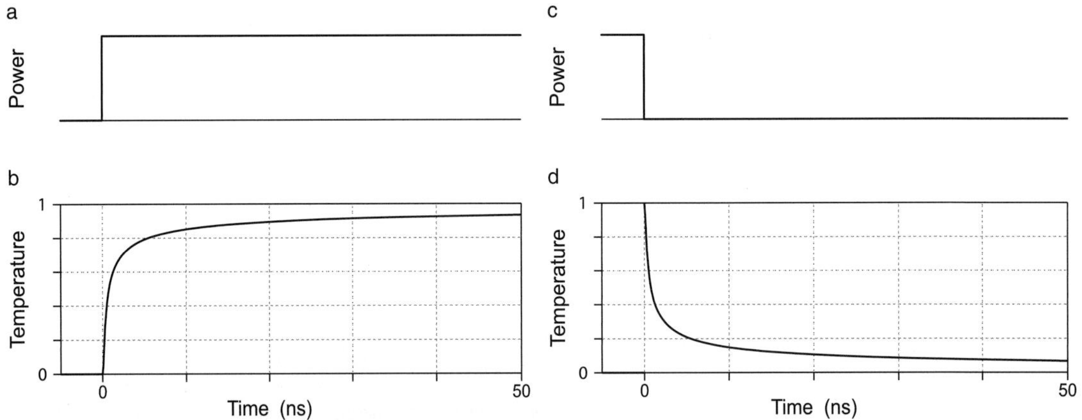

Fig. 2.12 (a,c) Heat power delivered by a point-like heat source in water as a function of time. (b,d) Resulting temperature evolution at 10 nm from the heat source.

Using the variable substitution $X = r/\sqrt{4D_s t'}$, one can show that $T(r,t)$ can be recast into[4]

$$T(r,t) = \frac{Q_0}{4\pi\kappa r}\left(1 - \mathrm{erf}\left(\frac{r}{\sqrt{4D_s t}}\right)\right). \tag{2.70}$$

This function is plotted in Figure 2.12b.

Let us consider now the opposite case where the illumination stops at $t = 0$ (see Figure 2.12c)

$$Q(t) = Q_0 \ \text{ for } t < 0 \tag{2.71}$$

$$Q(t) = 0 \quad \text{ for } t \geq 0. \tag{2.72}$$

One simply has the reverse evolution (see Figure 2.12d):

$$T(r,t) = \frac{Q_0}{4\pi\kappa r}\,\mathrm{erf}\left(\frac{r}{\sqrt{4D_s t}}\right). \tag{2.73}$$

Consequently, the transient temperature evolution of a nanoparticle (or several) after switching on or off the heating illumination *cannot be fitted with an exponential profile*, as one might be tempted to do, in order to retrieve the time constant. I personally had occasion to talk to researchers who observed, indeed, that a simple exponential fit did not work. They tried thus a double-exponential fit, involving the sum of two exponential functions and two time constants. The fit was naturally better, since involving two free parameters instead of one, but their conclusion that the systems was governed by two time constants was not relevant. It turned out that their measurements could be perfectly fitted using the error function, involving a single free parameter, *i.e.*, a single time constant.

---

[4] Note that such a formulation is not consistent with the fact that any information cannot travel faster than the speed of light, since Equations (2.69) and (2.131) imply that the temperature becomes nonzero everywhere in the universe at $t = 0^+$.

The expressions of the temperature evolution derived above answer an important question in thermoplasmonics: what is the timescale of the temperature evolution of a plasmonic system after switching on or off the illumination? According to this formalism, the time constant of the temperature evolution is to be found in the error functions. As $\mathrm{erf}(1) \approx 0.84$, the timescale

$$\tau = \frac{r^2}{4D_\mathrm{s}} \tag{2.74}$$

corresponds to the time when the temperature increase is 84% of the final temperature increase. With this notation, the error function reads $\mathrm{erf}(\sqrt{\tau/t})$. Thus, Equation (2.74) can be considered as the timescale of the temperature evolution. This means that the typical timescale of temperature variations depends linearly with the distance r from the heat source. The aforementioned question has thus no unique answer. It depends where you look.

To give an order of magnitude, 100 nm from a heat source located in water ($D_\mathrm{s} \approx 10^{-7}$ m$^2 \cdot$s^{-1}), according to Equation (2.74) the timescale of temperature variation equals $\tau = (10^{-7})^2/(4 \cdot 10^{-7}) \approx 20$ ns. One of the main advantages of using nanoparticles as nanosources of heat is the possibility of achieving very fast dynamics due to weak thermal inertia.

2.4 Harmonic Heating

Several applications of thermoplasmonics are based on a modulated heating of metal nanoparticles using a modulated laser intensity, for instance in photothermal imaging (see Section 6.7) [10, 13] or for the measurements of thermal diffusivity [16]. Such a system raises questions about the amplitude of the nanoparticle's temperature modulation, its phase delay compared to the light modulation and the spatiotemporal evolution of the temperature distribution in the surrounding medium. This section addresses all these questions, based on different systems: a point-like source, a spherical nanoparticle and multiple nanoparticles. A comprehensive study of this problem is proposed in Reference [9].

2.4.1 Thermal Point Source

To begin with, let us consider a system characterized by a uniform surrounding medium with a thermal conductivity κ_s, a specific heat capacity c_ps, a mass density ρ_s and a thermal diffusivity $D_\mathrm{s} = \kappa_\mathrm{s}/\rho_\mathrm{s} c_\mathrm{ps}$. Let us consider as well a localized heat source density

$$q(\mathbf{r}, t) = Q(t)\delta(\mathbf{r}). \tag{2.75}$$

In this section, consider a modulated heat source density at the angular frequency ω: $Q(t) = Q_0(1 + \cos \omega t)$. In this case, the differential equation that has to be solved anywhere in the system is:

$$\rho c_\mathrm{p}\, \partial_t T - \kappa \nabla^2 T = Q_0(1 + \cos \omega t)\delta(\mathbf{r}). \tag{2.76}$$

Noteworthily, the source term $Q(t)$ is not simply $Q_0 \cos \omega t$. An offset has been introduced because a heat source density cannot be negative (it would mean a heat sink). However, due to the linearity of the equation, the problem can be divided into two independent problems: a static problem led by the constant source term $Q_0 \delta(\mathbf{r})$ and a purely time-harmonic problem governed by the source term $Q_0 \cos \omega t \delta(\mathbf{r})$. The solutions of these two problems add up to form the solution of the initial problem. The static problem and its solution are already discussed in Section 2.2.1. We will thus focus on the time harmonic problem and naturally use a complex-number formalism such that

$$T(r,t) = \mathrm{Re}(\underline{T}^{\omega}(r,t)) \tag{2.77}$$

$$\text{where } \underline{T}^{\omega}(r,t) = \underline{T}e^{-i\omega t} \tag{2.78}$$

and $\underline{T}^{\omega}(r,t)$ solution of the equation

$$-i\omega \rho c_{\mathrm{p}} \underline{T}^{\omega} - \kappa \nabla^2 \underline{T}^{\omega} = Q_0 e^{-i\omega t} \delta(\mathbf{r}). \tag{2.79}$$

By looking at a solution of the homogeneous equation, in the form of a plane wave $\underline{T}^{\omega} = e^{ikr-i\omega t}$, a dispersion relation can be derived involving a complex wave vector amplitude:

$$\underline{k}_{\omega}^2 = \frac{i\omega}{D_{\mathrm{s}}} \tag{2.80}$$

$$\underline{k}_{\omega} = (1+i)\sqrt{\frac{\omega}{2D_{\mathrm{s}}}}. \tag{2.81}$$

The wave vector $\underline{k}_{\omega}$ is a complex number, which implies a wave attenuation, characteristic of a diffusion problem. By developing the expression $\exp(i\underline{k}_{\omega}r)$ using Equation (2.81), one can evidence a thermal wavelength $\lambda_{\mathrm{th}} = 2\pi/\mathrm{Re}(\underline{k}_{\omega})$:

$$\lambda_{\mathrm{th}} = 2\pi \sqrt{\frac{2D_{\mathrm{s}}}{\omega}} \tag{2.82}$$

and an attenuation length defined by $\delta_{\mathrm{th}} = 1/\mathrm{Im}(\underline{k}_{\omega})$:

$$\delta_{\mathrm{th}} = \sqrt{\frac{2D_{\mathrm{s}}}{\omega}}. \tag{2.83}$$

Importantly, $\lambda_{\mathrm{th}} = 2\pi \delta_{\mathrm{th}}$. Consequently, in thermodynamics, any thermal wave is damped over a distance that is 2π times shorter than the wavelength, irrespectively of the nature of the surrounding medium and the frequency [33]. In other words, a thermal wave is damped even before just one spatial oscillation. There is nothing to do against that and this is the reason why it does not make sense to speak about *thermal waves* in physics.

The solution of Equation (2.79) that fulfills the boundary conditions reads

$$\underline{T}^{\omega}(r,t) = Q_0 \, \underline{G}^{\omega}(r,t) \tag{2.84}$$

where $\underline{G}^{\omega}(r,t)$ is the Green's function of the problem:

$$\underline{G}^{\omega}(r,t) = \frac{e^{i\underline{k}_{\omega}r-i\omega t}}{4\pi \kappa_{\mathrm{s}} r}. \tag{2.85}$$

This is an important function for the dynamic study of thermal effects, which will be useful in the following.

Equation (2.77) gives the final general solution. Added with the solution of the steady state component, it gives

$$T(r,t) = \frac{Q_0}{4\pi\kappa_s r}\left[1 + e^{-r/\delta_{\text{th}}}\cos\left(\frac{2\pi}{\lambda_{\text{th}}}r - \omega t\right)\right].\tag{2.86}$$

2.4.2 Spherical Metal Nanoparticle under Harmonic Heating

Let us now increase by one step the complexity of the problem by considering a metal *sphere* as the nanosource of heat. Let us name κ, c_p and ρ the thermal conductivity, the heat capacity and the mass density of the metal constituting the nanosphere, and κ_s, c_{ps} and ρ_s those of the surroundings.

If one considers the inner temperature of the nanoparticle as uniform,[5] the physical quantities of interest in this problem are the temperature distribution in the surroundings $T(r,t)$ for $r > a$ and the nanoparticle temperature $T_{\text{NP}}(t) = T(a,t)$. The latter equality comes from temperature continuity considerations at the nanoparticle interface. As explained in the previous section, a temperature discontinuity occurs only if the nanoparticle is endowed with a finite surface conductivity, which may affect the nanoparticle dynamics by limiting the heat exchange rate. For the sake of simplicity, such an effect is neglected in this section, but its effect can be taken into account as shown in Reference [9].

The equations governing the temperature field read

$$\partial_t T = D_s \frac{1}{r^2}\partial_r(r^2\partial_r T)\tag{2.87}$$

$$V\rho c_p \frac{\mathrm{d}T_{\text{NP}}}{\mathrm{d}t} = \kappa_s\, 4\pi\, a^2 \partial_r T(a,t) + Q\tag{2.88}$$

where $V = 4\pi a^3/3$ is the volume of the nanoparticle. The second equation stems from an energy conservation statement. In our problem, the heat power Q absorbed (or equivalently delivered) by the nanoparticle equals

$$Q(t) = Q_0(1 + \cos\omega t).\tag{2.89}$$

Just like in the previous section, the static component of the heat source will be discarded, to focus on the time harmonic component, and the complex notation will be used for the heat source $\text{Re}(Q_0\, e^{-i\omega t})$. The temperature in the nanoparticle will be noted

$$T_{\text{NP}}(t) = \text{Re}(\underline{T}^{\omega}_{\text{NP}}(t))\tag{2.90}$$

$$\text{and}\quad \underline{T}^{\omega}_{\text{NP}}(t) = \underline{T}_{\text{NP}}e^{-i\omega t}\tag{2.91}$$

where $\underline{T}_{\text{NP}}$ is a constant complex number that contains the amplitude and the phase of the oscillation of the nanoparticle temperature. Finally, for the temperature field outside the nanoparticle, one can seek a solution in the form

$$T(r,t) = \text{Re}(\underline{T}^{\omega}(r,t)).\tag{2.92}$$

$$\text{where}\quad \underline{T}^{\omega}(r,t) = \underline{T}\frac{a}{r}e^{i(\underline{k}_{\omega}r - \omega t)}\tag{2.93}$$

---

[5] This is a common assumption in nanoplasmonics, as explained on page 47 (Equations (2.48)).

where $\underline{T}$ is a constant complex number and $\underline{k}_\omega$ is still the complex wave vector amplitude defined by Equation (2.81). It is worth recasting (2.87) and (2.88) using the complex notations, and new meaningful physical constants:

$$- i\omega\, \tau_s\, \underline{T}^\omega(r,t) = \frac{a^2}{r^2}\, \partial_r(r^2 \partial_r \underline{T}^\omega(r,t)) \tag{2.94}$$

$$-i\omega\, \tau_{NP}\, \underline{T}^\omega_{NP} = a\, \partial_r \underline{T}^\omega(a^+,t) + \frac{Q_0 e^{-i\omega t}}{4\pi\,\kappa_s\, a} \tag{2.95}$$

where two time constants have been introduced

$$\tau_s = \frac{a^2\, \rho_s c_{ps}}{\kappa_s} = \frac{a^2}{D_s} \tag{2.96}$$

$$\tau_{NP} = \frac{a^2\, \rho c_p}{3\,\kappa_s}. \tag{2.97}$$

Injecting Equation (2.93) into Equation (2.94) yields the same dispersion relation as Equation (2.81) defining $\underline{k}_\omega$, but $\underline{k}_\omega$ can be expressed differently as a function of the new parameters τ_s and D_s:

$$\underline{k}_\omega^2 a^2 = i\,\omega\,\tau_s \tag{2.98}$$

$$\underline{k}_\omega\, a = (1+i)\sqrt{\frac{\omega\,\tau_s}{2}} \tag{2.99}$$

$$\underline{k}_\omega = (1+i)\sqrt{\frac{\omega}{2 D_s}}. $$

All the physical quantities defined above, namely τ_s, τ_{NP}, κ_ω, λ_{th} and δ_{th} will be useful for the following discussions.

Injecting now the complex quantity (2.91) in Equation (2.95) yields

$$\underline{T}^\omega_{NP}(t) = \frac{Q_0 e^{-i\omega t}}{4\pi\,\kappa_s\, a\,(1 - i\underline{k}_\omega a - i\omega\,\tau_{NP})} \tag{2.100}$$

or equivalently, using Equation (2.99),

$$\underline{T}^\omega_{NP}(t) = \frac{Q_0 e^{-i\omega t}}{4\pi\,\kappa_s\, a\,\left(1 + (1-i)\sqrt{\frac{\omega\tau_s}{2}} - i\omega\,\tau_{NP}\right)}. \tag{2.101}$$

Interestingly, three timescales are involved in the response of the nanoparticle's temperature under harmonic heating: $\tau_\omega = 2\pi\omega^{-1}$ and the two time constants τ_s and τ_{NP} defined by Equations (2.96) and (2.97). Here is some more insight into their significations. (i) τ_ω is nothing but the period of the oscillations of the excitation. (ii) τ_s only involves the thermal characteristics of the surroundings. It is the timescale associated with any thermal process occurring in a uniform medium over a domain of typical size a (irrespectively of the presence of a nanoparticle). (iii) τ_{NP} is the timescale of the thermal variations within the nanoparticle immersed in a surrounding medium of conductivity κ_s. If c_m is too large, a heat exchange with the surroundings will not significantly vary the nanoparticle temperature, which will make any temperature variation last longer. In the limit case where c_m tends to infinity, the nanoparticle becomes a temperature reservoir that will no longer feature any temperature variation ($\tau_{NP} \to \infty$) despite the heating modulation. As the denominator of Equation (2.100) features three terms, three different regimes associated to different timescales are expected, depending on which term dominates. For most metals in water, however, the two frequencies frequencies are by chance very similar:

$$\frac{\omega \tau_{NP}}{\omega \tau_s} = \frac{\rho_m c_m}{3\, \rho_s c_{ps}} = \frac{\mu}{3} \tag{2.102}$$

where μ is an important dimensionless quantity defined as

$$\mu = \frac{\rho_m c_m}{\rho_s c_{ps}}. \tag{2.103}$$

For gold in water, $\mu \approx 0.6$. Consequently, there exist only two regimes. At sufficiently low frequency, the terms $\omega \tau_s$ and $\omega \tau_{NP}$ are much smaller than unity and thus negligible, which corresponds to the quasistatic regime: the temperature of the system faithfully follows the variations of the laser intensity in phase:

$$T_{NP}^{\omega}(t) = \frac{Q_0}{4\pi\,\kappa_s\,a}\cos(\omega t). \tag{2.104}$$

Another regime appears as soon as $\omega \tau_s$ or $\omega \tau_{NP}$ becomes on the order of 1. Since $\mu \approx 1$ for gold (and most metals) in water, the two terms $\omega \tau_s$ and $\omega \tau_{NP}$ becomes significant in Equation (2.101) from around the same frequency. This frequency can be estimated from the consideration $\omega \tau_s \sim 1$, which yields $a^2 \omega \sim D_s$. Let us define a dimensionless number

$$\xi_0 = \frac{\omega a^2}{D_s} = \omega\,\tau_s = |k_\omega|^2 R^2 = 2\frac{a^2}{\delta_{th}}. \tag{2.105}$$

The subscript "0" means *single* nanoparticle. The quasistatic regime is lost as soon as ξ_0 is larger than unity. The temperature oscillation becomes both delayed by $\pi/2$ and damped:

$$T_{NP}(t) = \frac{3}{\mu \xi_0}\frac{Q_0}{4\pi\,\kappa_s\,a}\cos(\omega t + \pi/2). \tag{2.106}$$

A delay of $\pi/2$ at high frequency is not conventional in physics. The fact that the delay is not π at high frequency evidences that a single nanoparticle under modulated illumination should not be considered as a resonator.

ξ_0 larger than unity, characterizing the high-frequency regime, means

$$a^2 \omega \geq D_s \approx 10^{-7}\ \mathrm{m^2 \cdot s^{-1}}. \tag{2.107}$$

This regime is usually *not achieved* in nanoplasmonics, for common nanoparticle sizes and for reasonable frequencies. Note that if the nanoparticle is not spherical, this expression can be used by replacing a by the characteristic nanoparticle size. Incidentally, in nanoplasmonics, when a single nanoparticle is illuminated by a modulated light, one can usually consider that its temperature faithfully follows the incident light variations, in phase and amplitude. We will see in the next section that this conclusion no longer applies when multiple nanoparticles are illuminated if collective effects appear. The amplitude and phase of $T_{NP}(t)$ are represented in Figure 2.13, demonstrating the two regimes.

Finally, from the expression of $\underline{T}_{NP}^{\omega}$ (see Equation (2.100)), one can express the complex amplitude of the temperature *outside* the nanoparticle. By noticing that $\underline{T}^{\omega}(a,t) = \underline{T}_{NP}^{\omega}(t)$ at any time, one can find the expression of $\underline{T}^{\omega}$:

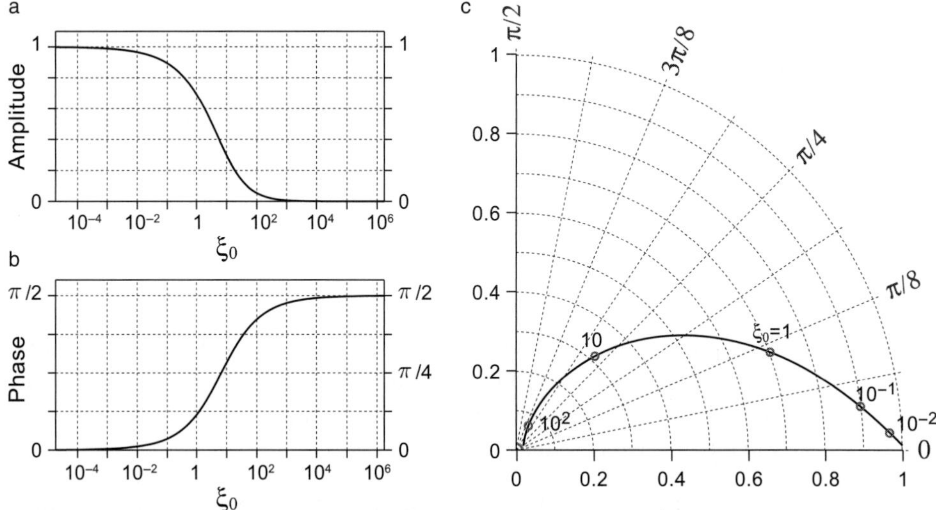

Fig. 2.13 Temperature response of a spherical nanoparticle in amplitude and phase submitted to harmonic heating as a function of the dimensionless number ξ_0 defined by Equation (2.105). (a) Amplitude. (b) Phase. (c) Polar plot of the response in amplitude and phase.

$$\underline{T}^{\omega}(r,t) = \frac{Q_0}{4\pi\,\kappa_s\,r}\frac{e^{i\underline{k}_{\omega}(r-a)-i\omega t}}{(1-i\underline{k}_{\omega}a-i\omega\,\tau_{\mathrm{NP}})}. \tag{2.108}$$

This time, for the temperature *outside* the nanoparticle, a delay naturally occurs compared with the excitation due to the term $e^{i\underline{k}_{\omega}r}$.

For $\underline{k}_{\omega}a \ll 1$, the usual case in nanoplasmonics, one ends up with the familiar expression that is commonly used, for instance, in photothermal imaging [10, 13] (see Section 6.7, on page 256) or for the measurements of thermal diffusivity [16]:

$$T(r,t) = \frac{Q_0}{4\pi\,\kappa_s\,r}\left[1+e^{-(r-a)/\delta_{\mathrm{th}}}\cos\left(\frac{2\pi}{\lambda_{\mathrm{th}}}(r-a)-\omega t\right)\right]. \tag{2.109}$$

Figure 2.14 plots the temperature as a function of the radial coordinate r at different times. For large enough nanoparticles, or slow enough light modulation ($\xi \ll 1$), the temperature distribution features large amplitudes and follows in phase the excitation. In the opposite case (here for $\xi = 10^2$), the temperature distribution tends toward a steady state. The heat source is fluctuating too rapidly compared to the excitation, and only its time-averaged value matters.

The presence of a (Kapitza) thermal resistivity at the interface between the nanoparticle and the surrounding medium can affect the thermal dynamics. This physics will not be described in this book because it is a marginal effect but the reader is invited to refer to Reference [9] for further details. The case of multiple nanoparticle illumination is much more common and very different from the single nanoparticle case. The next subsection is dedicated to this subject.

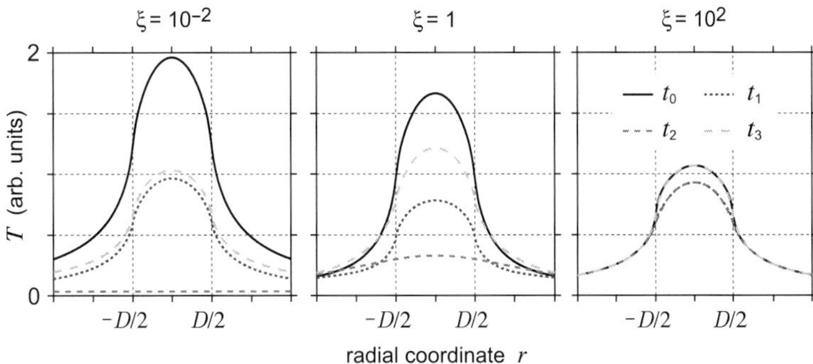

Fig. 2.14 Spatial profile of the temperature around a spherical gold nanoparticles illuminated by a modulated light intensity. Successive profiles have been plotted over a period $\tau = 2\pi/\omega$, at times $t_j = j\tau/4$. Figure reproduced with permission from [9]. Copyright 2014, American Physical Society.

2.4.3 Multiple Nanoparticles under Harmonic Heating

In this section, we consider the dynamic thermal response of an assembly of identical nanoparticles under the same illumination conditions, still modulated at ω. In this case, all the nanoparticles deliver the same average heat power Q_0 and the temperature increase anywhere in the system can be written as a linear sum of all the nanoparticle contributions:

$$\underline{T}^{\omega}(\mathbf{r}, t) = \sum_{j=1}^{N} \frac{Q_0}{4\pi \kappa_s r_j} \frac{e^{i\underline{k}_\omega(r_j - a) - i\omega t}}{(1 - i\underline{k}_\omega a - i\omega\tau_{\mathrm{NP}})} \tag{2.110}$$

where $r_j = |\mathbf{r} - \mathbf{r}_j|$.

In the following, the thermal homogenization regime will be considered, where the nanoparticles are close enough to generate a smooth temperature profile, delocalized throughout the nanoparticle distribution (see Section 2.2.5). We will not consider the other regime where the temperature increase is confined around each nanoparticle, which simply corresponds to the situation of the previous subsection.

Equation (2.110) can be simplified. We have seen in the previous section that the terms $\underline{k}_\omega a$ and $\omega\tau_{\mathrm{NP}}$ become dominant at very high frequencies. We will not consider the high frequency regime in the following. Indeed, we will see that the increased thermal inertia of an assembly of nanoparticles makes the temperature response completely damped at so high frequencies, in nanoplasmonics. Equation (2.110) can be simplified accordingly:

$$\underline{T}^{\omega}(\mathbf{r}, t) = T_{\mathrm{NP}}^0 \sum_{j=1}^{N} \frac{a\, e^{i\underline{k}_\omega r_j - i\omega t}}{4\pi \kappa_s r_j} \tag{2.111}$$

where $T_{\mathrm{NP}}^0 = Q_0/4\pi\kappa_s a$ would be the temperature of a single nanoparticle under cw illumination. Closed-form expressions of $\underline{T}^{\omega}(r, t)$ can be calculated in two and three dimensions if one considers a uniform and continuous circular or spherical heat source distribution and if one calculates the temperature response at the centre of the distribution.

With a circular heat source distribution of radius R featuring a heat source distribution q_0 (power per unit area), one gets

$$\underline{T}^{\omega}(0, t) = \frac{q_0}{2i\kappa_s \underline{k}_\omega} \left(e^{i\underline{k}_\omega R} - 1 \right) e^{-i\omega t}. \tag{2.112}$$

For a spherical heat source density of radius R with a heat source density q_0 (power per unit volume), one gets

$$\underline{T}^{\omega}(0, t) = \frac{q_0}{\underline{k}_\omega^2 \kappa_s} \left[e^{i\underline{k}_\omega R}(1 - i\underline{k}_\omega R) - 1 \right] e^{-i\omega t}. \tag{2.113}$$

Interestingly, regardless of the dimensionality, the temperature response is only dependent on the dimensionless parameter $\underline{k}_\omega R$, i.e., on the dimensionless real number $\xi = -i\underline{k}_\omega^2 R^2$ that can be recast as

$$\xi = \frac{\omega R^2}{D_s} = \frac{2R^2}{\delta_{th}}. \tag{2.114}$$

Thus, the dimensions that have to be compared are the thermal attenuation length δ_{th} and the size of the system R; or equivalently ωR^2 and the thermal diffusivity of the surroundings D_s. For low ξ values, the temperature of the nanoparticle distribution follows in phase the excitation. However, when $\xi \sim 1$, a delay occurs, along with a reduction of the oscillation amplitude. Interestingly, the limit of the phase delay at high ξ values depends on the dimensionality. When a $\pi/4$ phase shift is expected when heating a two-dimensional distribution of heat sources, a phase shift of $\pi/2$ is observed for a three-dimensional distribution of heat sources.

2.5 Pulsed Heating

This section is devoted to the physics of plasmonic nanoparticle heating under nanosecond- to femtosecond-pulsed illumination. Most of the physics discussed in this section is described in References [3, 27]. We consider in this section a pulsed laser illumination characterized by an average irradiance $\langle I \rangle$ (power per unit area) and a pulse repetition rate f. Under pulsed illumination, one often deals with the pulse fluence F instead of the time-averaged laser irradiance $\langle I \rangle$. It is defined as the energy density (energy per unit area) when considering one pulse. It can be expressed as a function of $\langle I \rangle$ and f:

$$F = \langle I \rangle / f. \tag{2.115}$$

The pulse duration τ_p is another important physical quantity characterizing the illumination. In most experiments the pulse duration ranges from around 100 femtoseconds to a few nanoseconds. The pulse profile can be described by a normalized[6] Gaussian profile:

$$q(t) = \frac{1}{\tau_p} \exp\left(-\pi \frac{(t - t_0)^2}{\tau_p^2} \right). \tag{2.116}$$

---

[6] This profile is normalized in the sense that $\int_{-\infty}^{\infty} q(t)\mathrm{d}t = 1$.

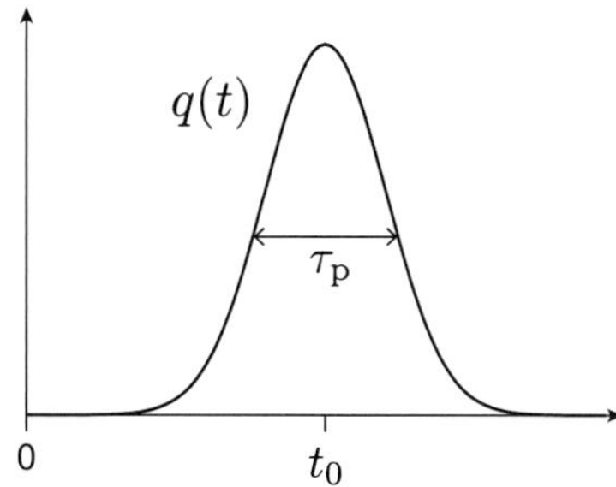

2.5.1 What's happening Inside the Nanoparticle

Absorption Process

The absorption of a short pulse of light by a metal nanoparticle can be described as a three-step process [15, 21] that involves different timescales:

1. **Electronic absorption.** During the interaction with the pulse, part of the incident pulse energy is absorbed by the gas of free electrons of the nanoparticle, much lighter and reactive than the ions of the lattice. The electronic gas thermalizes very fast to a Fermi–Dirac distribution over a timescale $\tau_{\mathrm{e}} \sim 100$ fs [21]. This leads to a state of non-equilibrium within the nanoparticle: the electronic temperature T_{e} of the electronic gas has increased while the temperature of the lattice (phonons) T_ℓ remains unchanged. The absorbed energy $\mathcal{E}_0$ reads:

$$\mathcal{E}_0 = \sigma_{\mathrm{abs}} \langle I \rangle / f = \sigma_{\mathrm{abs}} F. \tag{2.117}$$

2. **Electron–phonon thermalization.** Subsequently, this hot electronic gas relaxes (cools down) through internal electron–phonon interaction characterized by a timescale τ_{ep} to thermalize with the ions of the metal lattice. This timescale is not dependent on the size of gold nanoparticles except for nanoparticle smaller than 5 nm due to confinement effects [2]. Above this size and for moderate pulse energy, the timescale is constant and equals $\tau_{\mathrm{ep}} \sim 1.7$ ps [20, 17, 26]. With a good approximation, the power transfer from the electrons to phonons occurs according to an exponential decay function that reads:

$$p(t) = \frac{1}{\tau_{\mathrm{ep}}} \exp\left(-\frac{t}{\tau_{\mathrm{ep}}}\right). \tag{2.118}$$

At this point, the nanoparticle is in internal equilibrium at a uniform temperature ($T_e = T_\ell$), but is not in equilibrium with the surrounding medium that is still at the initial ambient temperature.

3. **External heat diffusion**. The energy diffusion to the surroundings usually occurs at higher characteristic timescale τ_d that reads (see further Equation (2.134))

$$\tau_d = \frac{\rho c_p a^2}{3\,\kappa_s} \tag{2.119}$$

where a is the radius of the nanoparticle, c_p is the volumetric heat capacity of the metal and κ_s is the thermal conductivity of the surrounding medium). For common nanoparticle sizes, it ranges from 1 ps to 1 ns (Figure 2.15).

For small nanoparticle ($<$ 20 nm), this third step can overlap in time with the electron–phonon thermalization [19] (as discussed hereinafter).

In the case of a *single* pulse illumination, from the physical quantities introduced so far, one can derive other useful physical quantities, namely the pulse intensity (or irradiance) $I(t)$ defined as a power per unit area for a single pulse:

$$I(t) = F\,q(t) \tag{2.120}$$

Note that $\int I(t)\mathrm{d}t = F$. Let us define also the power $P(t)$ transferred from the pulse to the electron gas, which reads

$$P(t) = \sigma_{abs}\,I(t) = \sigma_{abs}F\,q(t) \tag{2.121}$$

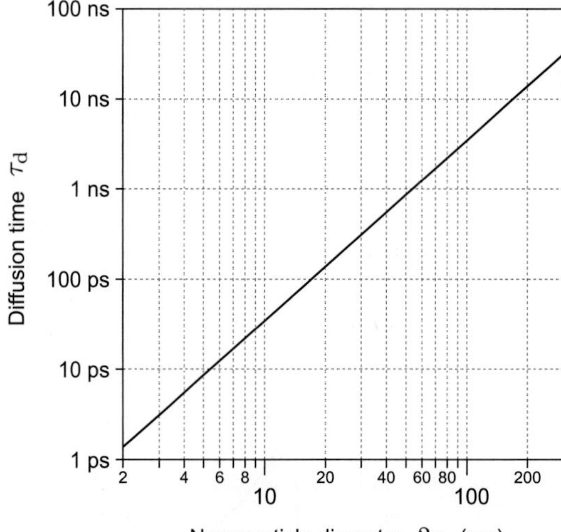

where σ_{abs} is the absorption cross section of the nanoparticle. The total energy absorbed by the nanoparticle reads thus

$$\mathcal{E}_0 = \int P(t)\mathrm{d}t = \sigma_{\mathrm{abs}}\, F. \tag{2.122}$$

A final and important physical quantity is the heat power transferred from the electron gas to the atomic lattice (*i.e.*, the phonons) of the nanoparticle:

$$Q(t) = P \otimes p\,(t) \tag{2.123}$$

$$= \sigma_{\mathrm{abs}}F\, q \otimes p\,(t) \tag{2.124}$$

where $A \otimes B$ denotes a function that equals the convolution between the functions A and B.

One can now discuss the occurrence of different regimes depending on the relative values of the three aforementioned time constants τ_{p}, τ_{ep} and τ_{d}. For very short pulses such that $\tau_{\mathrm{p}} \ll \tau_{\mathrm{ep}}$, one can consider the function q as a Dirac distribution compared to p, which yields $q \otimes p \approx p$. This is typically what happens in the femtosecond regime, since the typical pulse duration of 100 fs is shorter than $\tau_{\mathrm{ep}} = 1.7$ ps. On the contrary, if $\tau_{\mathrm{p}} \gg \tau_{\mathrm{ep}}$, typically with nanosecond pulses, $q \otimes p \approx g$. These approximations make the expression of $Q(t)$ much simpler: it features either an exponential decay in the femtosecond range $Q(t) = \sigma_{\mathrm{abs}}Fp(t)$, or a Gaussian shape in the nanosecond range $Q(t) = \sigma_{\mathrm{abs}}Fq(t)$. The *picosecond-pulse* range appears thus as an intermediate regime where $Q(t)$ cannot be approximated by a simple close-form expression, and has to be calculated using Equation (2.124).

Another approach to determine $Q(t)$ is the so-called two-temperature model (TTM) [2, 20, 18, 21, 25]. The TTM is a more sophisticated approach, which enables for instance the consideration of nonlinearities of the heat capacity of the metal nanoparticle as a function of temperature. It consists in numerically solving a set of two differential equations involving the electronic temperature T_{e} and the lattice temperature T_{ℓ}:

$$
\boxed{
\begin{aligned}
C_{\mathrm{e}}(T_{\mathrm{e}})\frac{\partial T_{\mathrm{e}}}{\partial t} &= -G \times (T_{\mathrm{e}} - T_{\ell}) + q(t) \\[2mm]
C_{\ell}\frac{\partial T_{\ell}}{\partial t} &= G \times (T_{\mathrm{e}} - T_{\ell})
\end{aligned}
}
\tag{2.125}
$$

where G reflect the electron–phonon coupling. The dependence of C_{e} on the temperature T_{e} is important and has to be taken into account to obtain quantitative modeling, especially because the electronic temperature can easily reach several 1000s of K. For gold, a good approximation is $C_{\mathrm{e}}(T_{\mathrm{e}}) \approx \gamma T_{\mathrm{e}}$ where $\gamma = 67.3$ J·m^{-3}·K^{-1} [21].

Subsequent Nanoparticle's Temperature Increase

An important question in pulsed heating is the maximum nanoparticle temperature increase δT_{max} achieved over time. This is the quantity that matters in most applications. For instance, if this quantity exceeds a given threshold, phase transitions or chemical reactions may occur in the surrounding medium. This maximum temperature increase depends on several parameters, and mainly on the pulse duration and the size of the nanoparticle. An

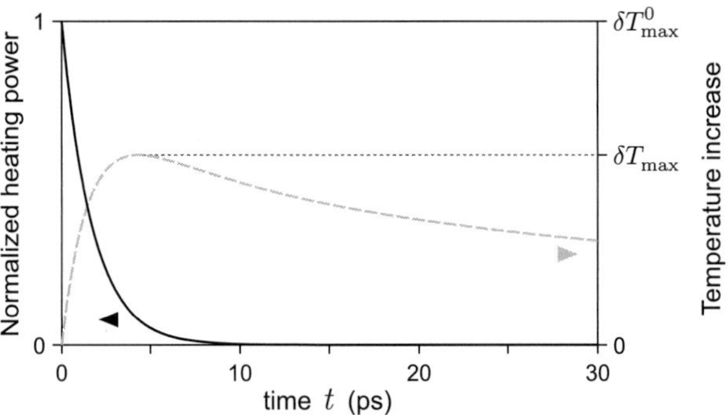

Fig. 2.16 Numerical simulations of the temperature evolution of a 20 nm gold nanosphere immersed in water under fs-pulsed illumination [27]. δT_{max} is defined as the maximum temperature increase achieved over time and $\delta T^0_{\mathrm{max}}$ is the ideal temperature increase, assuming approximations $\mathcal{A}1$ and $\mathcal{A}2$, as defined by Equation (2.129).

example of temperature evolution during a pulse absorption by a spherical gold nanoparticle is given in Figure 2.16. In this example, a femtosecond pulse is absorbed by the electron gas of the nanoparticle. This electron gas gives its energy to the nanoparticle ionic lattice over a typical timescale of 1.7 ps [20, 17, 26], which is the typical timescale of electron–phonon interaction in gold. This is why the heating profile of the nanoparticle displayed in Figure 2.16 is exponential and occurs over a 1-ps timescale, although the pulse duration is on the order of 100 fs (see Equation (2.118)). This plot defines δT_{max} and highlights the fact that the temperature rise is often fast (especially under *femtosecond* pulsed illumination) and may then decay more slowly depending on how fast heat can be drained away by the surrounding medium.

One can derive a first, simple estimation of this temperature increase δT_{max} if one assumes no heat release to the surrounding medium while the pulse is being absorbed. The energy absorbed by the nanoparticle can be directly expressed as a function of the nanoparticle temperature rise δT according to the relation:

$$\mathcal{E}_0 = V\rho c_{\mathrm{p}}\delta T \tag{2.126}$$

where V is the volume of the nanoparticle and $V\rho c_{\mathrm{p}}$ its total (lattice+electrons) heat capacity. Using Equation (2.117), this approximated nanoparticle temperature increase, coined the *ideal* temperature increase in the following, becomes

$$\delta T^0_{\mathrm{max}} = \frac{\sigma_{\mathrm{abs}}\langle I\rangle}{V\,\rho c_{\mathrm{p}}f} = \frac{\sigma_{\mathrm{abs}}F}{V\,\rho c_{\mathrm{p}}}. \tag{2.127}$$

As an example, for a gold nanorod, 50 nm long and 12 nm in diameter, at the plasmonic resonance ($\lambda_0 = 800$ nm), considering a random orientation, $f = 86$ MHz and $\langle I\rangle$=1.0 mW/μm^2, one obtains $\delta T^0_{\mathrm{max}} \approx 30°$C.

Interestingly, for small enough nanoparticles, typically for a gold nanoparticle diameter $2a$ below 60 nm, σ_{abs} is roughly proportional to the nanoparticle volume $V = 4\pi a^3/3$. One can thus define a constant ζ such that

$$\sigma_{\text{abs}} = \zeta a^3 \tag{2.128}$$

stands for a fitting function of $\sigma_{\text{abs}}(a)$ for small values of a. For spherical gold nanoparticles in water ($n = 1.33$), $\zeta = 0.430$ nm^{-1}. Using this approximation, Equation (2.127) can be recast into

$$\delta T_{\text{max}}^0 = \frac{3\zeta F}{4\pi \, \rho c_{\text{p}}}. \tag{2.129}$$

This expression is only dependent on the pulse fluence (power per unit area). The maximum temperature increase of a spherical gold nanoparticle is thus supposed to be independent of the nanoparticle size, at least for small enough nanoparticles.

To summarize, we have done two approximations to derive Equation (2.129):

Approximation $\mathcal{A}$1: The timescale of nanoparticle cooling/heating is much longer than the pulse duration. This approximation enabled us to derive a closed-form expression of the temperature increase (Equation (2.127)). When this approximation is not fulfilled, heat release occurs while the energy is being absorbed, which tends to reduce the maximum temperature increase. This typically occurs when the nanoparticle is sufficiently small or when the pulse duration is sufficiently long.

Approximation $\mathcal{A}$2: The absorption cross section is proportional to a^3. This assumption is only true for small nanoparticles (typically below a few tens of nanometers). For larger nanoparticles, the absorption cross section is damped compared to the a^3 law (see Figure 1.7 on page 9). Consequently, the temperature increase becomes weaker than the ideal temperature increase defined by Equation (2.129).

Consequently, in practice, the actual maximum temperature increase is always smaller than the ideal temperature increase defined by Equations (2.127) and (2.129).

Figure 2.17 plots the maximum temperature increase as a function of the nanoparticle diameter in order to further investigate the effect and validity of the two approximations. Four cases have been plotted, as labeled on the figure: the femtosecond-pulse case, the 1 ns pulse case, the case where Approximation $\mathcal{A}$1 has been made and the case where Approximations $\mathcal{A}$1 and $\mathcal{A}$2 have been made. Several comments can be made.

First, illuminating a nanoparticle under ns-pulsed illumination normally gives rise to a weaker temperature increase compared to fs-pulsed illumination, because nanoparticle cooling usually occurs over timescales shorter than 1 ns, which makes ns-pulsed heating too slow to achieve maximal temperature increase. However, this drawback did not prove to be prohibitive for applications in thermoplasmonics. Much investigation has been based on the use of nanosecond lasers, in particular for microbubble generation around gold nanoparticles.

Second, the temperature increase seriously decays for small nanoparticles. This decay comes from the faster energy release to the surrounding medium for small nanoparticles, which makes approximation $\mathcal{A}$1 no longer valid.

Third, the temperature increase seriously drops for large nanoparticles. This comes from the fact that the absorption cross section of a spherical nanoparticle no longer scales with the volume of the nanoparticle. This makes approximation $\mathcal{A}$2 no longer valid. As

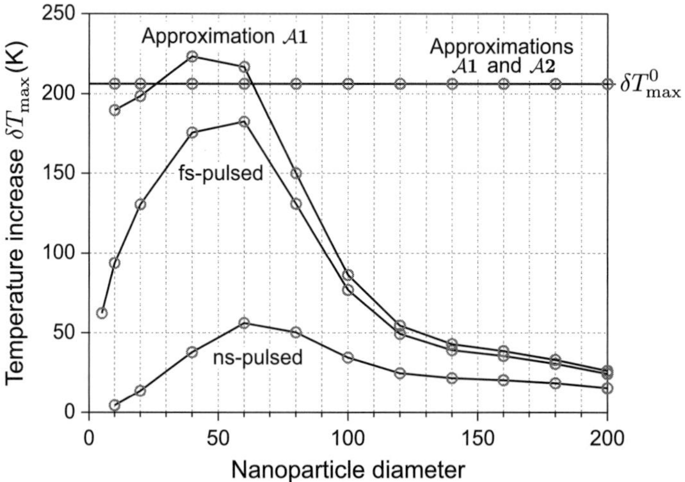

Fig. 2.17 Numerical simulations of the maximum temperature increase of a gold nanosphere immersed in water [27]. An arbitrary fluence of 5 J·m^{-2} was considered. Four cases have been represented, as labeled on the figure: the femtosecond-pulse case, the 1 ns pulse case, the case where Approximation $\mathcal{A}1$ has been made and the case where Approximations $\mathcal{A}1$ and $\mathcal{A}2$ have been made.

a result, there exists an optimum diameter for spherical gold nanoparticles for achieving large temperature rises, around 60 nm [23, 27].

Figure 2.18 plots the temporal evolution of the temperature of gold nanospheres of various diameters, and under different pulse regimes (nanosecond and femtosecond). One can see that the ideal temperature increase is never really reached, but one gets higher temperature increase in the case of femtosecond pulses, and for nanoparticle diameters close to 60 nm.

2.5.2 What's happening Outside the Nanoparticle

Thermal Point Source

In the ideal case of a thermal point source, the heat source density $q(t)$ can be described by a Dirac distribution $Q(t)\delta(\mathbf{r})$. And if the pulse duration is much faster than the heat diffusion in the surrounding medium ($\tau_\mathrm{p} \ll \tau_\mathrm{d}$), one can assume a $Q(t) = \mathcal{E}_0\delta(t)$. The heat diffusion equation reads thus

$$\rho c_\mathrm{p}\partial_t T(r,t) - \kappa\nabla^2 T(r,t) = \mathcal{E}_0\delta(\mathbf{r})\delta(t) \tag{2.130}$$

where $\mathcal{E}_0$ is the internal thermal energy of absorbed at time $t = 0$.

This ideal problem has a closed-form solution [3]:

$$T(r,t) = \frac{\mathcal{E}_0}{c_\mathrm{ps}\,\rho_\mathrm{s}}\frac{1}{(4\pi D_\mathrm{s}\,t)^{3/2}}\exp\left(-\frac{r^2}{4D_\mathrm{s}\,t}\right) \tag{2.131}$$

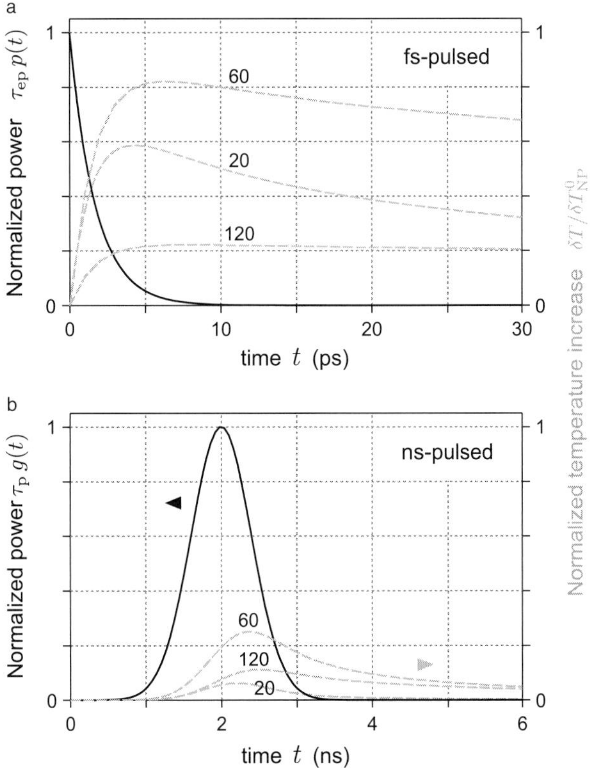

Fig. 2.18 Numerical simulations of the evolution of the temperature of gold nanospheres (dashed lines) following (a) a femtosecond pulse and (b) a 1-nanosecond pulse. Profiles of the nanoparticle heating power have been represented using solid lines. Three nanoparticle diameters have been considered: 20, 60 and 120 nm (as labeled in the figure). Reproduced from Reference [27]. Copyright 2015, American Chemical Society.

where r is the radial coordinate and $D_s = \kappa_s/\rho_s c_{ps}$ is the thermal diffusivity of the medium. Figure 2.19 plots the evolution of the temperature at $r = 10$ nm.

A physical quantity that is worth discussing in this context is the so-called *temperature envelope*, defined as the maximum temperature achieved over time at any location in the surrounding medium $T_{max}(r) = \max_t(T(r,t))$. The temperature envelope can be obtained from Equation (2.131) by calculating the time t for which $\partial_t T(r,t) = 0$ for any position r [3]. It yields a temperature envelope:

$$T_{max}(r) = \left(\frac{3}{2\pi\, e}\right)^{3/2} \frac{E_0}{c_{ps}\, \rho_s} \frac{1}{r^3}.$$

This physical quantity is important since the maximum temperature is usually what matters when investigating thermal induced processes. Interestingly, the thermal envelope features a $1/r^3$ spatial profile, which appears to be much more confined than the $1/r$ profile observed under cw illumination. This important conclusion is that pulsed illumination can

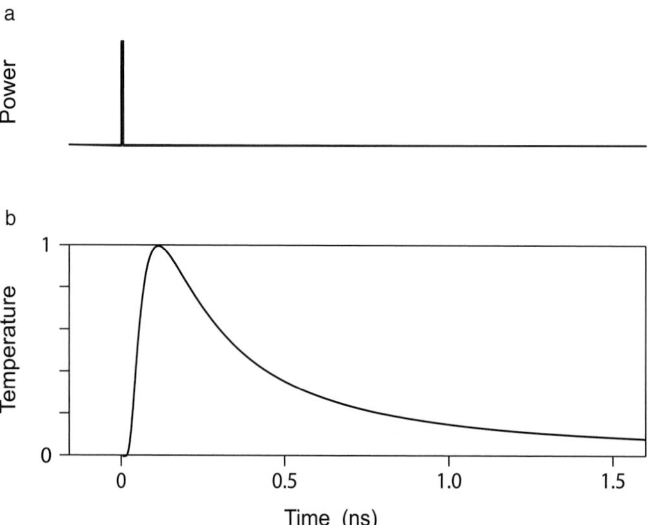

Fig. 2.19 (a) Heat power delivered by a point-like heat source in water as a function of time. (b) Resulting normalized temperature evolution at 10 nm from the heat source.

be used when a spatial confinement of the temperature increase is desired at the vicinity of the nanoparticle.

Note that the $1/r^3$ was derived here for an infinitely short pulse of light. This assumption is usually valid for femtosecond-pulsed illumination, sometimes for picosecond illumination, and never for nanosecond-pulsed illumination.

Spherical Nanoparticle

We consider now the more realistic case of a spherical metal nanoparticle, of radius a. In this case, no simple closed-form expression exists for the spatiotemporal evolution of the temperature of the system following the absorption of a pulse of light. Numerical simulations have to be conducted.

The problem can be simplified by considering the initial temperature profile $T(r, 0)$ as uniform (equals T_{NP}^0) inside the nanoparticle since the electron–phonon thermalization usually occurs much faster than the external heat diffusion. Then, one can also suppose that the nanoparticle temperature *remains* uniform during the evolution of the system because the thermal conductivity of the nanoparticle is much larger than the thermal conductivity of common surrounding media ($\kappa \gg \kappa_s$):

$$\forall\, t \text{ and } \forall\, r < a, \quad T(r, t) = T_{NP}(t). \tag{2.132}$$

The validity of these two approximations is discussed in Reference [3]. Under these assumptions, the system of equations governing the system reads

$$
\begin{cases}
\text{Diffusion equation:} \\[4pt]
\rho_{\mathrm{s}} c_{\mathrm{ps}} \partial_t T(r,t) = \kappa_{\mathrm{s}} \dfrac{1}{r^2} \partial_r \left(r^2 \partial_r T(r,t) \right) \quad \text{for } r > a \\[4pt]
\text{Boundary conditions:} \\[4pt]
V \rho c_{\mathrm{p}} \dfrac{\mathrm{d} T_{\mathrm{NP}}(t)}{\mathrm{d}t} = \kappa_{\mathrm{s}} 4\pi a^2 \partial_r T(a,t) \\[4pt]
\qquad\qquad\quad = -g\, 4\pi a^2 \Delta T(t).
\end{cases}
\tag{2.133}
$$

The first equation is the heat diffusion equation outside the nanoparticle. The two other equations come from considerations of energy conservation and will control the boundary condition at the nanoparticle interface ($r = a$). They involve a finite thermal interface conductivity g.

Interestingly, two new characteristic times arise from the boundary equations. They read:

$$
\tau_{\mathrm{NP}} = \frac{a^2 \, \rho c_{\mathrm{p}}}{3\, \kappa_{\mathrm{s}}},
\tag{2.134}
$$

$$
\tau_g = \frac{a \, \rho c_{\mathrm{p}}}{3\, g}.
\tag{2.135}
$$

If the surface conductivity g is small enough, the evolution of the nanoparticle temperature is governed by τ_g. If the surface resistivity is negligible, the evolution is governed by τ_{NP}.

The system of Equations (2.133) has to be numerically solved to address this problem. Figure 2.20 presents results of numerical simulations using normalized units of space and time ρ and τ defined by

$$
\rho = r/a
\tag{2.136}
$$

$$
\tau = t\, a^2 / D_{\mathrm{s}}.
\tag{2.137}
$$

This way, the line shapes in Figure 2.20 representing the temperature profile inside and outside the nanoparticles are universal and do not depend on the nanoparticle diameter a. For a given particle size a, the normalized coordinate ρ has to be multiplied by a and the normalized time τ by a^2/D_{s} to retrieve the actual coordinate r and time t. Hu and Hartland [19] have shown experimentally that the temporal evolution of the nanoparticle temperature can be conveniently fitted using a stretched exponential function:

$$
F(\tau) = \mathrm{e}^{-(\tau/\tau_0)^n}.
\tag{2.138}
$$

This function was used to fit the results of the numerical simulations regarding the evolution of the nanoparticle temperature as represented in the inset of Figure 2.20 (dashed line). The optimized fit parameters are $n = 0.39$ and $\tau_0 = 0.041$. This yields a useful formula giving the normalized nanoparticle inner temperature evolution for any particle radius a:

$$
F_a(t) = \exp\left[-\left(\frac{D_{\mathrm{s}}\, t}{0.041\, a^2} \right)^{0.39} \right].
\tag{2.139}
$$

Note that this useful formula assumes a zero interface resistivity $1/g$. For a finite value of g, the values of the fit parameters τ_0 and n are bound to be different.

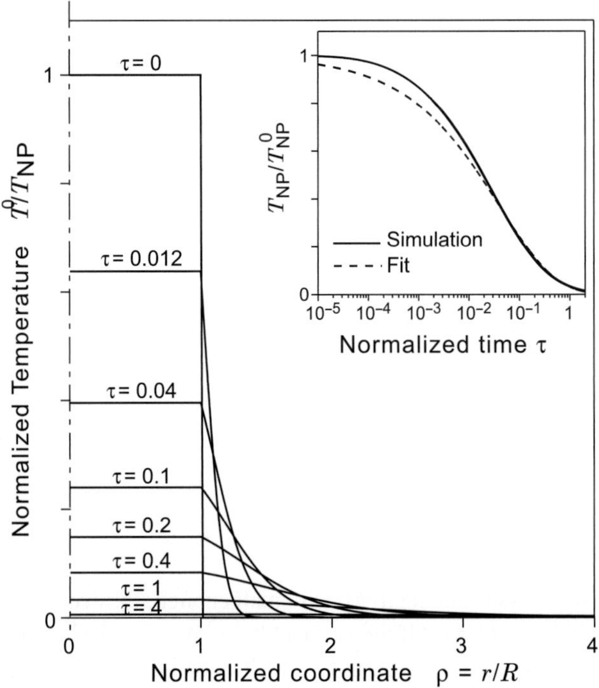

Fig. 2.20 Temperature profiles of a gold nanoparticle at various times as a function of the radial coordinates. Physical quantities have been normalized. The inset represents the evolution of the nanoparticle temperature, fitted using Equation (2.139). Figure reproduced with permission from [3]. Copyright 2011, American Physical Society.

Figure 2.21 aims at comparing the temperature profiles under pulsed and cw illuminations. Again, it plots the series of temperature profiles of Figure 2.20 along with the temperature envelope and the temperature profile of the steady state (cw illumination). The temperature envelope is a concept already discussed in Section 2.5.2 for a point source of heat. It is defined as the maximum temperature achieved over time at any location in the surrounding medium $T_{\max}(r) = \max_t(T(r,t))$. For a point source of heat, we had shown that the envelope profile scales as $1/r^3$. For the spherical source of heat we consider in this section, it appears that such a simple law does not exist. However, a stretched exponential function can be conveniently used to fit the envelope of the spatial temperature profile in the surrounding water (see inset of Figure 2.21):

$$F(\rho) = \exp\left[-\left(\frac{\rho - 1}{\rho_0}\right)^n\right]. \tag{2.140}$$

The fit parameters are $n = 0.45$ and $\rho_0 = 0.060$. These results illustrate to what extent pulsed illumination achieves a much higher degree of temperature confinement compared to cw illumination.

Another physical quantity under pulsed illumination that is worth discussing is the time-average temperature profile $\bar{T}(r)$, which can be defined as

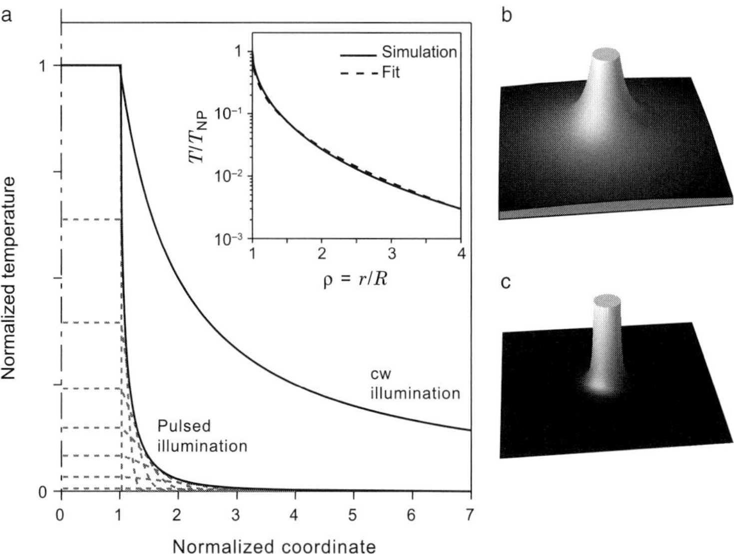

Fig. 2.21 (a) The temperature profiles at different times around a sphere heated by a pulse of light as a function of the normalized radial coordinate $\rho = r/a$ (dashed lines). The envelope of the temperature evolution along with a comparison with the profile under cw illumination are plotted in solid lines. The inset represents the envelope profile fitted using Equation (2.140). (b) Three-dimensional plot of the steady state temperature profile under cw illumination. (c) Three-dimensional plot of the temperature envelope under pulsed illumination. Figure reproduced with permission from [3]. Copyright 2011, American Physical Society.

$$\bar{T}(\mathbf{r}) = \langle T(\mathbf{r}, t)\rangle_t = f \int_0^{\frac{1}{f}} T(\mathbf{r}, t)\mathrm{d}t. \tag{2.141}$$

Unlike the temperature envelope $T_{\max}(\mathbf{r})$, the average temperature $\bar{T}(\mathbf{r})$ under pulsed illumination of a single nanoparticle always spreads out according to a $1/r$ law, just like in cw illumination, which is much broader than the $1/r^3$ spatial extension of the temperature envelope. Hence, it is important to note that the enhanced spatial confinement of the temperature usually put forward under pulsed illumination concerns only the *temperature envelope*, and not the time-average temperature.

Multiple Pulses and Repetition Rate

Experimentally, pulsed illumination rarely consists in sending a single pulse. It rather involves a series of pulses characterized by a repetition rate f. Importantly, all what we have discussed so far ceases to apply as soon as the nanoparticle does not have time to cool down between two successive pulses, *i.e.*, if the repetition rate f is too high. We have seen that the typical time-scale related to the dynamics of a nanoparticle following a pulse absorption is

$$\tau_{\mathrm{NP}} = \frac{a^2 \rho c_{\mathrm{p}}}{3 \kappa_{\mathrm{s}}}. \tag{2.142}$$

Hence, this timescale has to be compared with $1/f$ to figure out whether the system is in one regime or another. Let us define the dimensionless number[7]

$$\eta^{\text{self}} = (f\,\tau_{\text{NP}})^{-1}. \tag{2.143}$$

If $\eta^{\text{self}} \gg 1$, the nanoparticle will have time to cool down between two successive pulses and any discussion or numerical simulation can be based on the consideration of a single pulse. However, if $\eta^{\text{self}} \ll 1$, the nanoparticle temperature does not have the time to return to ambient temperature between two successive pulses, and a steady temperature offset within the nanoparticle appears. Both regimes are illustrated in Figure 2.22.

Let us discuss some orders of magnitude. For common Ti:Sapphire lasers, the repetition rate is around 80 MHz. For a gold nanoparticle in a dense medium such as water, $f^{-1} \sim \tau_{\text{NP}}$ yields a radius of around $a = 100$ nm. The important rule of thumb is thus the following: in nanoplasmonics, for nanostructures of the size of 100 nm or more, a pulsed illumination does not necessarily lead to fast transient temperature increases after each pulse. The temperature evolution can be steady. On the contrary, when the nanoparticle size is much less than 100 nm, typically colloidal nanoparticles, one can achieve

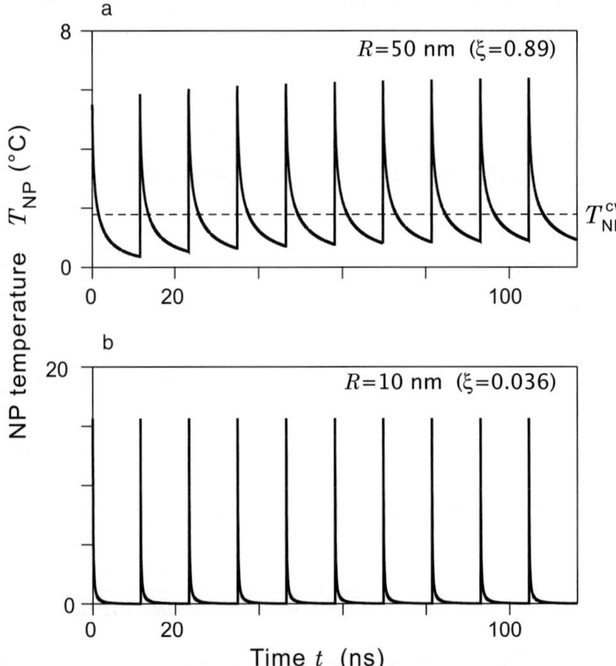

Fig. 2.22   Evolution of the nanoparticle temperature under pulsed illumination. (a) Regime where the temperature does not return to ambient conditions between successive pulses, characterized by a temperature offset. (c) Regime where temperatures rises are localized in time. Figure reproduced with permission from [3]. Copyright 2011, American Physical Society.

---

[7] The superscript "self" refers to a single particle. In the case of multiple nanoparticles under illumination, η^{self} becomes η^{ext} as defined in the next subsection.

strong transient temperature increases and regime of *time-localization*. However, note that this conclusion does not necessarily apply when multiple nanoparticles are illuminated even if they are all much smaller than 100 nm. This is explained in the following subsection.

Multiple Nanoparticles

The main benefit of using a pulsed illumination is to achieve a space and time confinement of a temperature increase on the nanoscale. However, such a confinement regime may not be systematic as soon as multiple successive pulses *or* multiple nanoparticles come into play. The effect of multiple successive pulses was discussed in the previous subsection. The purpose of this subsection is to discuss the influence of multiple nanoparticles under illumination.

When considering an assembly of N nanoparticles in proximity, we have seen in Subsection 2.2.5 that the temperature increase of any nanoparticle j of the system has two origins: a self contribution δT_j^{self} and a contribution δT_j^{ext} stemming for the $N - 1$ neighboring nanoparticles. In the case of pulsed illumination, δT_j^{self} is the temperature increase previously discussed in this chapter and its amplitude is given by Equation (2.127):

$$\boxed{\delta T^{\text{fs}} = \frac{\sigma_{\text{abs}} \langle I \rangle}{V \rho c_{\text{p}} f}.} \tag{2.144}$$

Hence, a visible temperature confinement will be achieved only when this temperature increase is significant compared to the collective temperature increase ΔT_j^{self} throughout the array. This statement leads to a condition that can ben expressed using another dimensionless number η^{ext} that quantifies the ratio $\Delta T^{\text{fs}}/\Delta T^{\text{self}}$. According to Reference [7], it yields for a linear chain of nanoparticles:

$$\eta_1^{\text{ext}} = \frac{3\,\bar{\kappa}}{2\,\rho c_{\text{p}} f\,a^3\,\ln N}\,\frac{p}{} \tag{2.145}$$

and for a 2D array of typical dimension L:

$$\eta_2^{\text{ext}} = \frac{\bar{\kappa}}{\rho c_{\text{p}} f\,a^3\,L}\,p^2. \tag{2.146}$$

In order to reach a confinement regime, the two conditions to be fulfilled are thus $\eta^{\text{self}} \gg 1$ and $\eta_m^{\text{ext}} \gg 1$, where m is the dimensionality of the system ($m = 1$ or 2 in the two above-mentioned cases). This amounts to saying that the instantaneous temperature rise after a pulse absorption by a nanoparticle has to be larger than both ΔT^{self} and ΔT^{ext}. Interestingly, the three dimensionless numbers defined in this section (see Equations (2.63), (2.145) and (2.146)) are linked by the relation:

$$\eta_m^{\text{ext}} = \eta^{\text{self}} \zeta_m. \tag{2.147}$$

2.6 Summary

Here is a summary of the most important take-home messages of this chapter.

Take-Home Messages of Chapter 2

- Heat generation in nanoplasmonics simply comes from a Joule effect generated by the electrons oscillating in the nanostructure.
- Thermal transfer in nanoplasmonics mostly occurs via heat conduction, rarely via convection or thermal radiation.
- Under steady state, the temperature spatial decay from a heat source scales as $1/r$ where r is the radial coordinate.
- Most of the time, the temperature inside plasmonic nanoparticles is uniform due to the much higher thermal conductivity of gold compared to common surrounding mediums.
- In the steady state, the presence of a thermal resistance between the nanoparticle and the surrounding medium does not affect the temperature in the surrounding medium. Only the temperature inside the nanoparticle is affected (it becomes larger).
- If many nanoparticles are illuminated at the same time, thermal collective effects are very likely to occur. In this situation, the temperature is smooth throughout the nanoparticle distribution despite the nanometrer dimension of the heat sources. One speaks about temperature homogenization.
- The temperature rise and decay of a plasmonic structure is not exponential and should not be fitted by an exponential law to retrieve any timescale. It is better described by the so-called error function $\mathrm{erf}(\sqrt{\tau/t})$, which varies much more slowly than an exponential profile $e^{-t/\tau}$.
- Under pulsed illumination, the temperature spatial decay from a heat source scales as $1/r^3$. It is therefore much more confined compared to the steady state, which varies as $1/r$.
- Thermal waves cannot be produced in physics, even when heating a nanoparticle using harmonic excitation, because a temperature field is governed by a diffusion equation.

References

[1] Adleman, J. R., Boyd, D. A., Goodwin, D. G., and Psaltis, D. 2009. Heterogenous Catalysis Mediated by Plasmon Heating. *Nano Lett.*, **9**(12), 4417–4423.

[2] Arbouet, A., Voisin, C., Christofilos, D., Langot, P., Del Fatti, N., Vallée, F., Lermé, J., Celep, G., Cottancin, E., Gaudry, M., Pellarin, M., Broyer, M., Maillard, M., Pileni, M. P., and Treguer, M. 2003. Electron–Phonon Scattering in Metal Clusters. *Phys. Rev. Lett.*, **90**(17), 177401.

[3] Baffou, G., and Rigneault, H. 2011. Femtosecond-Pulsed Optical Heating of Gold Nanoparticles. *Phys. Rev. B*, **84**, 035415.

[4] Baffou, G., Girard, C., and Quidant, R. 2010a. Mapping Heat Origin in Plasmonics Structures. *Phys. Rev. Lett.*, **104**, 136805.

[5] Baffou, G., Quidant, R., and García de Abajo, F. J. 2010b. Nanoscale Control of Optical Heating in Complex Plasmonic Systems. *ACS Nano*, **4**, 709.

[6] Baffou, G., Quidant, R., and Girard, C. 2010c. Thermoplasmonics Modeling: a Green's Function Approach. *Phys. Rev. B*, **82**, 165424.

[7] Baffou, G., Berto, P., Bermúdez Ureña, E., Quidant, R., Monneret, S., Polleux, J., and Rigneault, H. 2013. Photoinduced Heating of Nanoparticle Arrays. *ACS Nano*, **7**(8), 6478–6488.

[8] Baffou, G., Bermúdez Ureña, E., Berto, P., Monneret, S., Quidant, R., and Rigneault, H. 2014. Deterministic Temperature Shaping using Plasmonic Nanoparticle Assemblies. *Nanoscale*, **6**, 8984–8989.

[9] Berto, P., Mohamed, M. S. A., Rigneault, H., and Baffou, G. 2014. Time-Harmonic Optical Heating of Plasmonic Nanoparticles. *Phys. Rev. B*, **90**, 035439.

[10] Boyer, D., Tamarat, P., Maali, A., Lounis, B., and Orrit, M. 2002. Photothermal Imaging of Nanometer-Sized Metal Particles Among Scatterers. *Science*, **297**, 1160.

[11] Christopher, P., Xin, H., and Linic, S. 2011. Visible-Light-Enhanced Catalytic Oxidation Reactions on Plasmonic Silver Nanostructures. *Nature Chem.*, **3**, 467–472.

[12] Donner, J.S., Baffou, G., McCloskey, D., and Quidant, R. 2011. Plasmon-Assisted Optofluidics. *ACS Nano*, **5**, 5457–5462.

[13] Gaiduk, A., Ruijgrok, P. V., Yorulmaz, M., and Orrit, M. 2010. Detection Limits in Photothermal Microscopy. *Chem. Sci.*, **1**, 343.

[14] Govorov, A. O., Zhang, W., Skeini, T., Richardson, H., Lee, J., and Kotov, N. A. 2006. Gold Nanoparticle Ensembles as Heaters and Actuators: Melting and Collective Plasmon Resonances. *Nanoscale Res. Lett.*, **1**, 84.

[15] Grua, P., Morreeuw, J. P., Bercegol, H., Jonusauskas, G., and Vallée, F. 2003. Electron Kinetics and Emission for Metal Nanoparticles Exposed to Intense Laser Pulses. *Phys. Rev. B*, **68**, 035424.

[16] Heber, A., Selmke, M., and Cichos, F. 2015. Thermal Diffusivity Measured Using a Single Plasmonic Nanoparticle. *Phys. Chem. Chem. Phys.*, **17**, 20868.

[17] Hodak, J. H., Henglein, A., and Hartland, G. V. 1999. Size Dependent Properties of Au Particles: Coherent Excitation and Dephasing of Acoustic Vibrational Modes. *J. Chem. Phys.*, **111**, 8613.

[18] Hodak, J. H., Henglein, A., and Hartland, G. V. 2000. Electron–Phonon Coupling Dynamics in Very Small (between 2 and 8 nm Diameter) Au Nanoparticles. *J. Chem. Phys.*, **112**, 5942.

[19] Hu, M., and Hartland, G. V. 2002. Heat Dissipation for Au Particles in Aqueous Solution: Relaxation Time versus Size. *J. Phys. Chem. B*, **106**, 7029.

[20] Huang, W., Qian, W., El-Sayed, M. A., Ding, Y., and Wang, Z. L. 2007. Effect of Lattice Crystallinity on the Electron–Phonon Relaxation Rates in Gold Nanoparticles. *J. Phys. Chem. C*, **111**, 10751.

[21] Inouye, H., Tanaka, K., Tanahashi, I., and Hirao, K. 1998. Ultrafast Dynamics of Nonequilibirium Electrons in a Gold Nanoparticle System. *Phys. Rev. B*, **57**, 11334.

[22] Jackson, J. D. 1999. *Classical Electrodynamics*. Wiley.

[23] Katayama, T., Setoura, K., Werner, D., Miyasaka, H., and Hashimoto, S. 2014. Picosecond-to-Nanosecond Dynamics of Plasmonic Nanobubbles from Pump Probe Spectral Measurements of Aqueous Colloidal Gold Nanoparticles. *Langmuir*, **30**, 9504–9513.

[24] Keblinski, P., Cahill, D. G., Bodapati, A., Sullivan, C. R., and Taton, T. A. 2006. Limits of Localized Heating by Electromagnetically Excited Nanoparticles. *J. Appl. Phys.*, **100**, 054305.

[25] Letfullin, R. R., George, T. F., Duree, G. C., and Bollinger, B. M. 2008. Ultrashort Laser Pulse Heating of Nanoparticles: Comparison of Theoretical Approaches. *Adv. Opt. Technol.*, **2008**, 251718.

[26] Link, S., Burda, C., Wang, Z. L., and El-Sayed, M. A. 1999. Electron Dynamics in Gold and Gold–Silver Alloy Nanoparticles: the Influence of a Nonequilibrium Electron Distribution and the Size Dependence of the Electron–Phonon Relaxation. *J. Chem. Phys.*, **111**, 1255.

[27] Metwally, K., Mensah, S., and Baffou, G. 2015. Fluence Threshold for Photothermal Bubble Generation Using Plasmonic Nanoparticles. *J. Phys. Chem. C*, **119**, 28586–28596.

[28] Mukherjee, S., Zhou, L., Goodman, A. M., Large, N., Ayala-Orozco, C., Zhang, Y., Nordlander, P., and Halas, N. J. 2013. Hot-Electron-Induced Dissociation of H_2 on Gold Nanoparticles Supported on SiO_2. *J. Am. Chem. Soc.*, **136**, 64–67.

[29] Polleux, J., Rasp, M., Louban, I., Plath, N., Feldhoff, A., and Spatz, J. P. 2011. Benzyl Alcohol and Block Copolymer Micellar Lithography: A Versatile Route to Assembling Gold and in Situ Generated Titania Nanoparticles into Uniform Binary Nanoarrays. *ACS Nano*, **5**(8), 6355–6364.

[30] Richardson, H. H., Carlson, M. T., Tandler, P. J., Hernandez, P., and Govorov, A. O. 2009. Experimental and Theoretical Studies of Light-to-Heat Conversion and Collective Heating Effects in Metal Nanoparticle Solutions. *Nano Lett.*, **9**, 1139.

[31] Setoura, K., Okada, Y., Werner, D., and Hashimoto, S. 2013. Observation of Nanoscale Cooling Effects by Substrates and the Surrounding Media for Single Gold Nanoparticles under CW-Laser Illumination. *ACS Nano*, **7**(9), 7874–7885.

[32] Vásquez Vásquez, C., Vaz, B., Giannini, V., Pérez-Lorenzo, M., Alvarez-Puebla, R. A., and Correa-Duarte, M. A. 2013. Nanoreactors for Simultaneous Remote Thermal Activation and Optical Monitoring of Chemical Reactions. *J. Am. Chem. Soc.*, **135**, 13616–13619.

[33] Volz, S. (ed). 2007. *Microscale and Nanoscale Heat Transfer*. Springer.

Numerical Simulation Techniques

This chapter aims at describing simple numerical tools to determine the temperature distribution in arbitrary complex plasmonic systems. As the delivered heat power is nothing but the light power absorbed by the system, computing the absorption cross section is often a necessary step prior to any thermal calculation. For this reason, the first section of this chapter is devoted to explaining how absorption cross sections of nanoparticles can be calculated. Then, the second section is intended to describe some analytical and numerical tools suited to computing the temperature distribution inside and outside nanosources of heat.

3.1 Absorption Cross Section of a Nanoparticle

Computing absorption cross sections in plasmonics is no longer the purview of experienced theoreticians. Several simple tools have been developed over the last decade for this purpose, and not only commercial software. This section describes several simple numerical approaches to computing absorption cross sections: the physical quantities that form the basis of any photothermal effect in plasmonics.

3.1.1 Spherical Particles and Mie Theory

In this subsection, we consider a spherical particle of radius a, complex electric permittivity $\underline{\varepsilon} = \underline{n}^2$ embedded in a dielectric medium of permittivity $\varepsilon_s = n_s^2$. This particle is illuminated by a plane wave of angular frequency $\omega = 2\pi\, c/\lambda_0 = k\, c/n_s$.

As explained in Chapter 1 on page 4, in the quasistatic approximation, optical cross sections of spherical nanoparticles can be estimated using a simple closed-form expression of the polarizability $\underline{\alpha}$ [13]:

$$\begin{aligned}
\sigma_{\text{ext}} &= k\,\text{Im}(\underline{\alpha}) \\
\sigma_{\text{sca}} &= \frac{k^4}{6\pi}\,|\underline{\alpha}|^2 \\
\sigma_{\text{abs}} &= \sigma_{\text{ext}} - \sigma_{\text{sca}}
\end{aligned} \tag{3.1}$$

where

$$\underline{\alpha} = 3\,V\,\varepsilon_s\,\frac{\underline{\varepsilon}(\omega) - \varepsilon_s}{\underline{\varepsilon}(\omega) + 2\varepsilon_s}. \tag{3.2}$$

A refined closed-form expression of $\underline{\alpha}$ exists, which is a third-order development in ka [11]. It remains valid only for small values of ka:

$$\underline{\alpha} = \varepsilon_s V \frac{1 - \left(\frac{1}{10}\right)(\varepsilon + \varepsilon_s)x^2}{\left(\frac{1}{3} + \frac{\varepsilon_s}{\varepsilon - \varepsilon_s}\right) - \frac{1}{30}(\varepsilon + 10\varepsilon_s)x^2 - i\frac{16}{9}\varepsilon_s^{3/2}x^3} \tag{3.3}$$

where $V = \frac{4\pi}{3}a^3$ and $x = \frac{\pi a}{\lambda_0}$.

For arbitrary large spheres, the Mie theory has to be used. The reference document on this subject is the famous textbook by Bohren and Huffman, but the information is not easy to extract [7]. A self-sufficient description of the Mie theory is given below, along with MATLAB code.

Let us define a set of useful dimensionless parameters:

$$\underline{m} = \underline{n}/n_s$$

$$v = k\,a$$

$$\underline{w} = \underline{m}\,v.$$

In these conditions, the extinction, scattering and absorption cross sections are given by the formulae:

$$\sigma_{ext} = \frac{2\pi}{k^2} \sum_{j=1}^{\infty} (2j+1)\,\mathrm{Re}(\underline{\alpha}_j + \underline{\beta}_j)$$

$$\sigma_{sca} = \frac{2\pi}{k^2} \sum_{j=1}^{\infty} (2j+1)\,(|\underline{\alpha}_j|^2 + |\underline{\beta}_j|^2)$$

$$\sigma_{abs} = \sigma_{ext} - \sigma_{sca}$$

where

$$\underline{\alpha}_j = \frac{\underline{m}\,\psi_j(\underline{w})\,\psi_j'(v) - \psi_j(v)\,\psi_j'(\underline{w})}{\underline{m}\,\psi_j(\underline{w})\,\xi_j'(v) - \xi_j(v)\,\psi_j'(\underline{w})} \tag{3.4}$$

$$\underline{\beta}_j = \frac{\psi_j(\underline{w})\,\psi_j'(v) - \underline{m}\,\psi_j(v)\,\psi_j'(\underline{w})}{\psi_j(\underline{w})\,\xi_j'(v) - \underline{m}\,\xi_j(v)\,\psi_j'(\underline{w})}. \tag{3.5}$$

In these expressions, ψ_j and ξ_j are Ricatti–Bessel functions defined as:

$$\psi_j(x) = \sqrt{\frac{\pi x}{2}}\,J_{j+\frac{1}{2}}(x)$$

$$\xi_j(x) = \sqrt{\frac{\pi x}{2}}\left[J_{j+\frac{1}{2}}(x) + i\,Y_{j+\frac{1}{2}}(x).\right]$$

J_v and Y_v are Bessel functions of the first and second order, respectively. They are standard MATLAB functions, named respectively `besselj` and `bessely`. Note that these functions are solutions of the Bessel differential equation:

$$x^2\frac{d^2y}{dx^2} + 2x\frac{dy}{dx} + \left[x^2 - v(v+1)\right]y = 0$$

while ψ_j and ξ_j are solutions of the following differential equation:

$$x^2\frac{d^2y}{dx^2} + \left[x^2 - j(j+1)\right]y = 0.$$

ψ_j and ξ_j can be expressed as a sum of sines and cosines. For instance, the first terms read:

$$\psi_0(x) = \sin(x)$$
$$\xi_0(x) = \sin(x) - i\cos(x)$$
$$\psi_1(x) = \sin(x)/x - \cos(x)$$
$$\xi_1(x) = \sin(x)/x - i(\cos(x)/x + \sin(x)).$$

In Equations (3.4) and (3.5), the sum over j can be restricted to only a few terms, up to $j = N$. Bohren and Huffman [7] proposed the value $N = v + 4\,v^{1/3} + 2$.

In Equations (3.4) and (3.5), the primes indicate differentiation with respect to the argument in parentheses. The derivatives can be conveniently expressed as follows:

$$\psi_j'(x) = \psi_{j-1}(x) - \frac{j}{x}\,\psi_j(x)$$
$$\xi_j'(x) = \xi_{j-1}(x) - \frac{j}{x}\,\xi_j(x).$$

Here is a short piece of MATLAB code that reproduces this procedure.

MATLAB code: *Mie theory*

```
1    lambda0=532;           % wavelength in nm
2    a=20;                  % radius in nm
3    n_s=1.33;
4    n_Au=0.5144+2.2343*1i;% refractive index of gold
5                           % at lambda0
6    m=n_Au/n_s;
7    k=2*pi*n_s/lambda0;
8    x=k*a;
9    z=m*x;
10   N=round(2+x+4*x^(1/3))
11   j=(1:N);
12   sqr=sqrt(pi*x/2);
13   sqrm=sqrt(pi*z/2);
14   phi=sqr.*besselj(j+0.5,x);
15   xi=sqr.*(besselj(j+0.5,x)+i*bessely(j+0.5,x));
16   phim=sqrm.*besselj(j+0.5,z);
17   phi1=[sin(x),phi(1:N-1)];
18   phi1m=[sin(z),phim(1:N-1)];
19   y=sqr*bessely(j+0.5,x);
20   y1=[-cos(x), y(1:N-1)];
21   phip=(phi1-j/x.*phi);
22   phimp=(phi1m-j/z.*phim);
23   xip=(phi1+i*y1)-j/x.*(phi+i*y);
24   aj=(m*phim.*phip-phi.*phimp)./(m*phim.*xip-xi.*phimp);
25   bj=(phim.*phip-m*phi.*phimp)./(phim.*xip-m*xi.*phimp);
26   Qsca=sum((2*j+1).*(abs(aj).*abs(aj)+abs(bj).*abs(bj)));
27   Qext=sum((2*j+1).*real(aj+bj));
```

```
28   Qext=Qext*2*pi/(k*k);  % extinction cross sect. in nm^2
29   Qsca=Qsca*2*pi/(k*k);  % scattering cross sect. in nm^2
30   Qabs=Qext-Qsca;        % absorption cross sect. in nm^2
```

3.1.2 Ellipsoids (Mie–Gans Theory)

Simple expressions of the absorption cross section for nanoparticles of arbitrarily complicated morphologies are not usually available. However, one can still write a simple expression in the quasistatic regime when considering *ellipsoids* using the Mie–Gans formalism [13]. When the polarization of the incident light is along the axis j, the plasmonic enhancement factor (defined from the polarizability $\underline{\alpha}$ by $\underline{\alpha} = V\varepsilon_s\underline{\xi}$) reads

$$\underline{\xi}_j = \frac{1}{3}\frac{\underline{\varepsilon} - \varepsilon_s}{\varepsilon_s + L_j(\underline{\varepsilon} - \varepsilon_s)}, \tag{3.6}$$

where

$$L_j = \frac{R_1 R_2 R_3}{2} \int_0^\infty \frac{\mathrm{d}u}{(u + R_j^2)\sqrt{(u + R_1^2)(u + R_2^2)(u + R_3^2)}}, \tag{3.7}$$

where R_1, R_2 and R_3 denote the radii of the ellipsoid along the principal axes $j = 1, 2, 3$. Note that the depolarization factors L_j satisfy the relation $L_1 + L_2 + L_3 = 1$. $L_1 = L_2 = L_3 = 1/3$ for a sphere. L_j values are simple to compute with a few lines of code, for instance using MATLAB:

MATLAB code: *Calculation of the depolarization factors of ellipsoids*

```
1    % ellipsoid radii
2    R1=3; R2=1; R3=1;
3
4    % number of discretizatoin points
5    Nstep=1000;
6
7    % calculation
8    l1=0; l2=0; l3=0;
9    smax=max([R1*R1,R2*R2,R3*R3]);
10   step=smax/Nstep;
11
12   for s=0:step:100*smax;
13     sa=s;
14     sb=s+step;
15     roota=sqrt((sa+R1*R1)*(sa+R2*R2)*(sa+R3*R3));
16     rootb=sqrt((sb+R1*R1)*(sb+R2*R2)*(sb+R3*R3));
17     l1a=(sa+R1*R1)*roota;
18     l1b=(sb+R1*R1)*rootb;
19     l2a=(sa+R2*R2)*roota;
20     l2b=(sb+R2*R2)*rootb;
21     l3a=(sa+R3*R3)*roota;
22     l3b=(sb+R3*R3)*rootb;
23     l1=l1+(1/l1a+1/l1b)/2;
24     l2=l2+(1/l2a+1/l2b)/2;
25     l3=l3+(1/l3a+1/l3b)/2;
```

```
26   end
27
28   L1=step*R1*R2*R3/2*l1;
29   L2=step*R1*R2*R3/2*l2;
30   L3=step*R1*R2*R3/2*l3;
```

All the cross sections can be retrieved from $\underline{\xi}_j$ according to Equations 1.18, 1.19 and 1.20:

$$\sigma_{\text{ext},j} = k_0 n_s^2 V \text{Im}(\underline{\xi}_j) \tag{3.8}$$

$$\sigma_{\text{sca},j} = \frac{k_0^4 n_s^4}{6\pi} V^2 |\underline{\xi}_j|^2 \tag{3.9}$$

$$\sigma_{\text{abs},j} = \sigma_{\text{ext},j} - \sigma_{\text{sca},j}. \tag{3.10}$$

3.1.3 Nanoparticles of Arbitrary Morphology

General Methods

This part now addresses the most common case of a nanostructure of arbitrary morphology. In this case, the Mie theory no longer applies, and we cannot avoid a meshing of the nanostructure. At least three families of numerical approaches exist. They are all suited to computing cross sections, but also the electric field amplitude anywhere in the system.

The first family is based on a meshing of the whole system: nanostructure and surrounding medium. This implies the design of a bounded space surrounded by PMLs (perfect matching layers). This is a case for FDTD programs and commercial software such as Comsol [2]. Such an approach can be time-consuming since it requires a large mesh.

The second family is based on a meshing of solely the nanostructure (not the surrounding medium) and solves a self-consistent problem to compute the electric field amplitude anywhere in the system. This is the so-called Green Dyadic Tensor Technique [9].

Finally, the third approach is based on the use of a Boundary Element Method (BEM) [8, 13]. Such an approach only requires the meshing of the interface between the structure and the surrounding medium where fictitious inner and outer surface charges are calculated.

I will not linger over the details of these numerical approaches. I invite the reader to refer to the corresponding articles [9, 8, 13]. However, I will detail what I believe to be one of the most powerful procedures for carrying out numerical simulations in plasmonics: the parallel use of Blender and MNPBEM. I devote the whole of the following section to this subject.

Blender + MNPBEM

In 2012, two Austrian researchers, U. Hohenester and A. Trügler, published an article dedicated to describing their development of a MATLAB toolbox designed to carry out numerical simulations in plasmonics [10]. The underlying theory is the BEM and they name their toolbox MNPBEM. Thanks to this MATLAB package, the use of the BEM method for plasmonics is no longer restricted to a couple of groups in the world, and to experienced theoreticians. Any MATLAB user can now easily conduct numerical simulations in plasmonics using the MNPBEM package.

I propose to make this approach even more powerful using Blender [1], a free, professional 3D modeler. This very powerful free software makes it possible to simply design arbitrarily sophisticated meshed surfaces and interfaces though a user-friendly interface. To make it work, the structure has to be meshed using triangular meshing, and it has to be designed and modified in the *Edit mode*, not in the *Object mode*. Once the structure has been designed, the corresponding meshing can be exported as an ASCII file using this command in the Blender command window:

Blender code: *To export a 3D triangular mesh as an ASCII file*

```
bpy.ops.export_mesh.ply(filepath="/nanoparticle.txt",
check_existing=True,filter_glob="*.ply",
use_normals=True,use_uv_coords=True,use_colors=True)
```

This command has been tested in version 2.77a of Blender (April 2016 version). It may differ for other versions. Finally, the exported ASCII file has to be read using an appropriate MATLAB program.

MATLAB function: *To import into MATLAB a 3D triangular mesh made by Blender*

```
function p=blenderImport(fileName,taille)

data=fileread(fileName);
%% Find number of vertices
pos=findstr(data,'element_vertex');
comp=0;
j=0;
valString=findstr('a','e');
while comp==0
    numString=data(pos+15+j);
    num=findstr('0123456789',numString);
    comp=isempty(num);
    if comp==0
        valString=[valString numString];
    end
    j=j+1;
end
eval(['Np=' valString ';'])
fprintf('model_with:\n__%d_vertices\n',Np)

%% Find number of faces
pos=findstr(data,'element_face');
comp=0;
j=0;
valString=findstr('a','e');
while comp==0
    numString=data(pos+13+j);
    num=findstr('0123456789',numString);
    comp=isempty(num);
    if comp==0
        valString=[valString numString];
    end
    end
```

```matlab
        j=j+1;
end
eval(['Nf=' valString ';'])
fprintf('  %d faces\n',Nf)

%% Find mesh point coordinates and triangle links
pos=findstr(data,'end_header');
j=0;
m=1;
comp=0;
res=zeros(1,6*Np);
while j < 6*Np  + 4*Nf
    valString=findstr('a','e');
    while comp==0
        numString=data(9+pos+m);
        num=findstr('0123456789.-e',numString);
        comp=isempty(num);
        if comp==0
            valString=[valString numString];
        end
        m=m+1;
    end
    comp=0;
    if ~isempty(valString)
        j=j+1;
        eval(['res(j)=' valString ';'])
    end
end
res2=reshape(res(1:6*Np),6,Np);

% points coordinates x, y, z
verts=zeros(Np,3);
for j=1:Np
    verts(j,1)=res2(1,j);
    verts(j,2)=res2(2,j);
    verts(j,3)=res2(3,j);
end

% points triangles n1,n2,n3
res3=reshape(res(6*Np+1:6*Np+4*Nf),4,Nf)';
faces=[res3(:,2:4),NaN*zeros(Nf,1)]+1-min(res3(:));

%% computation of the coord. of the centres of the faces
f.x=zeros(1,Nf);
f.x=zeros(1,Nf);
f.x=zeros(1,Nf);
for j=1:Nf
    n1=faces(j,1);
    n2=faces(j,2);
    n3=faces(j,3);
    f.x(j)=(verts(n1,1)+verts(n2,1)+verts(n3,1))/3;
    f.y(j)=(verts(n1,2)+verts(n2,2)+verts(n3,2))/3;
    f.z(j)=(verts(n1,3)+verts(n2,3)+verts(n3,3))/3;
end

%% computation of Rj=|sj|
```

```
f.R=zeros(1,Np);
for j=1:Np
    f.R(j)=sqrt( verts(j,1)*verts(j,1)+...
                 verts(j,2)*verts(j,2)+...
                 verts(j,3)*verts(j,3) );
end

%% orientation of the surface elements
jf=1;
n1=faces(jf,1);
n2=faces(jf,2);
n3=faces(jf,3);
ori=verts(n3,:)*cross(verts(n1,:),verts(n2,:))';
if ori>0
    for jf=1:Nf
        n1=faces(jf,1);
        n2=faces(jf,2);
        faces(jf,1)=n2;
        faces(jf,2)=n1;
    end
end
verts( :, 1 ) = verts( :, 1 ) * taille(1);
verts( :, 2 ) = verts( :, 2 ) * taille(2);
verts( :, 3 ) = verts( :, 3 ) * taille(3);
p = particle( verts, faces );
```

Once the mesh has been properly imported into MATLAB, the MNPBEM package can be used in a standard way, as depicted in the documentation. As an example, Figure 3.1 presents the absorption cross section spectrum computed with the MNPBEM package for a gold nanoparticle morphology designed using Blender. Interestingly, the same 3D mesh can be used both to perform the numerical simulation and to render realistic 3D images, the original purpose of Blender. The 3D structure in Figure 3.1a therefore *rigorously* represents the morphology used in the numerical simulations to obtain the results of Figure 3.1b,c.

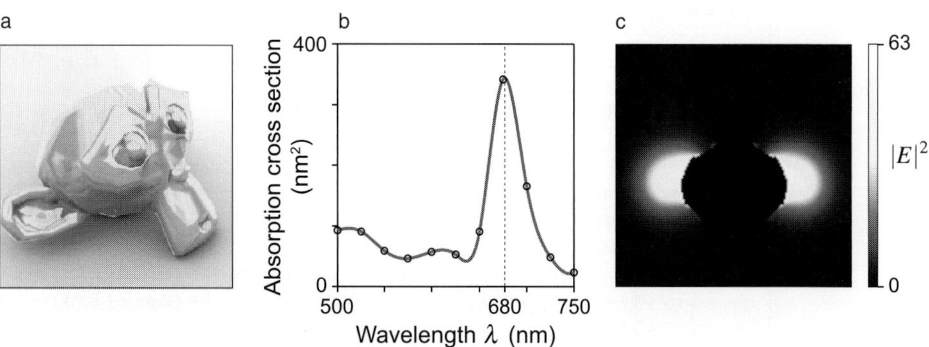

Fig. 3.1 (a) Three-dimensional view of the nanoparticle, rendered using Blender. (b) Absorption cross section obtained using the MNPBEM package from the same 3d mesh created by Blender. (c) Near-field map of the same nanoparticle.

3.2 Temperature of a Nanoparticle in the Steady State

3.2.1 Spherical Particle

The absorption cross section σ_abs is often the important first physical quantity to determine prior to any temperature calculation. For instance, for a *spherical* particle or radius a, once σ_abs is determined (for instance, using Mie theory as detailed in the previous section), the temperature increase in the steady state within the nanoparticle simply reads

$$\delta T = \frac{\sigma_\text{abs} I}{4\pi \kappa_\text{s} a} \tag{3.11}$$

as derived in Subsection 2.2.2 on page 47. In this expression, a is the radius of the nanoparticle, I the light irradiance (power per unit area) and κ_s the thermal conductivity of the surrounding medium. In this equation, one assumes that the nanoparticle temperature is uniform. This assumption is valid most of the time in nanoplasmonics, as explained in Subsection 2.2.2, and we will make this approximation in the rest of this chapter.

Thus, determining the temperature of a spherical nanoparticle does not require numerical simulations. But this is an exception: most of the time, numerical simulations are required. The other exception is the case of an ellipsoid, as described in the next subsection.

3.2.2 Ellipsoids and Spheroids

An ellipsoid is the three-dimensional analogue of an ellipse. The standard equation of an ellipsoid centred at the origin of a Cartesian coordinate system and aligned with the axes is

$$(x/a_x)^2 + (y/a_y)^2 + (z/a_z)^2 = 1. \tag{3.12}$$

a_x, a_y, a_z are the semi-axis lengths of the ellipsoid along the axes x, y, z.

An ellipsoid is called a *spheroid* when two of the semi-axes are equal, for instance $a_x = a_y$.

- $a_x = a_y > a_z$ corresponds to an *oblate* spheroid.

- $a_x = a_y < a_z$ corresponds to an *prolate* spheroid.

For spheroids ($a_x = a_y \hat{=} a_\rho$), one usually defines the aspect ratio as

$$r = a_z/a_\rho. \tag{3.13}$$

An oblate spheroid corresponds to $r < 1$ and a prolate spheroid to $r > 1$. One can also define the equivalent radius a_0 of the ellipsoid, which is the radius of a sphere of equal volume:

$$a_0 = (a_\rho^2 a_z)^{1/3}. \tag{3.14}$$

Although analytical expressions for the temperature increase δT exist for ellipsoids, only the case of spheroids is detailed in this section. The expressions for ellipsoids are more complicated, involving Riemann functions, and are not really of interest.

In practice, nanoparticles rarely resemble spheroids. But such morphologies can be conveniently used to conduct numerical simulation aiming at investigating the effect of a nanoparticle asymmetry or elongation.

The thermodynamic problem consisting in calculating the temperature increase δT of an ellipsoid has formal analogy with an electrostatic problem [4]. Temperature and electrostatic potential are both governed by the Poisson equation, which becomes the Laplace equation in the absence of sources: $\nabla^2 A = 0$. The temperature, the thermal conductivity and the heat source density in thermodynamics become the electric potential, the permittivity and the charge density in electrostatics. In particular, the problem of an electric conductor of charge Q is equivalent to the problem of an object of infinite thermal conductivity, heated by a power P. In electrostatics, one has $U = Q/C_{el}$ where U is the electrostatic potential of the object and C_{el} its electric capacitance. In thermodynamics, the equivalent relation reads [4]

$$\delta T = \frac{P}{C_{th}} \tag{3.15}$$

where C_{th} is similar to a thermal capacitance. For instance, for a sphere of radius a, $C_{el} = 4\pi\epsilon_0 a$ in electrostatics, while in thermodynamics:

$$C_{th} = 4\pi\kappa a. \tag{3.16}$$

Determining the electric capacitances of *spheroids* is a problem that was solved a long time ago. Fortunately, there exist closed-form expressions. They are different depending on the nature of the spheroid (oblate of prolate). Let us define the Laplace radius a_L such that the thermal capacitance can be written $C_{th} = 4\pi\kappa a_L$ [5] and the temperature increase of the spheroid:

$$\delta T = \frac{\sigma_{abs} I}{4\pi\kappa a_L}. \tag{3.17}$$

The concept of the Laplace radius was introduced in Reference [5]. For prolate spheroids ($r > 1$), the Laplace radius reads

$$a_L = a_z \frac{2\sqrt{1 - r^{-2}}}{\ln \dfrac{1 + \sqrt{1 - r^{-2}}}{1 - \sqrt{1 - r^{-2}}}}. \tag{3.18}$$

For oblate spheroids ($r < 1$), the Laplace radius reads

$$a_L = a_\rho \frac{\sqrt{1 - r^2}}{\arcsin\sqrt{1 - r^2}}. \tag{3.19}$$

3.2.3 Discs and Rods

Discs and rods are common nanoparticle morphologies in plasmonics. Discs are often fabricated using e-beam lithography while rods can easily be synthesized chemically.

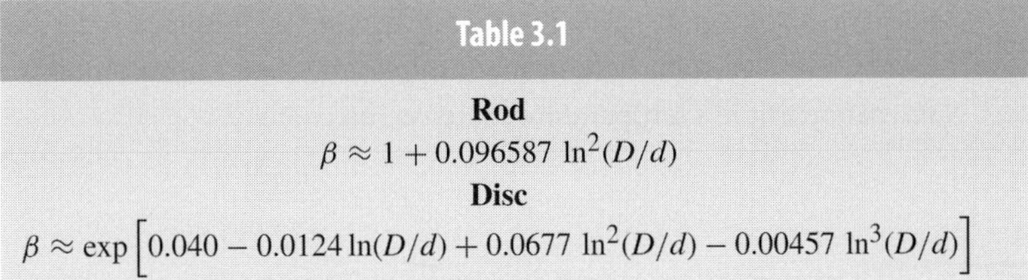

Table 3.1

Rod
$$\beta \approx 1 + 0.096587 \, \ln^2(D/d)$$
Disc
$$\beta \approx \exp\left[0.040 - 0.0124 \ln(D/d) + 0.0677 \ln^2(D/d) - 0.00457 \ln^3(D/d)\right]$$

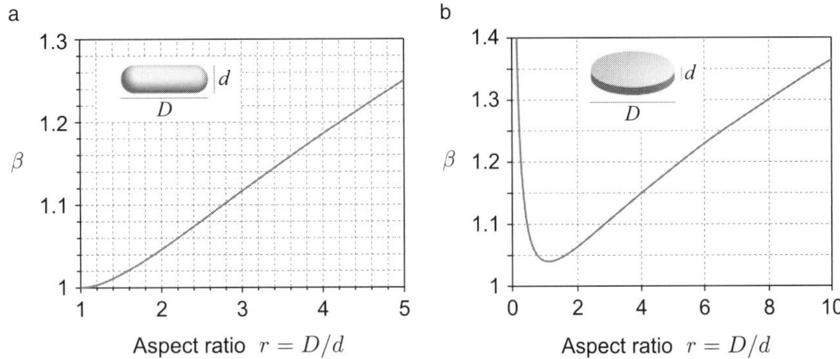

Fig. 3.2 Correction factor β to determine the temperature increase of nanoparticles as a function of their aspect ratio $r = D/d$. (a) Case of a rod (with hemispherical caps). (b) Case of a disc.

Unlike ellipsoids, there exist no simple closed-form expressions of the temperature for these morphologies. In Reference [4], thermal capacitances C_{th} have been calculated for discs and rods for various aspects ratios, and the results have been fitted using approximate closed-form expressions. Let us define a correction factor β such that [4]:

$$\delta T = \frac{\sigma_{\text{abs}} I}{4\pi \kappa a_0 \beta} \tag{3.20}$$

where a_0 is the equivalent radius, *i.e.*, the radius of a sphere of equal volume. Fitting formulas of numerical simulations are given below as a function of the aspect ratio D/d [4]:

For a given absorption cross section, any deviation of the nanoparticle from a sphere at constant volume leads to a decrease of its temperature due to a larger surface-to-volume ratio leading to a more efficient release of the energy in the surroundings. Consequently, β is a dimensionless number that is supposed to be larger than unity. Interestingly, β is only dependent on the shape of the nanoparticle, not on the size. For instance, for a nanorod, β is only dependent on the aspect ratio. Figure 3.2 plots the fitting expressions for rods and discs.

3.2.4 Nanoparticles of Arbitrary Morphology

Calculation of the Laplace Radius a_{L} by Laplace Matrix Inversion

For other more sophisticated geometries (triangles, dimers, stars, etc.), numerical simulations have to be conducted to determine the nanoparticle's temperature increase. A simple numerical algorithm exists that can determine the Laplace radius a_L [5] defined by Equation (3.17). For a nanoparticle of any morphology, the Laplace radius a_L is defined such that the nanoparticle's temperature increase reads

$$\delta T = \frac{\sigma_{abs} I}{4\pi \kappa a_L}. \tag{3.21}$$

The numerical procedure developed in Reference [5] is the following. The structure has to be meshed following a regular distribution of vertices. One can then define what is named the Laplace matrix $\mathbb{A}$ defined by its elements

$$A_{ij} = 1/r_{ij} \tag{3.22}$$

where r_{ij} is the distance between the vertices i and j, and $A_{ii} = 2/p$ for a cubic meshing with lattice parameter p. The Laplace radius simply reads

$$a_L = \sum_{i,j} (\mathbb{A}^{-1})_{ij}. \tag{3.23}$$

Example MATLAB code related to a cubic particle is provided below:

MATLAB code: *To compute the Laplace radius of nanoparticles*

```
1    L=40;           %cube edge
2    p=2;            %mesh step
3    N=(L/p+1)^3;    %number of points
4    S=zeros(N,3);
5
6    n=0;
7    L2=L/2;
8    for ix=-L2:p:L2
9         for iy=-L2:p:L2
10             for iz=-L2:p:L2
11                 n=n+1;
12                 S(n,1)=ix;
13                 S(n,2)=iy;
14                 S(n,3)=iz;
15             end
16         end
17    end
18
19    A=zeros(N,N); %Laplace matrix
20    for i=1:N
21         for j=1:N
22             if i==j
23                 A(i,j)=2/p;
24             else
25                 dx=S(i,1)-S(j,1);
26                 dy=S(i,2)-S(j,2);
27                 dz=S(i,3)-S(j,3);
28                 A(i,j)=1/sqrt(dx*dx+dy*dy+dz*dz);
29             end
30         end
```

```
31   end
32   Ainv=inv(A);
33   aL=sum(Ainv(:))  %Laplace radius
```

In this particular example, the calculated Laplace radius for a 40 nm cube is $a_{eq} = 27$ nm, a value that is larger than the equivalent radius $a_0 = 24$ nm. This means that the temperature increase of a cube will be smaller than the temperature increase of a sphere, for a constant heat power and nanoparticle volume. This is consistent with the fact that a sphere leads to the most efficient temperature increase due to the smallest surface/volume ratio.

A meshing *in volume* as presented in Reference [5] is sufficient but not necessary. The same results could be obtained by meshing only the surface of the nanoparticle. Indeed, assuming the temperature is uniform, for a given delivered power $\sigma_{abs}I$, the nanoparticle temperature increase would be the same if the heat generation were located only at the nanoparticle interface. To create meshings of surfaces, MNPBEM [10] or Blender [1] can conveniently be used. Here, for instance, is MATLAB code that shows how to deal with the mesh and compute a Laplace radius using MNPBEM:

MATLAB code: *Computing a nanoparticle temperature using MNPBEM*

```
1    p = trisphere( 1444, 10 );
2    tri=p.faces(:,1:3);    % number of the 3 vertices of the
     triangles
3    pos_v=p.verts;         % coordinates of the vertices
4    Nt=size(tri)*[1;0];    % number of triangles
5    pos_t=zeros(Nt,3);     % coordinates of the centres of the
     triangles
6    aire_t=zeros(Nt,1);    % areas of the triangles
7    for it=1:Nt
8        ax=pos_v(tri(it,1),1);
9        ay=pos_v(tri(it,1),2);
10       az=pos_v(tri(it,1),3);
11       bx=pos_v(tri(it,2),1);
12       by=pos_v(tri(it,2),2);
13       bz=pos_v(tri(it,2),3);
14       cx=  pos_v(tri(it,3),1);
15       cy=pos_v(tri(it,3),2);
16       cz=pos_v(tri(it,3),3);
17       pos_t(it,1)=(ax+bx+cx)/3;
18       pos_t(it,2)=(ay+by+cy)/3;
19       pos_t(it,3)=(az+bz+cz)/3;
20       aire_t(it)=0.5*norm(cross([(ay-ax);(by-bx);(cy-cx)
     ],...
21           [(az-ax);(bz-bx);(cz-cx)]));
22   end
23   A=zeros(Nt,Nt);        % Laplace matrix
24   for ii=1:Nt
25       for jj=1:Nt
26           dx=pos_t(ii,1)-pos_t(jj,1);
27           dy=pos_t(ii,2)-pos_t(jj,2);
28           dz=pos_t(ii,3)-pos_t(jj,3);
29           A(ii,jj)=sqrt(dx*dx+dy*dy+dz*dz);
30       end
```

```
31  end
32  Amin=sqrt(sum(aire_t)/Nt);   % Average neighbor interdistance
33  for ii=1:Nt
34      A(ii,ii)=Amin/2;
35  end
36  A=1./A;
37  Ainv=inv(A);
38  aL=sum(sum(Ainv));   % Laplace radius
```

This code computes the Laplace radius of a sphere of radius 5 nm. The result gives $a_L = 5.08$ nm (instead of 5). For other morphologies, line 1 of the code defining a particle p has to be modified accordingly.

3.3 Temperature Anywhere, and Not Necessarily in the Steady State

This section is devoted to explaining what kinds of simple computation techniques can be used to investigate more than merely the temperature of nanospheres, or nanoparticles temperature in the steady state.

3.3.1 Spheres in Transient Regime

Studying the dynamics of transient temperature variations is a common requirement in thermoplasmonics, especially in the context of pulsed illumination. This problem is more complex, but sometimes it is sufficient to consider a simple sphere to address the problem, which makes it possible to conduct numerical simulations in one dimension and to markedly reduce the computation time. Considering spheres can be interesting when the size of the nanoparticle matters and not really the morphology.

Such an approach has been used in Reference [12], where the aim was to study the influence of the size of the nanoparticle on the bubble generation threshold. This article used MATLAB code based on a Runge–Kutta 4 method (previously developed and used in Reference [3]). This code enables the computation of the spatiotemporal evolution of the temperature inside and outside the nanoparticle $T(r, t)$ following the absorption of a pulse of light. This code is too long to be reproduced here, but the full MATLAB code was provided in Supplementary Information of Reference [12] and the reader is invited to refer to this source to use this approach.

3.3.2 Temperature in the Surrounding Medium for a Nanoparticle of Arbitrary Morphology

Section 3.2 focused on the temperature increase within the nanoparticle. This section now focuses on the temperature outside the nanoparticle. The main difference is that the temperature outside the nanoparticles is nonuniform. Thus, the question is no longer one of determining the temporal evolution but rather its spatiotemporal evolution, which makes the problem more complicated.

…Using a Green's Function Approach

At first glance, one may think about a Green's function approach consisting in discretizing the nanoparticle volume and considering each unit cell as a source of heat power Q_i. Under this assumption, the resulting temperature increase anywhere is the system would stem for these N unit cells according to the relation

$$\delta T(\mathbf{r}) = \sum_{i=1}^{N} \frac{Q_i}{4\pi \kappa_s |\mathbf{r}_i - \mathbf{r}|}. \tag{3.24}$$

This may seem natural, but it would lead to an incorrect temperature distribution as it does not take into account the fact that the thermal conductivity of the heat source (the nanoparticle) κ is different from that of the surroundings κ_s. In other words, using Equation (3.24) amounts to considering that the nanoparticle conductivity equals the conductivity of the surroundings.

However, this approach can be adjusted to yield valid calculations, provided a fictitious heat source distribution $(\tilde{Q}_i)_{i \in [1,N]}$ is used in Equation (3.24), as explained in Reference [5]. Once this fictitious heat source density is determined using a matrix inversion, it can be used to calculate the temperature distribution both inside and outside the nanoparticle using Equation (3.24). Note that such an approach is based on the assumption that the temperature is uniform within the nanoparticle, which is justified in most cases in nanoplasmonics. The algorithm is not detailed in this book, and the reader is invited to refer to Reference [5] for more information.

Temperature Calculations Using Comsol Multiphysics

Comsol Multiphysics can be conveniently used to compute a temperature distribution inside and outside a nanoparticle from a specified heat source distribution. Note that the source distribution within the nanoparticle can also be computed using the RF module of Comsol Multiphysics. Comsol is therefore a standalone approach to address the full problem.

The CAD (Computer-Aided Design) tool is not especially user-friendly, and much less powerful than Blender [1]. It would be difficult to use it to design complicated morphologies. However, the computations are usually very fast, even with a three-dimensional mesh, as a temperature field is scalar.

Let us first draw attention to an important caution. One should always conduct thermodynamical numerical simulations in *three dimensions*, not two, even in the case of a very elongated object. For instance, a metal wire cannot be considered as being infinitely long. Indeed, the temperature profile around an infinitely long heated wire scales as $\ln(r)$, not as $1/r$. Consequently, the temperature cannot tend toward a finite temperature (the ambient temperature) when r tends to infinity, and any numerical simulation using this boundary condition would be diverging for this reason, or would give a wrong temperature profile around the structure, strongly dependent on the boundary conditions.

Regarding the boundary conditions, it is important to design a box that is much larger than the largest dimension of the heat source. The temperature on the boundaries of the box can be the ambient temperature if the box is sufficiently large. But the slow $1/r$ decay of temperature can be problematic. For this reason, I encourage considering a spherical box of radius R (let's say ten times larger than the largest dimension of the heat source) with a temperature that equals $Q/4\pi\kappa_s R$, where Q is the heat power delivered by the heat source. Far from the heat source, the temperature is indeed bound to match this value everywhere on the sphere.

Finally, the morphology of the heat source (*i.e.*, of the plasmonic nanostructure) will have to be designed using the CAD tool of Comsol. The corresponding volume will have to be endowed with a thermal conductivity κ (and possibly with a thermal capacity and mass density, but only in the case of a transient evolution). The volume will have also to be endowed with a heat source density q which will be nothing other than the absorption cross section (see previous section) multiplied by the light irradiance (power per unit area) and divided by the volume of the nanoparticle:

$$q = \frac{\sigma_{\mathrm{abs}} I}{V} = \frac{Q}{V}. \tag{3.25}$$

This amounts to considering a uniform heat source distribution in the metal. Note that this is not in accordance with reality (see Figure 2.2 on page 39), but such a simplification is not supposed to affect the temperature profile as soon, as the thermal conductivity of the heat source is much larger than that of the surroundings.

3.3.3 Multiple Nanoparticles

It is common in nanoplasmonics to consider many nanoparticles under illumination: thousands, millions or billions of them. In this case, it is not realistic (and, as we will see, not necessary) to mesh the whole system to compute the temperature distribution.

Do Thermal Collective Effects Occur?

When considering several nanoparticles, as we have shown in the previous chapter, in Subsection 2.2.5 (page 50), the temperature increase δT_j of the nanoparticle j originates from two contributions [6], a self contribution and a contribution stemming for the other $N-1$ neighboring nanoparticles. These contributions were named, respectively, $\delta T_j^{\mathrm{self}}$ and $\delta T_j^{\mathrm{ext}}$ so that

$$\delta T_j = \delta T_j^{\mathrm{self}} + \delta T_j^{\mathrm{ext}}. \tag{3.26}$$

Depending on the relative significance of these two terms, two regimes may occur. First, if $\delta T_j^{\mathrm{self}}$ is dominant for any NP j, the system is said to be in the temperature confinement regime: temperature increases occur only in the vicinity of the nanoparticles. On the contrary, if $\delta T_j^{\mathrm{ext}}$ is dominant, a so-called thermal collective effect occurs, characterized by a smoothly varying temperature distribution on the micrometric scale, despite the nanometric nature of the heat sources. In order to easily predict the occurrence of one regime or another, a simple dimensionless parameter has to be estimated, which reads

$$\zeta_m = \frac{p}{a\,N^{(m-1)/m}} \quad \text{if } m \geq 2 \tag{3.27}$$

$$\zeta_1 = \frac{p}{a\,\ln N} \quad \text{if } m = 1, \tag{3.28}$$

where p is the average neighboring nanoparticle interdistance, a the nanoparticle radius, or the typical size of the nanoparticle in the case of a non-spherical nanoparticle, N the number of nanoparticles in the system and m the dimensionality ($m = 1$ for a chain of nanoparticles, $m = 2$ for an array, etc.). ζ_m can be conveniently used to predict the occurrence of one regime or another, but it can also be used to give an estimation of δT_j^{ext}:

$$\delta T_j^{\text{ext}} = \zeta_m^{-1}\,\delta T_j^{\text{self}}. \tag{3.29}$$

If $\zeta_m \gg 1$, numerical simulations can be conducted upon considering a single thermally isolated nanoparticle.

If $\zeta_m \sim 1$, collective effects cannot be neglected, and all the nanoparticles have to be taken into account to compute the temperature anywhere in the system. But this can easily be done numerically, considering all the N nanoparticles as point-like heat sources:

$$\delta T(\mathbf{r}) = \sum_{i=1}^{N} \frac{Q_i}{4\pi\kappa_s|\mathbf{r}_i - \mathbf{r}|}, \tag{3.30}$$

where Q_i is the heat power delivered by the nanoparticle i and $\mathbf{r}_i$, its position vector.

If $\zeta_m \ll 1$, thermal collective effects dominate and a homogenization effect occurs. One can replace the distribution of N nanoparticles with a continuous heat source density. This is made easier if the nanoparticle distribution is uniform (like in Figure 3.3). The continuous heat source density is thus uniform. Irrespectively of the geometry of the system (nanoparticles distributed in two or three dimensions), calculations can be carried out considering *continuous* 2D or 3D heat source densities. In some cases (uniform circular heat source, Gaussian beam, etc.), some closed-form expressions can be derived, as explained in Section 2.2.6.

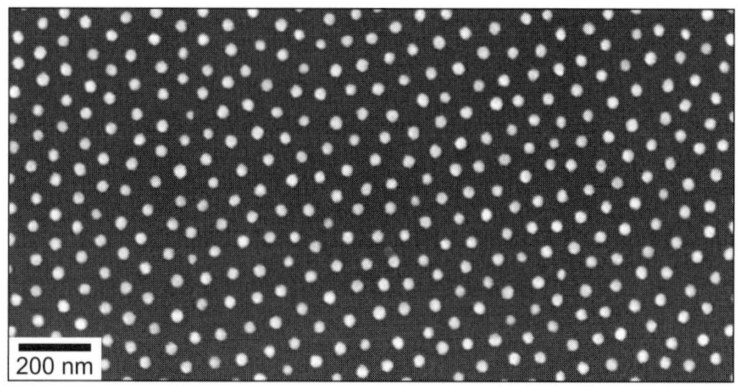

Fig. 3.3 Example of a uniform nanoparticle distribution obtained by block copolymer micellar lithography (courtesy of Julien Polleux).

Two-Dimensional Heat Sources

Quite often in thermoplasmonics, heat originates from a distribution of nanoparticles on a planar substrate (like in Figure 3.3). If thermal collective effects dominate, the heat source density can be considered continuous, despite the nanoscale and discrete nature of the heat sources, as explained above. If no simple closed-form expression exists, one can use a simple formalism based on the use of Green's functions to compute the resulting temperature distribution in three dimensions.

As explained on page 95, Equation (3.24) cannot be used to compute the temperature distribution created by a set of heat sources Q_i if the sources are connected, forming a single piece of metal, because the high thermal conductivity of the metal plays a fundamental role and is not taken into account with this approach. However, when the sources of heat Q_i are disconnected, like in Figure 3.3, one can use this equation:

$$\delta T(\mathbf{r}) = \sum_{i=1}^{N} \frac{Q_i}{4\pi\kappa_s|\mathbf{r}_i - \mathbf{r}|}. \tag{3.31}$$

When thermal collective effects occur, the heat can even be considered continuous. For a uniform, two-dimensional distribution of identical particles, one has thus:

$$\delta T(\mathbf{r}) = \iint_{S} \frac{q(\mathbf{r}')}{4\pi\kappa_s|\mathbf{r}' - \mathbf{r}|}\mathrm{d}\mathbf{r}' \tag{3.32}$$

where $q(\mathbf{r}') = AI(\mathbf{r}')$, A is the absorbance of the nanoparticle layer and $I(\mathbf{r}')$ is the light beam profile in power per unit area. This expression is nothing but the convolution between the function $q(\mathbf{r}')$ and the thermal Green's function (2.34) defined on page 45:

$$\delta T = q \otimes G. \tag{3.33}$$

Figure 3.4 gives a visual illustration of this convolution.

Numerically, a convolution can be conveniently performed using a Fourier transform formalism. By defining $\mathcal{F}f$ as the Fourier transform of any function f, one has a simple multiplication

$$\mathcal{F}\delta T = \mathcal{F}q \times \mathcal{F}G \tag{3.34}$$

$$\delta T = \mathcal{F}^{-1}(\mathcal{F}q \times \mathcal{F}G). \tag{3.35}$$

Let us attempt a practical example. Imagine you want to compute the temperature distribution upon illuminating a BCML sample (like in Figure 3.3) using a uniform laser beam

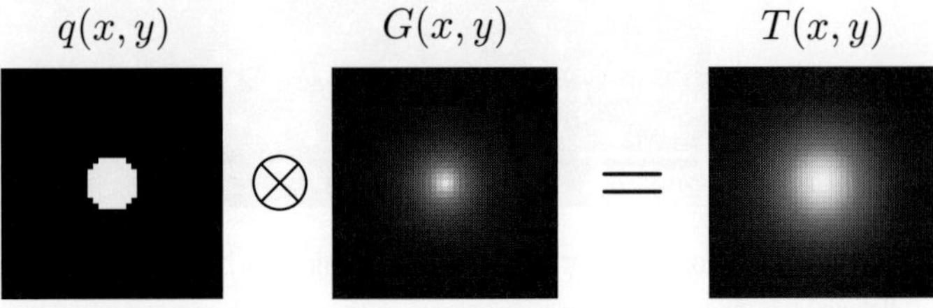

Fig. 3.4   Convolution between a heat source density and the thermal Green's function that gives the temperature profile. Low resolution 41 $\times$ 41px images have been used. The corresponding MATLAB code is given on the next page.

of diameter D. We will consider the temperature profile over a plane located at a distance h from the layer of nanoparticles and parallel to it. The nanoparticles have an absorption cross section σ_{abs} and an areal density n (in m^{-2}). The temperature distribution reads:

$$\delta T(x, y) = \frac{\sigma_{\text{abs}} n}{4\pi \kappa_s} \iint_{\mathcal{S}} \frac{1}{\sqrt{(x' - x)^2 + (y' - y)^2 + h^2}} \mathrm{d}x' \mathrm{d}y'. \tag{3.36}$$

Numerically, to compute this integral, one has to define two matrices, the first one reproducing the heat source with 1 values over the domain $\mathcal{S}$ and the second one corresponding to the Green's function $1/r$. Then, one has to compute the Fourier transforms (using the function `fft` if using MATLAB), make the matrix product element by element and compute the inverse Fourier transform (using the function `ifft` if using MATLAB).[1]

Here is MATLAB code corresponding to the case depicted in Figure 3.4.

MATLAB code: *Computing a temperature profile by convolution with a Green's function*

```
1    % Green function
2    Nx=20;
3    Ny=20;
4    x=-Nx:Nx;
5    y=-Ny:Ny;
6    h=3;
7    xx=ones(2*Ny+1,1)*x;
8    yy=y'*ones(1,2*Nx+1);
9    Green=1./sqrt(xx.^2+yy.^2+h);
10
11   % Heat source
12   source=zeros(2*Ny+1,2*Nx+1);
13   a=5;
14   for ix=-Nx:Nx
15       for iy=-Ny:Ny
16           if ix*ix+iy*iy<a*a
17               source(iy+Ny+1,ix+Nx+1)=1;
18           end
19       end
20   end
21
22   % Convolution
23   dT=ifftshift(ifft2(fft2(Green).*fft2(source)));
```

References

[1] www.blender.org

[2] www.comsol.fr

[3] Baffou, G., and Rigneault, H. 2011. Femtosecond-Pulsed Optical Heating of Gold Nanoparticles. *Phys. Rev. B*, **84**, 035415.

---

[1] The image used on the cover of this book was computed this way.

[4] Baffou, G., Quidant, R., and García de Abajo, F. J. 2010a. Nanoscale Control of Optical Heating in Complex Plasmonic Systems. *ACS Nano*, **4**, 709.

[5] Baffou, G., Quidant, R., and Girard, C. 2010b. Thermoplasmonics Modeling: A Green's Function Approach. *Phys. Rev. B*, **82**, 165424.

[6] Baffou, G., Berto, P., Bermúdez Ureña, E., Quidant, R., Monneret, S., Polleux, J., and Rigneault, H. 2013. Photoinduced Heating of Nanoparticle Arrays. *ACS Nano*, **7**(8), 6478–6488.

[7] Bohren, C. F., and Huffman, D. R. 1983. *Absorption and Scattering of Light by Small Particles*. Wiley Interscience.

[8] García de Abajo, F. J., and Howie, A. 2002. Retarded Field Calculation of Electron Energy Loss in Inhomogeneous Dielectrics. *Phys. Rev. B*, **65**, 115418.

[9] Girard, C. 2005. Near Fields in Nanostructures. *Rep. Prog. Phys.*, **68**, 1883.

[10] Hohenester, U., and Trügler, A. 2012. MNPBEM – A Matlab Toolbox For the Simulation of Plasmonic Nanoparticles. *Comput. Phys. Commun.*, **183**, 370–381.

[11] Mayer, S. A. 2007. *Plasmonics*. Springer.

[12] Metwally, K., Mensah, S., and Baffou, G. 2015. Fluence Threshold for Photothermal Bubble Generation Using Plasmonic Nanoparticles. *J. Phys. Chem. C*, **119**, 28586–28596.

[13] Myroshnychenko, V., Rodríguez-Fernández, J., Pastoriza-Santos, I., Funston, A. M., Novo, C., Mulvaney, P., Liz-Marzán, L. M., and García de Abajo, F. J. 2008. Modeling the Optical Response of Gold Nanoparticles. *Chem. Soc. Rev.*, **37**, 1792–1805.

4

Thermal Microscopy Techniques

This chapter introduces the thermal microscopy techniques that have been developed and used to probe local temperature in plasmonic structures under illumination. Reliably imaging a temperature distribution with a sub-micrometric resolution is not trivial, which certainly explains why the first temperature measurement in plasmonics dates only from 2009.

To date, ten families of techniques have been developed for this purpose. Interestingly, most of them rely on far-field *optical* microscopy techniques, which explains why part of the optics community is now tackling problems of thermodynamics. Half of the techniques are based on *fluorescence* measurements of molecular probes and are gathered in the first section. Then, the other techniques are assigned to specific sections, namely nanodiamond spectroscopy, wavefront sensing, Raman scattering spectroscopy, X-ray absorption spectroscopy, and scanning thermal microscopy. Each section of this chapter begins with a rather detailed explanation of the underlying physics in play. For this reason, the interest of this chapter is also to enter the physics of concepts such as fluorescence emission, surface-enhanced Raman scattering, X-ray absorption or nanodiamond's photophysics.

4.1 Introduction

Probing or mapping temperature on the submicrometric scale is not an easy task, even in the twenty-first century. Standard IR thermal radiation measurements, which are usually done on the macro scale, no longer apply at such small dimensions since the involved wavelength of the radiations (more than 10 μm) would lead to a very poor spatial resolution [39]. Moreover, most optical components are not transparent in this wavelength range (lenses, glass coverslips, etc.). Unlike light (which is endowed with a propagative nature), heat just diffuses. This makes any temperature distribution arising from a nanosource of heat confined at its vicinity, and not propagating to the far field. For these reasons, first attempts to probe a temperature field at small scales were based on the use of local probes consisting of a small composite tip acting as a nanoscale thermocouple or bolometer. This is the so-called SThM (scanning thermal microscopy) technique, invented in 1986 [80]. It enables a spatial resolution of around 50 nm. This technique would therefore have been ideal for temperature measurements in nanoplasmonics ... if only it were not so invasive.

Magnetic resonance imaging (MRI) was one of the first thermal imaging techniques used in thermoplasmonics. The groups of West and Halas used MRI measurement to retrieve the temperature in living animals in the context of photothermal therapies [33]. But once again, just like infrared imaging, this technique was not developed in the research context of plasmonics and remains dedicated to macroscopic measurements. For this reason, it will not be described in this chapter. The chapter rather focuses of new temperature microscopy techniques suited to image the temperature surrounding single, or a collection of, nanoparticles.

It was necessary to wait till 2009 for a first temperature microscopy technique suited for plasmonics experiments [2]. The trick was to scatter fluorescent molecules in the surroundings of the plasmonic nanoparticles of interest and perform standard fluorescent microscopy measurements. As many fluorescence properties are temperature-dependent, mapping fluorescence naturally leads to a temperature map, provided a calibration curve was previously established. Fundamentally, the trick was therefore to turn thermal information into optical information in the visible range in order to make it propagative and noninvasively detectable from the far field. Most thermal imaging techniques subsequently developed were based on this thermo-optical transduction along with the use of an optical microscope, which also provides the ability to focus a laser light over a diffraction limited spot and achieve submicrometric heating with precise positioning.

Two typical sample geometries have been used to develop thermal microscopy techniques in plasmonics and illustrate their capabilities. In one configuration (Figure 4.1a), colloidal gold nanoparticles are dispersed in a liquid (sometimes injected in a microfluidic channel) and a laser is heating a portion of the fluid, over the whole thickness of the liquid layer. A second approach (Figure 4.1b) instead consists in using a two-dimensional distribution of plasmonic nanostructures lying on a solid, planar substrate (typically lithographic structures on a glass coverslip). All the techniques that will be introduced in this chapter are based on one or other of these geometries.

Three different and interesting reading-templates of the literature can be drawn. First, one can categorize the different temperature measurement techniques in terms of temperature-dependent *physical quantities* that have been used for temperature probing and imaging:

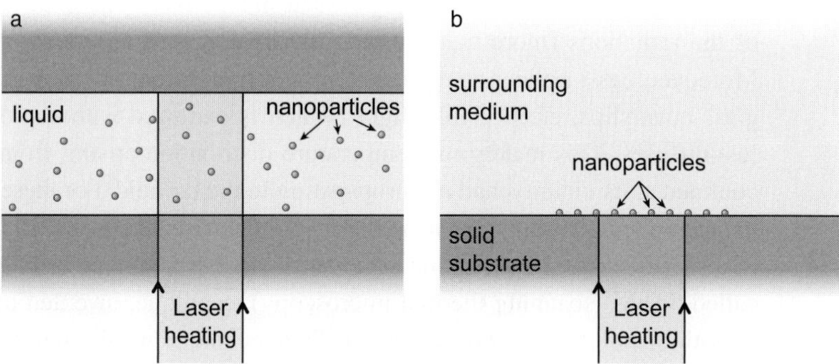

Fig. 4.1 Side view of the two typical sample geometries for investigation of temperature microscopy techniques in plasmonics: (a) Colloidal nanoparticles dispersed in a solvent and (b) single or multiple nanoparticles lying on a solid substrate.

- fluorescence
- Raman effect
- electronic spin resonance
- refractive index
- X-ray absorption

Second, one can obtain a different overview of the literature by reviewing the temperature sensitive *probes* for temperature probing and mapping:

- fluorescent molecules (fluorescein, GFP, fluorescent polymer, ...)
- nanodiamonds
- thermocouples
- hybrid organic-inorganic nanoparticles
- quantum dots
- no probe (label-free)

Finally, one can instead describe the techniques themselves, and this is how the following sections of the chapter will be organized:

- Fluorescence microscopy
- Optical and microwave spectroscopy of nanodiamonds
- Optical wavefront sensing
- Raman scattering spectroscopy
- X-ray absorption spectroscopy
- Scanning thermal microscopy

To give a comprehensive overview before detailing each temperature microscopy technique, Figure 4.2 summarizes seven years of investigation in this field of research in a chronological manner using a timeline.

4.2 Fluorescence Microscopy

Fluorescence properties of molecules such as fluorescence intensity, fluorescence spectrum or fluorescence polarization anisotropy depend on temperature. Thus, it suffices to disperse fluorescent molecules in the surrounding medium of heated nanoparticles (Figure 4.1b) and use common fluorescence microscopy techniques to achieve temperature measurements with a diffraction-limited spatial resolution, provided a calibration curve was previously determined. Such an approach therefore requires two light beams: typically one laser light to heat the plasmonic structures and one laser light to excite fluorescence. The "heating" wavelength has to be distinct from both the fluorescence excitation wavelength and the fluorescence emission wavelength range. In common situations, the fluorescence excitation is performed using a blue laser, the fluorescence emission is collected in the green region

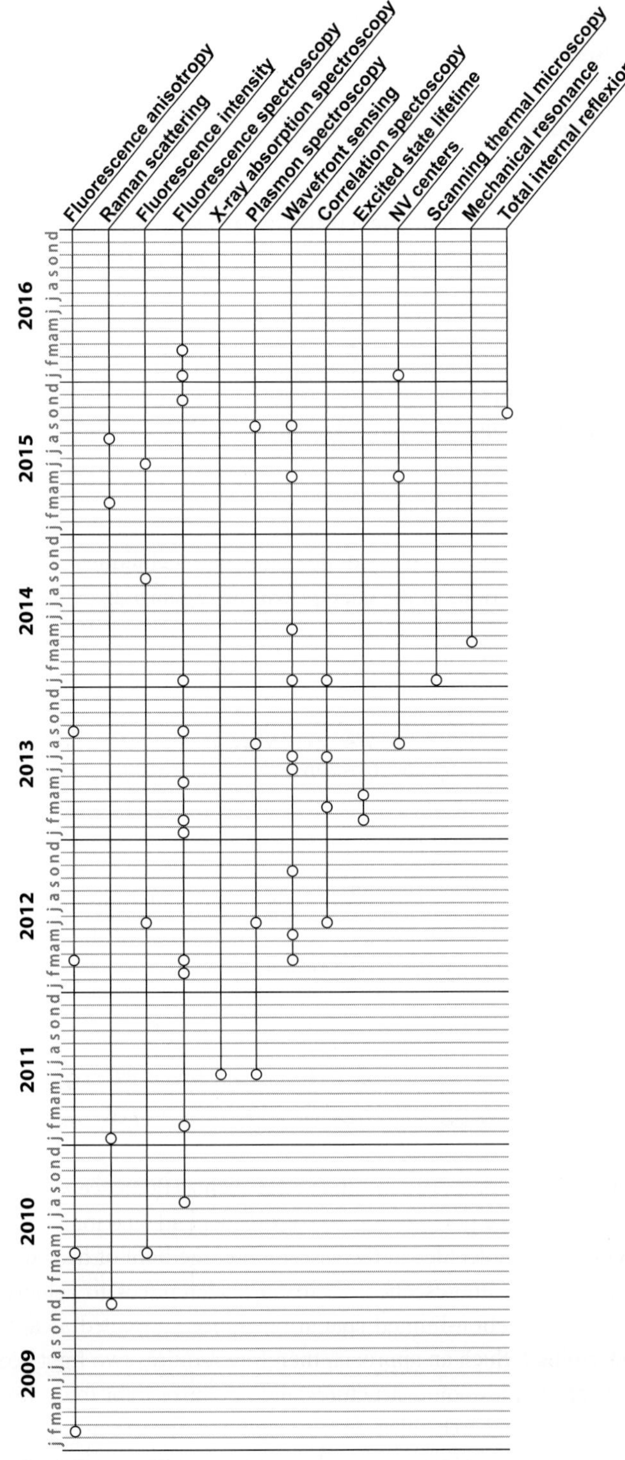

Fig. 4.2   Chronology of the articles reporting on temperature mapping for investigation in plasmonics.

and heating is performed in the near-IR range. This way, the low laser intensity for fluorescence excitation is not supposed to heat the nanoparticles and the low photon energy of the heating laser is not supposed to bleach the molecules even at high laser power.

Among all the temperature-dependent fluorescence properties, the less reliable certainly is fluorescence *intensity*. Indeed, fluorescence intensity depends on many other factors besides temperature, such as fluctuations of laser power, nonuniformity of fluorophore concentration, and photo-bleaching effects. For this reason, the community rather considered physical quantities such as lifetimes, spectra or polarization anisotropy.

Note that the proximity of a metal nanoparticle is supposed to modify the properties of fluorescent molecules, such as the fluorescence intensity, the excited state lifetime or the emitted polarization, which could in principle affect the measurements and yield artifacts. However, such effects affect only a small part of the molecules within the fluorescence excitation laser spot, those located at the nanometric distance of the metal surface. The proximity of the metal has not been evidenced as being a problem for this reason.

Importantly, there exist other parameters of the surrounding medium that can affect fluorescence properties and yield flawed measurement especially if the surrounding medium is as complex as the cytoplasm of a living cell. Indeed, fluorescence is known to depend on ion concentrations, viscosity and pH, and these physical quantities are known to vary if a living organism experiences a temperature variation. Hence, caution has to be used when measuring temperature in living cells using fluorescence means, as recently highlighted in References [6, 9].

So far four different physical quantities related to fluorescence have been used in plasmonics to conceive temperature measurement techniques. These related techniques are based on

1. Fluorescence anisotropy
2. Fluorescence intensity
3. Excited state lifetime
4. Fluorescence spectra
5. Fluorescence correlation spectroscopy

The next five subsections aim at reviewing these techniques.

4.2.1 Fluorescence Polarization Anisotropy

In general, a population of fluorophores illuminated by a linearly polarized incident light re-emits *partially* polarized fluorescence due to random molecular orientation [74]. In this context, one usually defines the fluorescence polarization anisotropy as

$$r = \frac{I_\parallel - I_\perp}{I_\parallel + 2I_\perp} \tag{4.1}$$

where $I_\parallel$ and $I_\perp$ are the intensities of the fluorescence polarized parallel and perpendicular with respect to the incident polarization.[1] For randomly oriented fluorophores

---

[1] The factor 2 is a convention. It comes from the fact that the total emitted fluorescence intensity reads $I = I_x + I_y + I_z = I_\parallel + 2I_\perp$ with $I_\parallel = I_x$ and $I_\perp = I_y = I_z$.

whose absorption and emission dipolar moments are parallel, the theoretical polarization anisotropy r_0 in the absence of any molecular motion is maximal and reaches a value of 0.4. The measured value of r is closely related to molecular rotation arising from Brownian dynamics according to Perrin's equation [74]:

$$\frac{1}{r} = \frac{1}{r_0}\left(1 + \frac{\tau_s}{\tau_r}\right)$$
(4.2)

where τ_r is the rotational correlation time and τ_s the excited state lifetime. This equation means that substantial molecular rotation (induced by its Brownian dynamics) during the lifetime of the excited state leads to a fluorescence depolarization, $i.e.$, a lower value of r. An increase of the temperature contributes to lower the fluorescence polarization anisotropy r since it gives rise to a faster rotation of the molecules, $i.e.$, a lower value of τ_r (see Figure 4.3c). This happens according to the Debye/Stokes/Einstein equation:

$$\tau_r = \frac{V\eta(T)}{k_B T}$$
(4.3)

where T is the temperature, $\eta(T)$ the dynamic viscosity of the medium, V the hydrodynamic volume and k_B the Boltzmann constant. The maximum temperature sensitivity of r is achieved when τ_r is on the order of magnitude of the fluorescence lifetime τ_s.

The first successful measurement of a temperature field created around plasmonic nanoparticles was introduced by the group of Romain Quidant in 2009 and was based on

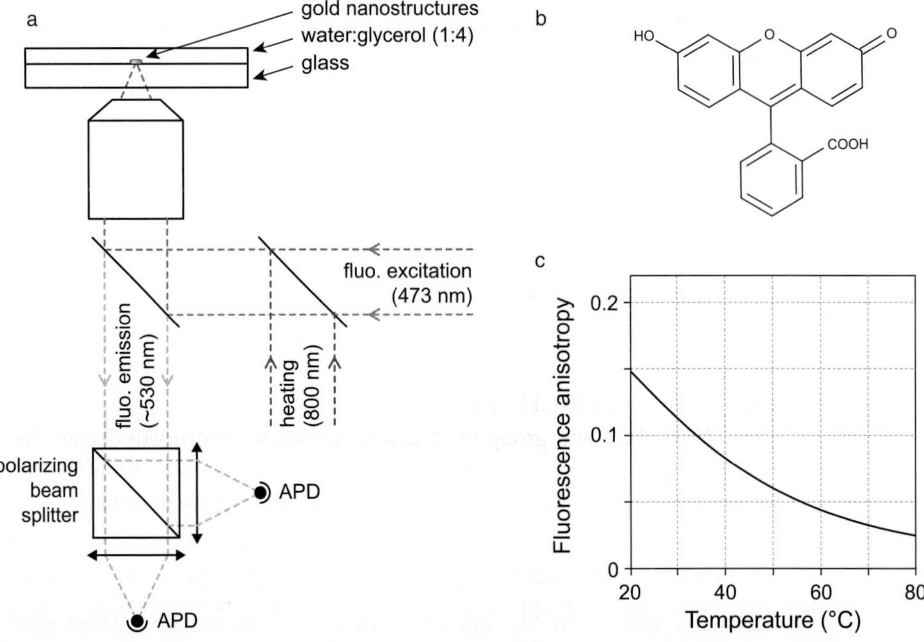

Fig. 4.3 (a) Experimental set up configuration used in References [3, 2, 27, 26] to measure a temperature increase using fluorescence polarization anisotropy. (b) Representation of a fluorescein molecule. (c) Fluorescence polarization anisotropy of fluorescein molecules as a function of temperature.

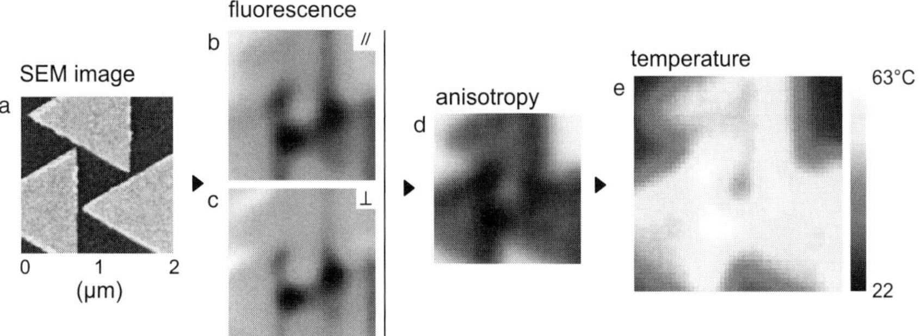

Fig. 4.4 Temperature measurements performed on gold microstructures by fluorescence polarization anisotropy imaging. (a) Scanning electron microscope image of gold microstructures. (b,c) Two cross-polarization maps of the emitted fluorescent of surrounding fluorescein molecules. (d) Fluorescence polarization anisotropy calculated from images (b,c). (e) Temperature map calculated from image (d).

fluorescence polarization anisotropy measurements [2]. Fluorescein molecules (Figure 4.3b) were dispersed in a liquid surrounding gold nanoparticles deposited on a glass coverslip. According to Equation (4.3), fluorescein molecules (around 1 nm big) spin too fast in water, making r equal to zero, no matter the temperature. In order to make the approach efficient for temperature measurements, the authors used a 4:1 glycerol–water mixture as the surrounding medium in order to increase the viscosity η and decrease τ_r. The fluorescence polarization anisotropy was measured by raster scanning confocal microscopy and using a detection path composed of a cubic polarizing beam splitter used to split the two cross polarization components onto two avalanche photodiodes (APDs), as represented in Figure 4.3a. Finally, the temperature map was retrieved from the fluorescence anisotropy image using the calibration curve displayed in Figure 4.3c. A temperature sensitivity of 1°C was achieved, along with a diffraction limited resolution (around 250 nm). These good figures are counterbalanced by the complexity of the detection part, where two avalanche photodiodes have to be aligned with each other. Figure 4.4 presents some temperature maps of gold nanoparticles obtained using this technique. Note that in their approach, the authors overlapped the heating beam and the fluorescence excitation beam and raster scan the sample. This creates temperature maps that are a bit more complicated to interpret than steady state temperature distributions that could have been obtained by raster scanning the excitation laser beam with a fixed heating laser.

In 2010, the same group used this technique to investigate deeper the physics of heat generation in plasmonic structures, and evidenced a localization mismatch between optical hot spots in the near fields and thermal hot spots within plasmonic structures [3].

The use of a viscous surrounding medium such as glycerol can be prohibitive for some applications, in particular in chemistry or biology, where water would be preferred. In 2012, the group of Quidant circumvented this problem and managed to perform temperature measurements in water by adjusting the size of the molecules [27]. Indeed, enlarging the size of a molecule (instead of increasing the viscosity of the solvent) is another means of slowing down its rotational Brownian motion. Proteins are good candidates for this purpose as they are much bigger than common fluorescent molecules, and the authors have shown that Green Fluorescent Proteins (GFP) in water by chance are endowed with a rotational correlation time on the order of their fluorescent lifetime, making them suitable temperature probes for biological applications. Successful temperature mapping was conducted on HeLa and U-87 MG cancer cell lines with a temperature sensitivity of *ca.* 0.4°. Later on, in 2014, the same group managed to map for the first time the temperature in living organisms (C-elegans worms) using this technique [26].

4.2.2 Fluorescence Intensity

Temperature estimations in plasmonics have also been performed by fluorescence *intensity* measurements. Three totally different mechanisms have been reported.

The first mechanism relies on the natural dimming of fluorescence intensity of most fluorophores as the temperature increases. This approach was reported for the first time in plasmonics by Chiu and Chu only recently, in 2015 [18]. The well-known issue with this approach is that fluorescence intensity depends on many physical parameters besides temperature, such as fluorophore concentration, laser intensity or molecular bleaching. For these reasons, measurement using this mechanism are known to be less reliable. This is certainly the reason why this technique was introduced only in 2015 in a single article and was not one of the first techniques reported in this context, despite its simplicity.

A second, more subtle mechanism was introduced much earlier, in 2010, by the Group of Lene B. Oddershede [11, 45]. This group has long-standing expertise in nanoparticle optical trapping. In particular, this group succeeded in trapping metal nanoparticles. As nanoparticle optical trapping usually requires significant laser power, the natural question they faced was how to know the subsequent temperature of the metal nanoparticle during trapping. For this purpose, the group introduced a clever approach to determine the temperature of single 80 nm to 200 nm gold nanoparticles. Fluorescent molecules were incorporated in a lipid bilayer (a biologically relevant matrix) in which a metal nanoparticle was trapped (Figure 4.5). These fluorescent molecules had a preference for the gel phase of the lipid bilayer. A gel-to-liquid phase transition to the liquid phase was induced by heating the nanoparticle. The spatial spreading of the liquid phase around the nanoparticle was dependent on the temperature increase of the nanoparticle. The larger the temperature, the wider the spatial distribution of the liquid phase. As the fluorescent molecules tend to escape from the liquid phase, nanoparticle heating came along with a decrease of fluorescence over an area, whose diameter was dependent on the temperature increase. This was directly monitored to indirectly retrieve the nanoparticle temperature, knowing the nanoparticle size and the melting temperature of the liquid bilayer ($T_\mathrm{m} = 33°$C).

In 2014, Ebrahimi and coworkers introduced a third technique based on fluorescence intensity measurements [28]. This technique is singular as it uses gold nanoparticle as temperature sensors. The principle is based on the quenching of fluorescent molecules in the vicinity of gold nanoparticle (Figure 4.6). DNA strands were attached at one end to spherical gold nanoparticles and labeled at the other end with a specific fluorophore. At low temperature, the structure remains in a collapsed form and the fluorescence is quenched due to the proximity of the metal with the fluorescent molecule. At higher temperature, above the melting point of the DNA strands, the structure opens and the fluorescence is no longer

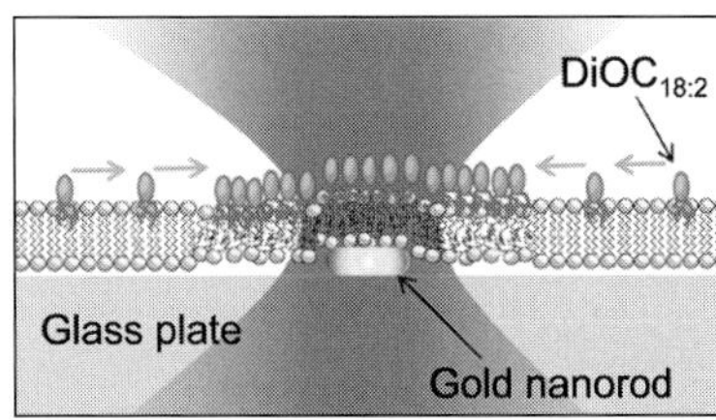

Fig. 4.5 Sample chamber used in References [11, 45] where a nanorod is immobilized on a glass surface and is covered by a gel phase lipid bilayer containing a low molar fraction of DiOC$_{18:2}$ fluorophores. Reproduced with permission from Reference [45]. Copyright 2012, American Chemical Society.

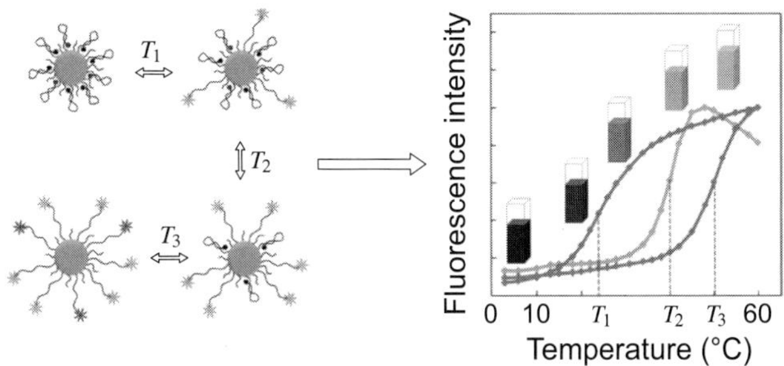

Fig. 4.6 Principle of the nanothermometers reported by Ebrahimi *et al.* Reproduced with permission from Reference [28]. Copyright 2014, American Chemical Society.

quenched. The authors refined the degree of complexity of their nanothermometers by functionalizing on every gold nanoparticle three kinds of fluorophores respectively to three kinds of DNA strands featuring different and successive melting temperatures (22.4, 38.7 and 48.7°C). Using this three-color method, precise determination of the temperature could be achieved over a biologically relevant temperature range (15–60°C) with a precision of ~ 0.5°C. The principle of this technique is summarized in Figure 4.6.

4.2.3 Fluorescence Spectroscopy

Fluorescence spectral line shapes of most chemical compounds are likely to be temperature-dependent. This effect is at the basis of some thermal microscopy approaches in thermoplasmonics. Two common situations [37] are encountered in the literature: (i) a temperature increase induces a measurable spectral shift of a fluorescence line (Figure 4.7a), or (ii) the temperature increase does not induce a spectral shift but a distortion of the spectral line shape associated with a modification of the relative intensities at two different wavelengths (Figure 4.7b). The benefit of such approaches is that they are not dependent on the fluorescence intensity. Thus, they are free from artifacts coming from laser power fluctuations, nonuniformity of fluorophore concentration, molecular blinking, etc.

Such an approach was introduced by Clarke *et al.* in 2010 [20]. The authors monitored temperature using Indo-1 fluorescent molecules and quantum dots. Indo-1 was well known to be a Ca^{2+} sensor in cell biology, but it was also shown to be sensitive to temperature

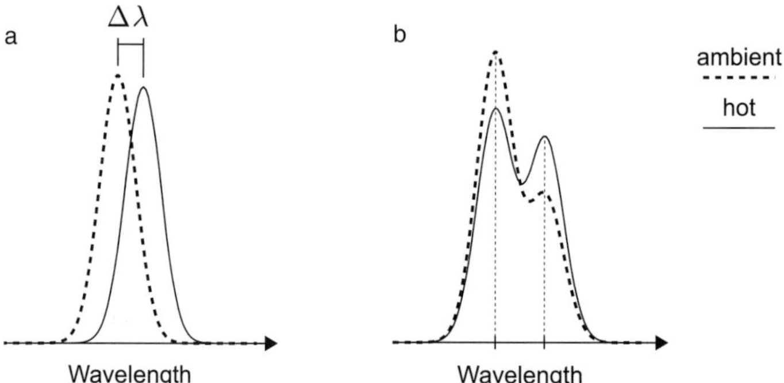

Fig. 4.7   (a) Fluorescence line shift upon heating. (b) Fluorescence spectrum distortion upon heating. In both cases, dashed lines refer to spectra at ambient temperature and solid lines correspond to spectra at higher temperature.

via a spectral blue-shift of a fluorescence emission line shape. On the contrary, quantum dots exhibit a red-shift of their fluorescence spectrum as the temperature increases. Experiments were conducted on agglomerated nanorods on a glass substrate in a liquid environment. Instead of measuring the spectral shift $\Delta\lambda$ for each pixel of the image (very time consuming), the authors measured fluorescence intensity images under wide field illumination using an EM-CCD (electron multiplying charged coupled device) and considered the ratio between two images acquired under separate collection wavelength ranges. Such an approach is fast, but suffers from a poor signal-to-noise ratio and requires the acquisition of two successive images. Some inconsistencies can be noticed in this work, such as the extremely slow timescale of heating (a few minutes) while the authors are heating a micrometric area. Then, they observed strong fluid convection and even bubble generation for temperature changes supposed to be only a few degrees. Such effects are supposed to occur at much higher temperatures (see Chapter 5).

In 2011, Carlson, Khan and Richardson [13] proposed to embed fluorescent ionic centres directly within the substrate, instead of dispersing fluorescent compounds in a surrounding fluid. They deposited a thin solid AlGaN film doped with Er^{3+} ions on a glass slide as represented in Figure 4.8a. The photoluminescence of Er^{3+} ionic centres is known to be temperature-dependent via a spectral line shape distortion, as schematized in Figure 4.7b. The microscopy approach consisted therefore in acquiring one spectrum per pixel and process it by taking the intensity ratio at two separate wavelengths. Such a procedure is time consuming and sometimes requires up to one hour to get an image. This way, the authors managed to measure the temperature distribution when heating single lithographic 120 nm nanodots and single colloidal 40 nm nanoparticles, in steady conditions, as represented in Figure 4.8b.

Some interesting and honest discussions have been conducted in this article. First, the authors noticed that the nanoparticle itself can modify the fluorescence spectrum of the ionic centre, and lead to an apparent negative temperature variation at zero laser power. Second, the authors showed that when heating such small nanoparticles, caution has to be used as most of the heat generation is likely to come from the underlying sample (that is also weakly absorbing), and not from the nanoparticle itself. Finally, as such nanoparticles

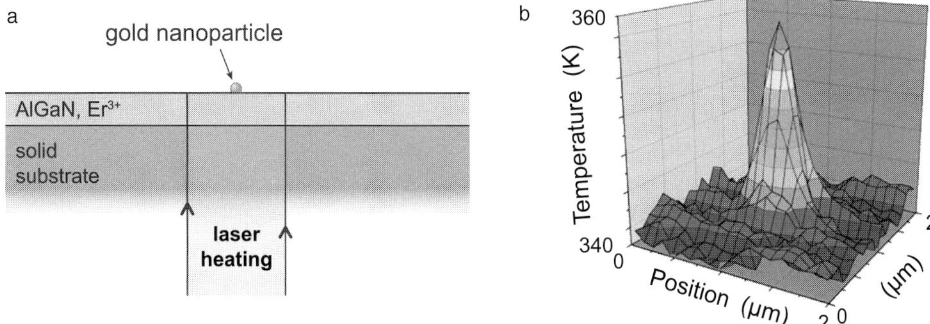

Fig. 4.8 (a) Side view of the sample geometry used in Reference [13]. The thermosensitive part consisted of Er^{3+} ionic centres embedded in a thin AlGaN layer. (b) Temperature image of a 40 nm gold nanoparticle. Reproduced with permission from Reference [13]. Copyright 2011, American Chemical Society.

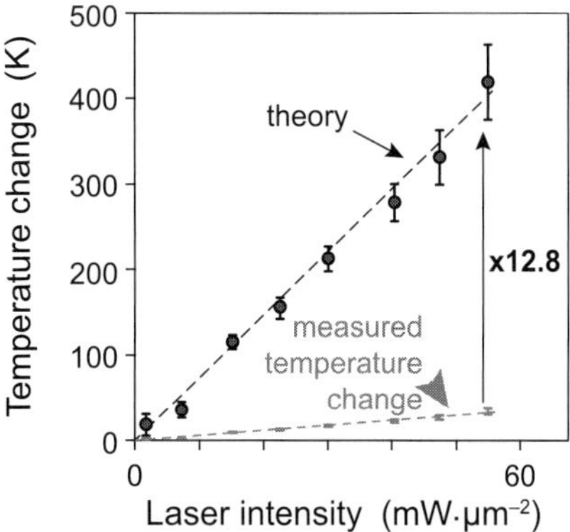

Fig. 4.9 Temperature increase of a 40 nm gold nanoparticle as a function of laser intensity reported in Reference [13]. Experimental measurements are multiplied by a factor of 12.8 to circumvent the spatial limited resolution and match theoretical predictions. Reproduced with permission for Reference [13]. Copyright 2011, American Chemical Society

are smaller than the diffraction limited spot (approx. 500 nm), the measured maximum temperatures were necessarily underestimated. In order to retrieve the actual temperature increase, the authors computed the expected maximum temperature increase by numerical simulation to derive a correction factor of 12.8, and multiplied the experimental measurements by this factor, as represented in Figure 4.9. Unfortunately, because of this correction procedure, this technique that is supposed to be experimental becomes theoretical to some extent, since a correction factor is applied to the measurements in order to match them with numerical simulations.

Although this thermal microscopy technique is slow and requires some pre-knowledge of the sample, especially when nanometric objects are investigated, this technique remains particularly convenient and represents significant progress in the field because it is one of the only approaches that does not necessitate a surrounding medium above the substrate. Moreover, ionic centres do not thermo-bleach and can be used for arbitrarily large temperature increases, while most fluorescent molecules bleach above 70°C. Finally, such fluorescent nanoprobes are not submitted to any thermal-induced fluid convection or variations of molecular probe concentration [79, 20].

Such advantages enabled the authors to conduct subsequent experiments involving other problematics such as measuring thermal conductivity and absorption cross section of nanowires [14] and investigate water superheating [15].

The use of Er^{3+} ionic centres for temperature measurements, as proposed by the group of Richardson, belongs to a family of temperature measurement techniques that involves the use of ions from the lanthanide series. Lanthanide-based compounds are well-known analytical tools, commonly used in biology or chemistry, as sensors or labels. Their fluorescence is most intense when lanthanide ions have a third degree of oxidation. Although all lanthanides proved to be efficient, the most common are Sm(III), Er(III), Eu(III), Tb(III), and Dy(III). The most singular properties of lanthanide-based compounds are their long lifetimes, in the microsecond to millisecond range, which allows the use of time resolved spectroscopy. Fluorescence intensity, lifetime and spectra of lanthanides are dependent on pH, temperature, chemical environment, etc. Reference [64] provides an exhaustive review of the lanthanide-based compounds that have been used as temperature sensors (not in relation with plasmonics, though).

In 2012 Maestro *et al.* [46] reported on the use of quantum dots as temperature nanoprobes in a solution containing gold nanorods. The fluorescence spectral line shape of Cd-Se quantum dots exhibits a red shift upon heating, like in Figure 4.7a, at a rate of 0.1 ± 0.05 nm·K^{-1}. The calibration curve measured by the authors is displayed in Figure 4.10a. Using this approach, the authors performed point measurements of temperature in a gold nanorod solution within a microfluidic channel, 200 µm in height. They evidenced temperature increases of a few tens of kelvins and managed to quantify the absorption efficiency (absorption over extinction) of the gold nanorods. In their experimental approach, the authors chose to focus both laser beams using separate microscope objectives located on both sides of the microfluidic channel (see Figure 4.10a). Subsequently [47, 48], the same authors applied their experimental technique to investigate the photothermal properties of a larger variety of nanoparticles, such as carbon nanotubes, nanocages, nanoshells and nanostars. Note that in all these works, the aim was not to *image* a temperature field, but rather to perform point measurements in order to investigate the photothermal properties of nano-objects. The fact that the authors didn't implement a raster scanning system on their device illustrates a limitation of the approaches based on spectrum acquisition: acquiring an image requires measuring and processing one spectrum per pixel, which is rather time-consuming.

In 2013, the group of Daniel Jaque [59] pursued the initiative of Richardson by using lanthanide-based compounds. They used Nd^{3+}-doped LaF_3 (core) nanoparticles shielded by an undoped LaF_3 shell, excited at 808 nm and emitting around 885 and 1060 nm. The approach was, however, slightly different for several reasons. First, they used Nd^{3+}:LaF_3 nanoparticles within the medium of interest, instead of Er^{3+} ionic centres within the substrate. Second, the context of this work was to use these nanoparticles as a means to measure the temperature for biomedical applications, directly into tissues, thanks to the red-IR absorption and emission range of their compounds overlapping the transparency

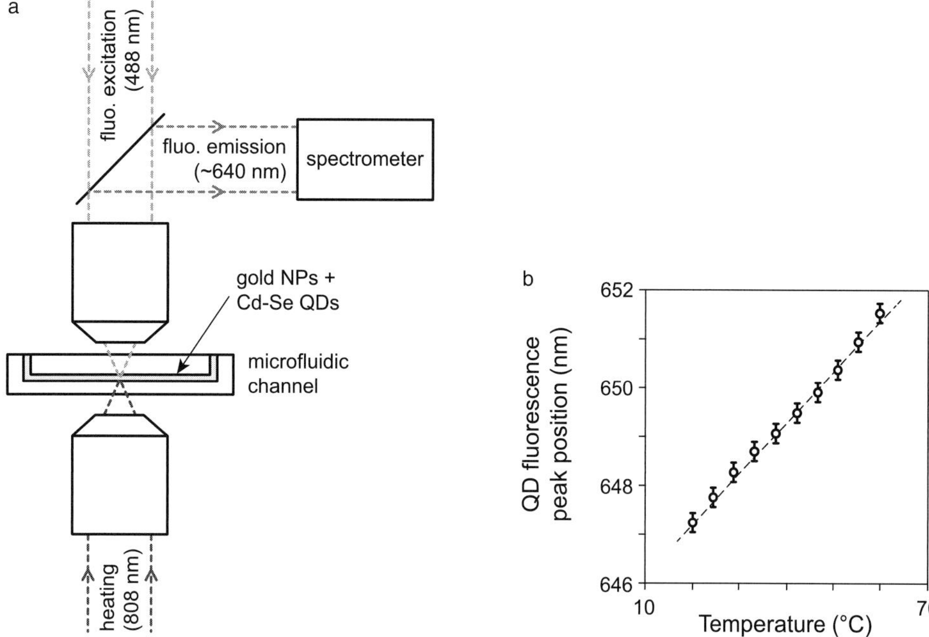

Fig. 4.10 (a) Experimental set up configuration used in Reference [46] to measure a temperature increase in a gold nanorod solution using Cd-Se quantum dots. (b) Wavelength corresponding to the fluorescence emission maximum of Cd-Se quantum dots as a function of temperature, measured in Reference [46].

window of human tissues. The thermal sensitivity was $\pm 2°C$. The authors compared the brightness of their novel Nd^{3+}:LaF_3 probes to their previous quantum dots under similar conditions (nanoparticle concentration, excitation intensity, exciting collecting optics). The emission intensity generated by the Nd^{3+}:LaF_3 nanoparticle colloidal solution was found to be over one order of magnitude higher than that generated by the CdTe quantum dot colloidal solution. This larger efficiency was attributed to the fact that Nd^{3+}:LaF_3 nanoparticles are excited by means of one-photon excitation (a down-conversion process), whereas for CdTe quantum dots, a two-photon excitation process through virtual states is needed [46, 49]. In parallel, the same group also reported the used of other lanthanide nanoparticles consisting of Er/Yb co-doped $NaYF_4$ nanocrystals [38], and compared them with Cd-Te quantum dots. Here again, the thermal sensitivity of the lanthanide particles was given by relative intensity changes between two emission spectral peaks, while the sensitivity of the quantum dots was related to a spectral shift.

In 2013, Debasu and coworkers [22] also used lanthanide based compounds as nanothermometers. They used $(Gd,Yb,Er)_2O_3$ directly attached to gold nanorods in solution. The excitation mode was, however, different from previous studies. While Richardson or Jaque were simply measuring a one-photon fluorescence process, the mechanism involved in this study was a so-called *up-conversion* process, associated with an excitation at 980 nm and an emission in the visible range [85]. Up-conversion, that results in the emission at higher energy, relies on two successive single photon absorptions, enabled *even using a cw laser*

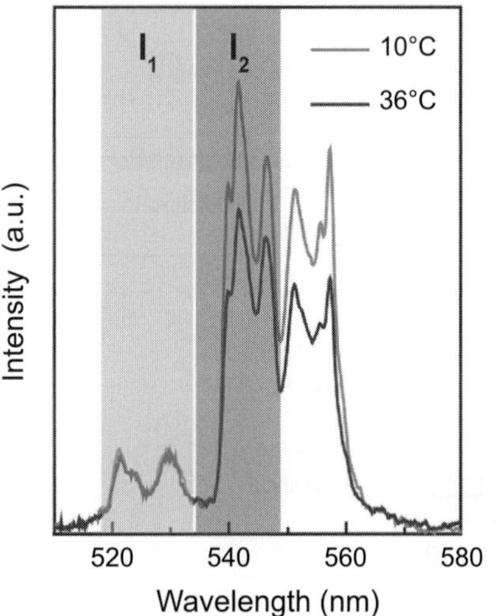

Fig. 4.11 Emission spectra obtained from a single optically trapped NaYF$_4$:Er^{3+},Yb^{3+} nanoparticles at two different temperatures. The intensity ratio I1/I2 is highly temperature-sensitive. Reproduced from Reference [60]. Copyright 2016, Wiley-VCH Verlag GmbH & Co. KGaA.

illumination by the long lifetime of the first excited state of lanthanide-based compounds (in the microsecond to millisecond range), which makes the probability of subsequent photon absorption before de-excitation non-negligible. The authors achieved a resolution of 0.3–2.0 K in the range up to 800°C.

In 2015, the group of Fiorenzo Vetrone [61] used hybrid mixture of luminescent lanthanide-based nanoparticles (NaGdF4: Yb$_3^+$ 20%, Er$_3^+$ 2%) with gold nanorods (20 nm × 100 nm in average). The localized plasmon resonance (980 nm) of the gold nanorods matched with the absorption wavelength of the first excited state of Yb$_3^+$, which led to a large plasmonic resonant enhancement on the absorption, which was dependent on the distance between the nanorod and the luminescent nanoparticles. The distance was adjusted by the thickness of a SiO$_2$ layer encapsulating the luminescent nanoparticles. Here again, the luminescence of the Er$_3^+$/Yb$_3^+$ doped nanoparticles relied on an up-conversion process, and could be processed to retrieve temperature. In this study the role of the nanorods were twofold: collective optical heating and luminescence enhancement.

In 2016, concomitantly, the groups of Jaque [60] and the group of Richardson [10] managed to optically trap temperature sensitive 2 μm particles in a fluid medium to measure a temperature field in three dimensions. Jaque used colloidal NaYF$_4$:Er^{3+},Yb^{3+} particles as temperature microprobes in living cells heated by gold nanoparticles [60]. The microparticles were trapped with a 980 nm laser. The same laser was responsible of a bright anti-Stokes emission in the visible range, characterized by two thin peaks at 525 and 550 nm, whose intensity ratio was highly temperature-dependent. The authors already used this kind of particles in 2010 for temperature mapping in living cells [77], but they were not optically trapped at that time. The group of Richardson reported on a very similar work using optically trapped 150 nm Er$_2$O$_3$ erbium oxide nanoparticles [10]. The

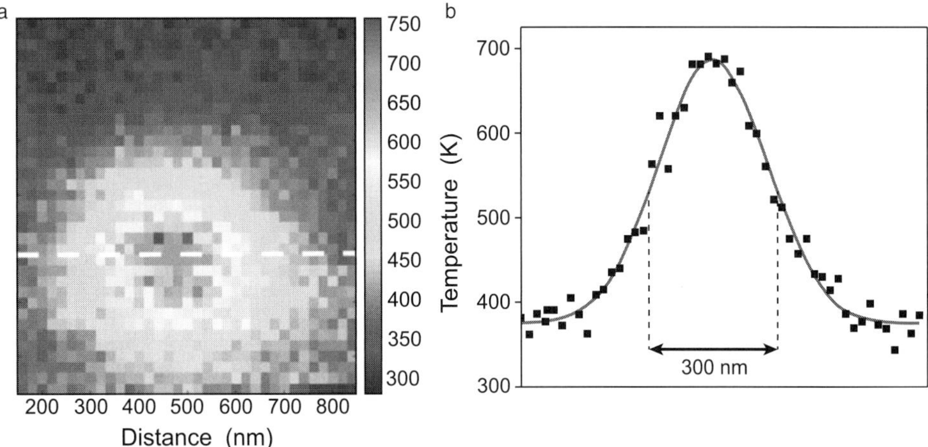

Fig. 4.12 (a) Thermal image of a nanostructure with the laser-trapped Er_2O_3 nanoparticle. The white dotted line represents the place where the temperature profile shown in (b) is located. The solid line is a Gaussian fit (FWHM of 300 nm) to the temperature profile. Reproduced with permission from Reference [10]. Copyright 2016, Springer-Verlag Berlin Heidelberg.

probe nanoparticles were much smaller than the microparticles of Jaque and the message of Richardson was that they could achieve a subdiffraction spatial resolution using such small temperature probes. However, several criticisms can be raised regarding this work. First, the authors are using the same laser beam (at 532 nm) to heat and probe the temperature. This feature is likely to turn into an issue for arbitrary systems because the ideal laser intensity for trapping has no reason to be of the order of the laser intensity required for heating at a desired temperature. The aforementioned two-beam approach of Jaque is more natural and convenient in that sense. Second, the authors claim a subdiffraction resolution of 165 nm that is not evident by looking at their measurements (see Figure 4.12) because the heating of such large nanoparticles was not supposed to generate a subdiffraction temperature field. The authors had to perform a deconvolution of the image. The article would have been more convincing with a direct measurement of a subdiffraction-limit FWHM.

In 2016, other trapping experiments were conducting, but not with the same configuration. Šiler and coworkers aimed to trap a gold nanoparticle in water and map the surrounding temperature [78]. This time, not the probe but the source of heat was trapped. Fluorescent molecules were dispersed in the fluid surrounding the trapped nanoparticle and the temperature was measured using a blue laser exciting the fluorescence over a diffraction-limited spot. Two types of fluorescent molecules were mixed, Rhodamine B and Rhodamine 110, and the fluorescence spectra of the mixture was measured to retrieve the temperature. Figure 4.13 explains the overall procedure.

4.2.4 Excited State Lifetime Imaging

When a chemical compound absorbs a photon, it generally remains in the excited state for a while before relaxing by spontaneously emitting a photon or through a non-radiative

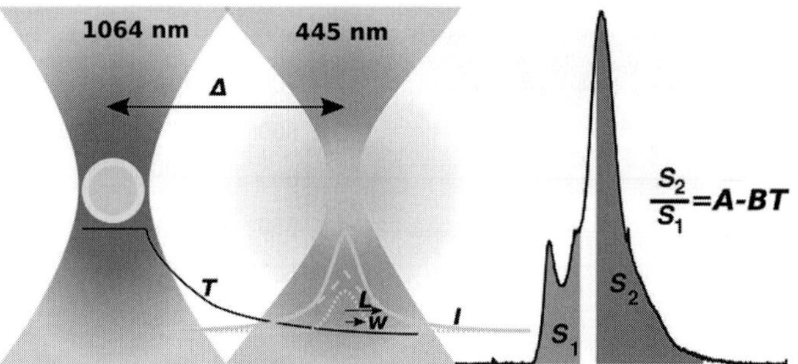

Fig. 4.13   Principle of proposed method. A core–shell PMMA-gold nanoparticle is trapped and serves as a heat source. The increased temperature changes the fluorescence of the dyes diluted in the surroundings. The fluorescence is excited by the second laser in a spot separated by distance Δ from the particle. Reproduced with permission from Reference [78]. Copyright 2016, The Optical Society.

process. The average delay τ_s between absorption and de-excitation is called excited state lifetime (ESL) and generally is on the order of a few nanoseconds [44, 74] in the case of fluorescent compounds. The delay between absorption and relaxation is stochastic by nature, *i.e.*, it is not deterministic and has to be described by a probability distribution. For a simple system, the probability of de-excitation over time obeys an exponential law. Let N_0 be the number of exited molecules at time $t = 0$, the evolution of the number of excited molecules over time reads

$$N(t) = N_0 \exp(-t/\tau_s). \tag{4.4}$$

Although de-excitation does not necessarily occur radiatively, the luminescence intensity from a population of fluorophores is proportional to the number of excited molecules $N(t)$ at any time. Hence, a common approach to measure an ESL consists in following the decay of luminescence intensity of the system over time, subsequent to a short laser pulse excitation, much shorter than the ESL. However, as the de-excitation is usually extremely fast, one has to use a time-correlated-single-photon technique where one measures the time between sample excitation by a pulsed laser and the emission of a photon (or rather the arrival of a photon on the detector). The measurement of this time delay is repeated many times to reconstruct the exponential decay line shape and retrieve the ESL. ESL *images* can be obtained using a technique called FLIM (fluorescence lifetime imaging). The ESL of a fluorophore is sensitive to its microenvironment and can be used to probe physical quantities such as ionic concentrations, pH, microviscosity or local temperature. For this reason, ESL imaging is widely used in cellular biology.

Although FLIM could have been used in nanoplasmonics to image a temperature distribution, it was never done this way. The few articles that reported ESL measurements for temperature probing in nanoplasmonics only conducted point measurements.

In 2013, Coppens *et al.* [21] reported for the first time on ESL measurements to probe the temperature increase created by a plasmonic system. Instead of dispersing luminescent molecular probes in a surrounding liquid, the authors chose to conduct their experiments on a substrate that was luminescent itself: the substrate consisted of a 100 nm thick ruby

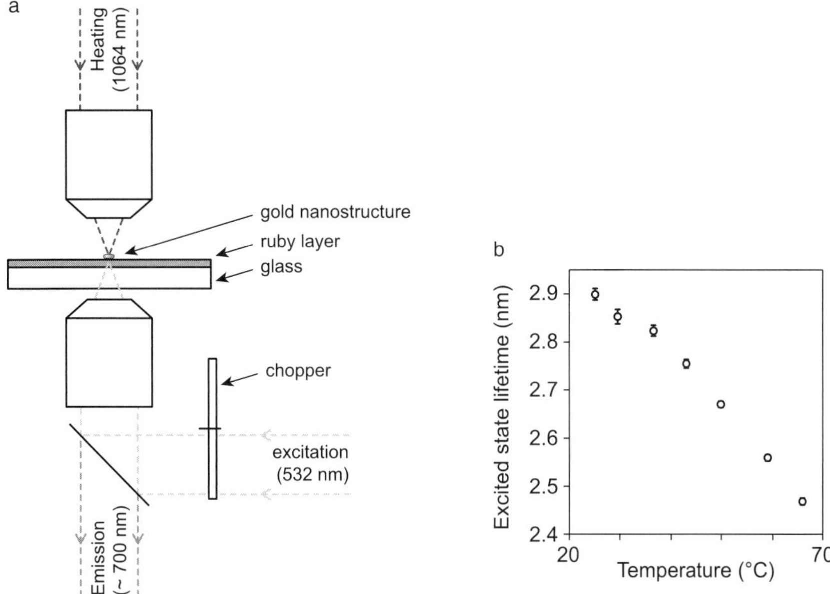

Fig. 4.14 (a) Experimental set up configuration used in Reference [21] to measure a temperature increase of gold nanostructures on a ruby luminescent layer. (b) Excited state lifetime of the ruby layer as a function of temperature, measured in Reference [21]

layer (Al_2O_3:Cr^{3+}) deposited on a glass coverslip. The particularity of ruby is that its ESL is on the order of a few milliseconds (see Figure 4.14b), which makes any ESL estimation simpler and less expensive than the aforementioned pump probe approach. In particular, for ESL in the millisecond range, pulses of light can be generated using a chopper (see Figure 4.14a). Ruby's phosphorescence emission lies around 700 nm and its ESL is known to be temperature sensitive over a large temperature range (room temperature up to *ca.* 550°C). In order to make their measurements possible, the authors had to choose metal structures with a plasmonic resonance wavelength (1064 nm) far from the phosphorescence emission range. Figure 4.14b represents the calibration curve relating temperature and ESL of the Ruby layer. The systems under study were bow-tie and diabolo gold nanostructures made by electron beam lithography. The authors aimed at comparing their relative efficiency as nanosources of heat.

At the end of their article, although they only have conducted point measurements, the authors mentioned that such an approach could readily be extended to 2D mapping of temperature distributions with a resolution given by the diffraction limit of the luminescence excitation beam. In their current configuration, the mapping speed would be limited by the millisecond lifetime of ruby. However, using fluorescent probes with a lifetime on the order of a few nanoseconds, standard FLIM could be used to acquire an image where the mapping speed would be only limited by the laser repetition rate. According to the authors, with a 1 kHz repetition rate, a 125 × 125 pixel image could be acquired in *ca.* 40 s.

In parallel, Freddi *et al.* also reported on temperature probing using ESL measurements [29]. The difference was that they used Rhodamine molecules, which are fluorescent and feature an ESL on the nanosecond scale. Rhodamine molecules were neither dispersed in the surrounding liquid nor embedded in the substrate, they were electrostatically bound to the plasmonic nanoparticles themselves via a 4.4 nm thick molecular coating acting as a spacer to avoid fluorescence quenching by the metal proximity. The authors evidenced that the ESL was dependent on the environment of the molecules, and in particular on the proximity of the nanoparticle metal (which made the ESL drop from 2.4 ns to 1.6 ns). However, this is no problem as long as the calibration curve relating temperature and ESL is acquired from these decorated nanoparticles, not from free Rhodamine molecules in solution. A temperature sensitivity of 0.028 ± 0.004 ns·K^{-1} was measured. The authors also mentioned the possibility to extend their approach to temperature imaging. In this case, around 2000 photons per pixel would be required, which should correspond to about 10 to 50 ms exposure time per pixel.

4.2.5 Fluorescence Correlation Spectroscopy

When exciting solvated fluorescent molecules over a small region (generally by focusing a laser beam in a solution of fluorophores), the collected fluorescence may feature temporal fluctuations due to molecules entering and leaving the observation volume because of molecular Brownian motion. These fluctuations contain valuable information on molecular dynamics, in particular diffusion coefficients and binding constants. These fluorescence fluctuations are all the more significant if the average number of fluorophores in the illuminated volume is reduced, *i.e.*, if the molecular concentration is low.

In order to extract valuable information from the time trace $I(t)$ of these fluorescence intensity fluctuations, one generally computes the associated autocorrelation function $G(\tau)$ defined as

$$G(\tau) = \frac{\langle I(t)\, I(t+\tau)\rangle_t}{\langle I(t)\rangle_t^2} \tag{4.5}$$

where $\langle \ldots \rangle_t$ denotes a time average over the time variable t. This is the basic principle of a well-known technique named fluorescence correlation spectroscopy (FCS).

As molecular Brownian dynamics is strongly temperature-dependent, temperature variations can be evidenced by FCS measurements [36]. Indeed, the diffusion constant D of an object of hydrodynamic radius R in a liquid of dynamic viscosity η is given by

$$D = \frac{k_{\mathrm{B}}T}{6\pi\,\eta(T)R}. \tag{4.6}$$

Increasing temperature contributes not only to increase $k_{\mathrm{B}}T$ but also to a marked decrease of the viscosity $\eta(T)$, in most solvents. For instance, the viscosity of water decreases by a factor of 4.6 from ambient temperature to 100°C (see Appendix B). The effect is even stronger when working in glycerol as a drop of three orders of magnitude occurs over the same temperature range. This is the reason why performing FCS measurements is usually a very sensitive way to measure temperature in solvents such as glycerol.

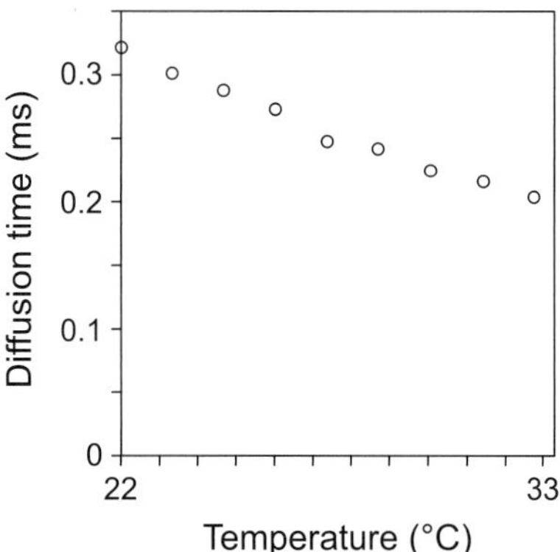

Fig. 4.15 Diffusion time of Rhodamine molecules as a function of temperature measured in Reference [71].

In 2012, Toshimitsu *et al.* [71] reported on optical trapping of polymer chains in liquid assisted by plasmonic nanostructures. In this context, they discuss the effect of a local optical induced temperature increase that affects the trapping efficiency by inducing a repulsive force coming from Soret effect (see Section 5.3). The work was the occasion for the authors to introduce a temperature measurement technique based on FCS. They dispersed Rhodamine molecules in ethylene glycol solution surrounding the plasmonic nanostructures under illumination and performed FCS measurements to retrieve the local temperature as a function of laser power. The calibration relating diffusion times measured by FCS and temperature is displayed in Figure 4.15. In 2013, the same group published a related article where the idea was to further focus on this FCS approach for temperature probing in nanoplasmonics [82].

In 2013, Hao Vu *et al.* [31] introduced a technique to measure temperature on the microscale based on rotational scattering correlation spectroscopy (rSCS).[2] The authors used asymmetric 140 nm spheres that were coated with a gold layer on only one side, in such a way that the angular scattering profile of such an object was strongly non-isotropic. The same type of correlation measurements can be done by illuminating nano-objects. However, in this case, the measured signal does not come from fluorescence but from simple light scattering, and the intensity fluctuations do not come from molecules traveling through the observation volume, but from rotational Brownian motion of the objects diluted in glycerol. The originality of the work is that the plasmonic nanoparticles simultaneously act as nanoheaters and thermal nanosensors. Given the involved timescale of the rotational Brownian motion (submicrosecond), long time traces (presumably several seconds) have to be recorded in order to retrieve a nanoparticle temperature with a good precision. Figure 4.16 plots the calibration curve used in this work, relating rotational correlation time and temperature.

---

[2] As this technique is not based on fluorescence measurements, it should not appear within this section. However, as the principle is similar to FCS, I chose to mention this work in this subsection instead of dedicating a new section in this chapter for a single article.

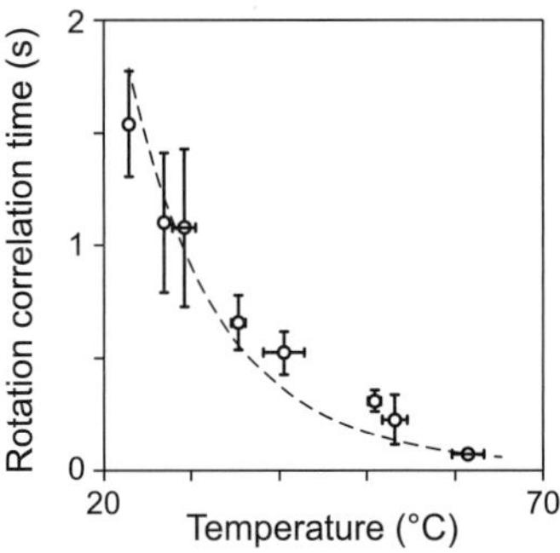

Fig. 4.16 Rotational correlation time of asymmetric nanoparticles as a function of temperature measured in Reference [31].

In 2014, Hormeño and coworkers performed another type of correlation spectroscopy to retrieve the temperature of nanoparticles [35],[3] but the physics was different. The authors designed hybrid nanoparticles made of a gold core and a pNIPAM polymer shell. Around 35°C, pNIPAM undergoes a phase transition from an expanded phase to a collapsed phase, which was intended to modify the Brownian motion the nanoparticles due to a variation of their hydrodynamic volume. This volume was probed by dynamic light scattering (DLS), a technique suited to measure particle size in a solution via autocorrelation of the scattered light signal collected from the sample. The temperature of the nanoparticles can be retrieved this way as far as the relation between the DLS signal and the temperature is determined.

4.3 Microwave Spectroscopy of Nanodiamonds via Fluorescence Measurements

Diamond nanoparticles (or nanodiamonds) endowed with one or several negatively charged nitrogen vacancy (NV) defects in their carbon atomic lattice are widespread and increasingly being used in many areas of research due to their unique properties of photo-luminescence, spin resonance in the microwave range, magnetosensitivity, photostability, chemical inertia and biocompatibility. They led to numerous applications from biology to quantum physics [66, 24, 73, 43] and have proven to behave as efficient probes of temperature [1, 16, 72, 43, 50, 56], magnetic field [62] and electric field [25].

A NV centre consists of a triplet state formed by two unpaired electrons. The electronic energy diagram, sketched in Figure 4.17a, displays two types of energy transitions: transitions in the *microwave range*, and transitions in the *optical range* associated with fluorescence emission.

Microwave transitions occur between three sublevels of the ground triplet state level. Two of the sublevels are generally degenerated ($|m_s = +1\rangle$ and $|m_s = -1\rangle$). The third

---

[3] This technique is not based on fluorescence either.

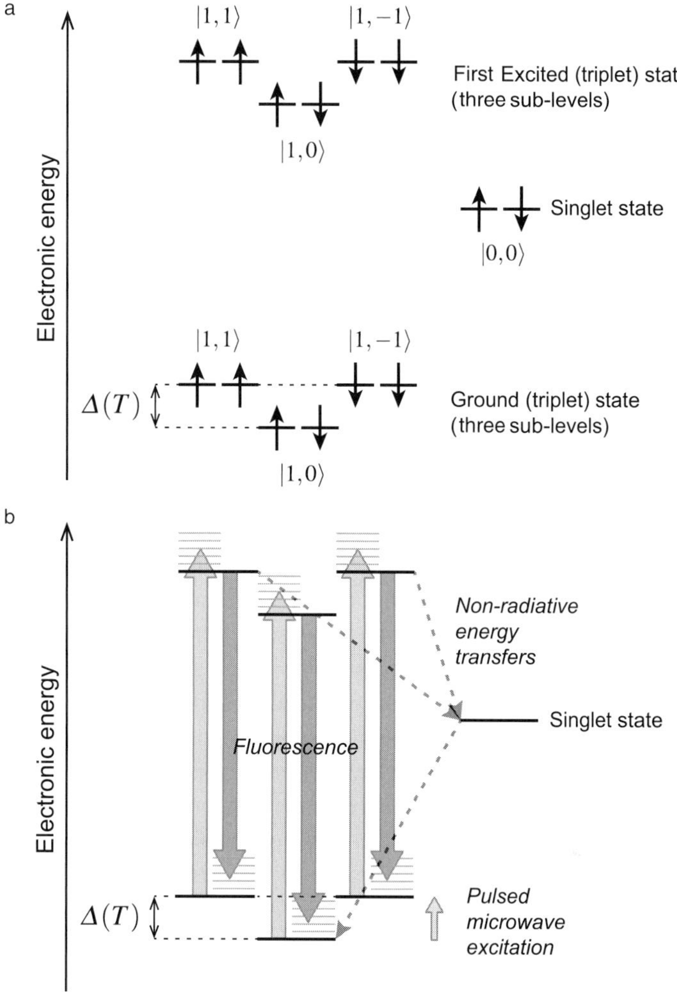

Fig. 4.17 (a) Electronic energy diagram of a NV center. The splitting energy Δ between the $|1, 0\rangle$ and $|1, \pm1\rangle$ states is temperature-dependent. (b) Same diagram where the energy transfer pathways induced by optical and microwave excitation have been indicated. Under green illumination (from 460 nm to 580 nm), a red fluorescence is emitted (from 637 nm to 800 nm) from the first excited triplet state. Spin transitions between the $|1, 0\rangle$ and $|1, \pm1\rangle$ states can be controlled by pulsed microwave irradiation. Excitation from $|1, \pm1\rangle$ has a finite branching ratio into the metastable singlet state $|0, 0\rangle$ that subsequently de-excites to the ground state over a timescale of 300 ns (dashed lines). This de-excitation pathway via the singlet state is non-radiative.

sublevel ($|m_{\mathrm{s}} = 0\rangle$), lying at a lower energy, reveals an energy splitting arising from the presence of a local crystal asymmetry. This splitting energy Δ is called the axial zero-field splitting parameter and it corresponds to an angular frequency of around $(2\pi) \times 2.87$ GHz. It is the main physical quantity of the energy diagram that is temperature-dependent. This dependence is presumably due to local thermal expansion of the diamond atomic lattice.

In parallel, *optical* transitions occur between the ground level and the first excited level. Such transitions are efficiently excited using green light and reemits in the red region of the spectrum, between 637 nm and 800 nm (see Figure 4.17b). While optical excitation from the $|m_s = 0\rangle$ sublevel is spin preserving, excitation from the sublevels $|m_s = +1\rangle$ or $|m_s = -1\rangle$ enables a non-radiative decay via a singlet state with a long lifetime (of the order of 300 ns), which subsequently necessarily decays toward the $|m_s = 0\rangle$ sublevel of the ground state. Consequently, exciting a NV center using a green light rapidly clears out the populations of the states $|m_s = \pm 1\rangle$ and the NV centers becomes spin polarized in the $|m_s = 0\rangle$ sublevel. In other words, this singlet state decay allows for an efficient initialization of the electronic spin using light.

Microwave transitions and optical transitions are intimately related. For instance, by illuminating the system using a microwave frequency matching Δ, transitions occur from $|m_s = 0\rangle$ to $|m_s = \pm 1\rangle$, which tends to deem the emitted fluorescence because populating the $|m_s = \pm 1\rangle$ states yields non-radiative pathways, as explained above. This fluorescence deeming can be used to measure Δ via a technique named ODMR for "optically detected magnetic resonance" or equivalently FDMR for "fluorescence detected magnetic resonance". The principle consists in spanning the incoming microwave frequency and measuring in parallel the fluorescence intensity. Δ corresponds to the microwave frequency that induces a deeming of the fluorescence emission, via the above-mentioned mechanism.

As $\Delta(T)$ is temperature-dependent, ODMR can be conveniently used to measure the temperature of NV centres, provided a calibration curve was previously established [55]. Note that the calibration curve is supposed to vary from one NV centre to another. Thus, the energy splitting Δ of the given NV centre, or of the given nanodiamond, must be measured at room temperature prior to conducting temperature measurements.

In 2010, Acosta *et al.* studied for the first time the temperature dependence of the microwave resonance spectra of NV centres using ODMR [1]. In this study, the authors investigated ensembles of NV centres in single-crystal bulk diamond (they did not use nanodiamonds). The authors recorded full ODMR spectra around the resonance (Figure 4.18a) and they evidenced that both the resonance energy Δ and the splitting energy E were temperature-dependent.[4] They measured a sensitivity $dD/dT = -74.2$ kHz·K^{-1}. This figure has to be compared with the width of the resonance which is on the order of 10 MHz. This gives a sensitivity of around 10^{-2} K^{-1} (Figure 4.18b). The splitting energy E was shown to be much less sensitive to temperature: $|dE/(EdT)| \approx 10^{-4}$ K^{-1} (Figure 4.18c).

In 2013, the Lukin's group was able to measure the temperature at specific locations in living cells using nanodiamonds dispersed in the cytoplasm. Heating was induced inside living cells using gold nanoparticles under illumination [43] and temperature measurement were performed using ODMR of the nanodiamonds. The authors claimed a temperature sensitivity as small as 1.8 mK (a sensitivity of 9 mK·Hz$^{-1/2}$). However, this figure is related to measurements of single NV centres in *bulk* diamond, not using diamond nanoparticles. When using diamond nanoparticles contained hundreds of NV centres, the

---

[4] This energy E corresponds to the small energy difference between the states $|m_s = -1\rangle$ and $|m_s = +1\rangle$, arising from the local magnetic field of the diamond lattice. E can be further increased by applying an external magnetic field.

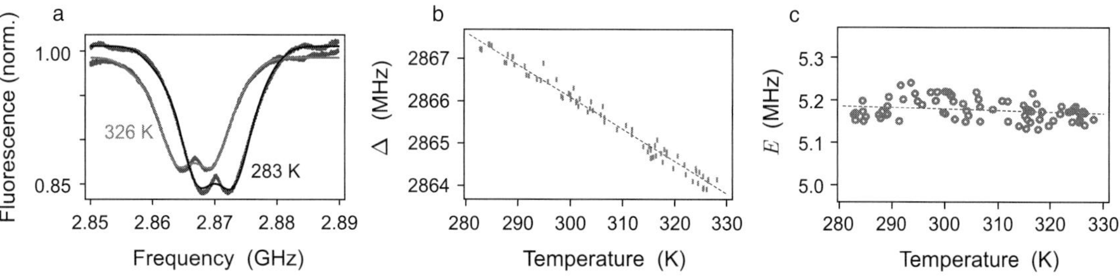

Fig. 4.18 (a) ODMR spectra at 283 and 326 K. (b) Magnetic resonance energy Δ retrieved from ODMR spectra, as a function of temperature, with linear fit. (c) Values of E retrieved from ODMR spectra, as a function of temperature, with linear fit. Reproduced with permission from Reference [1]. Copyright 2010, American Physical Society.

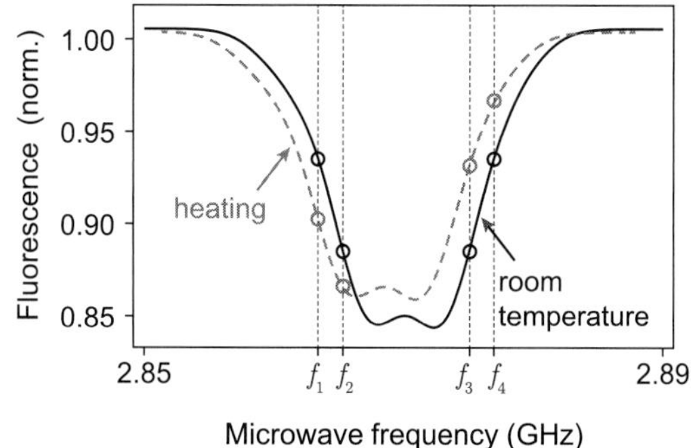

Fig. 4.19 Schematic of the method of the four-frequency measurement used by the Lukin's group to perform optically detected magnetic resonance (ODMR) in living cells, as in Reference [43].

sensitivity is rather on the order of $0.2\ \mathrm{K\cdot Hz}^{-1/2}$, as evidenced by their measurements (see Figure 4.20). In their experiments, the authors did not measure the full microwave spectra. Spanning the microwave wavelength over the full range of interest would have been very time-consuming. The authors only sampled four points of the ODMR spectrum, *i.e.*, four frequencies, to extract the thermal-induced shift of the spectra, as represented in Figure 4.19. Figure 4.20a represents a fluorescence map from nanodiamonds. In this figure, five nanodiamond locations are spotted by circles and a heated gold nanoparticles is indicated using a cross. The temperature increase retrieved from the fluorescence of all these nanodiamonds as a function of the distance from the heat source is in good agreement with the $1/r$ decay law of the temperature profile, as represented in Figure 4.20b.

In 2015, Tzeng *et al.* refined the use of NV centres as temperature probes in plasmonics [72]. The authors designed a method that allowed probing of the temperature changes over a temperature range of 100 K, with a precision of 1 K and a temporal resolution of better than 10 μs. This much faster acquisition time compared to what was previously reported (a few seconds) is due to the use a three-point method coupled with a pump–probe-type measurement. Three frequencies were measured. Two frequencies were positioned at the edge of the resonance to improve the sensitivity to any shift of resonance, and an additional third frequency corresponded to a far-off resonant frequency at which the microwave has no

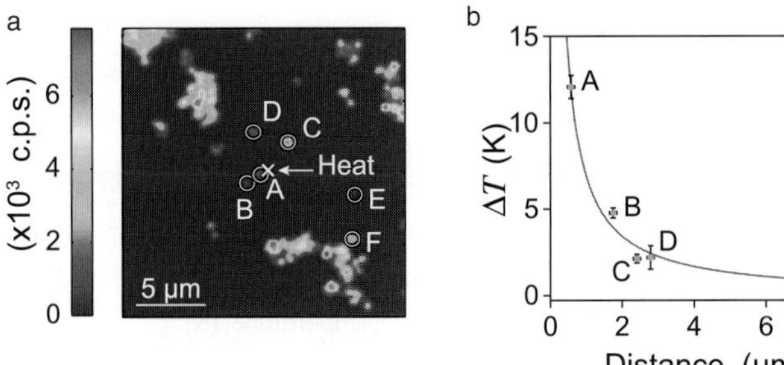

Fig. 4.20   (a) Two-dimensional confocal scan of nanodiamonds (circles) and a gold nanoparticle (cross) spin-coated onto a glass coverslip. The color bar represents fluorescence intensity, expressed in counts per second (c.p.s.). (b) Temperature changes measured (red points) at the six nanodiamond locations in (a) as a function of distance from the illuminated gold nanoparticle (cross). The blue curve is the theoretical $1/r$ temperature profile based on the steady state solution of the heat equation. All data in this figure were obtained on a glass coverslip, and all error bars correspond to 1 s exposure time. Reproduced with permission from Reference [43]. Copyright 2013, Nature Publishing Group.

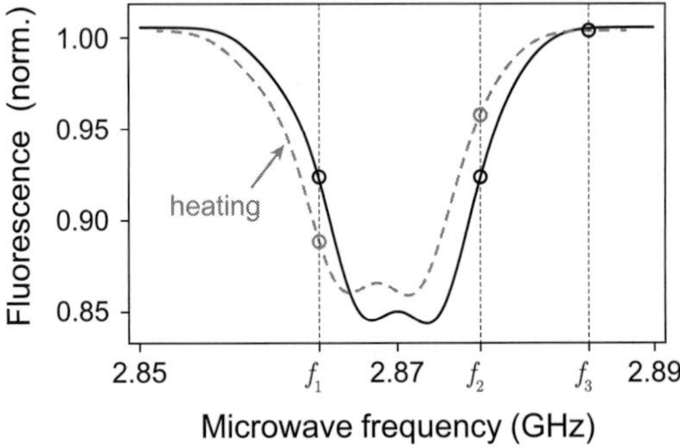

Fig. 4.21   Schematic of the method of the three-frequency measurement used by Tzeng *et al.* to perform fast ODMR, as in Reference [72].

observable effect on the ground state spins of the NV centres (see Figure 4.21). An advantage of this three-point method is that it possesses a built-in ability for self-correction of signal fluctuation. The authors compared their three-frequency approach to the four-frequency measurement of the Lukin's group, mentioned above [43]. With the Lukin's approach, the working range is limited to about ± 15 K due to the linear temperature dependence assumption. In comparison, the Tzeng's method is about three times lower in sensitivity but has a much wider (10-fold) working range, and the analytical relationship between the frequency shift and the measured fluorescence intensities is available even at large temperature changes. Moreover, the method is extendable to time-domain measurement by performing pump–probe-type experiments using a pulsed laser as the heating source.

In 2016, Tétienne and coworkers managed to functionalized the tip of the scanning probe of an AFM with a nanodiamond endowed with multiple NV centres [70]. With this device, the authors could map the photoinduced heating generated by laser irradiation of a single gold nanoparticle in a fluid environment. This device also enabled the authors to retrieve

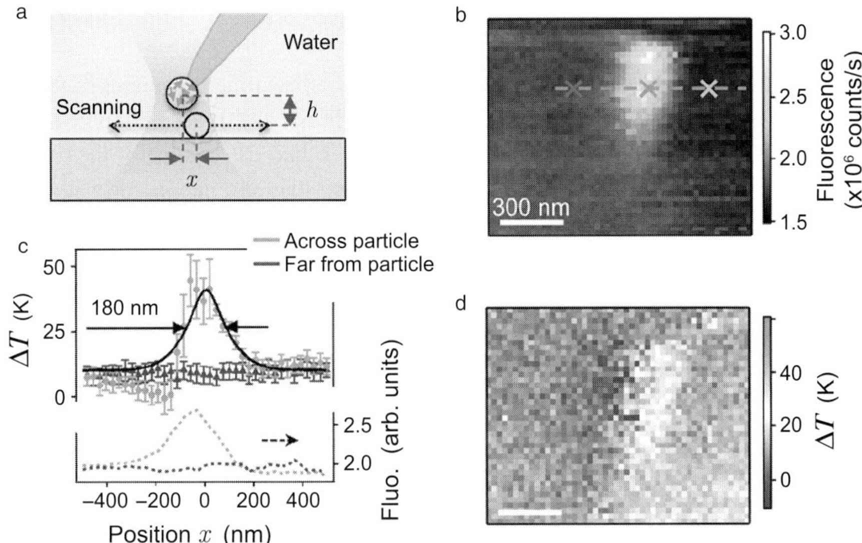

Fig. 4.22 Thermal imaging of a photoheated gold nanoparticle as conducted in Reference [70]. (a) Schematic of the experiment. (b) Fluorescence image obtained by scanning a nanodiamond on top of a 40 nm gold particle. The integration time is 1.5 s per pixel and thus 50 min for the whole frame. (c) Line cuts extracted from (d), which is the temperature map calculated from the image (b). Reproduced with permission from Reference [70]. Copyright 2016, American Chemical Society.

topographic information of the sample with a spatial resolution of 100 nm. The temperature accuracy was on the order of a few 10s of Kelvin (see Figure 4.22). The origin of such poor sensitivity is not clear. The reason may come from the fact that the authors were aiming at making an image, forcing them to reduce the exposure time per pixel to make an image within a reasonable amount of time, which was still on the order of one hour. Previous studies that achieved a better sensitivity were just doing single point measurement with exposure times on the order of a few seconds. The authors underlined the fact that the use of a NV *ensemble* is crucial to this particular experiment. Indeed, the signal from a *single* NV centre would be overwhelmed by the background fluorescence produced by the gold particle, making measurements in plasmonics highly challenging using the conventional single NV approach.

4.4 Techniques Sensitive to Refractive Index Variations

Any liquid features a variation (usually a decrease) of its refractive index n upon heating. When locally heating a nanoparticle in a solvent, this variation naturally occurs over a volume confined in the vicinity of the nanoparticle and provides a way to optically probe a local temperature variation. Investigating a temperature variation by probing a refractive index modification is a strong benefit compared to the aforementioned techniques as it is *label-free*.

In 2002, Boyer *et al.* leveraged for the first time this effect in the context of plasmonics. When heating a nanoparticle in a liquid, since the portion of the fluid that features a refraction index variation is much bigger than the volume of the nanoparticle, the system becomes suddenly much more scattering than the nanoparticle itself. In other words, by heating a nanoparticle, it becomes detectable under dark field imaging. This trick was used, for instance, to detect nanoparticles as small as a few nanometers and perform tracking experiments in living cells. The benefit compared to tracking with fluorescent molecules is that a nanoparticle never bleaches. However, this technique is not considered as a thermal imaging technique in this chapter as it is not suited to *quantitative* temperature measurement. The purpose of this technique was more to *detect* nanoparticles rather than to measure a temperature in plasmonics. This technique, named *photothermal imaging*, is however discussed in detail in Chapter 6, listing the applications of thermoplasmonics, on page 256.

In 2012 and 2013, Setoura *et al.* [68, 69] proposed to measure the shift of the scattering spectra of nanoparticles under heating due to this thermal-induced modification of the surrounding refractive index. As the effect is weak, large temperature increases had to be reached. This approach allowed the estimation of nanoparticle temperature up to 400°C with an accuracy of 20°C on the basis of a spectral calculation using Mie theory. However, some pre-knowledge on the nanoparticle geometry is required for quantitative temperature retrieval.

In 2011, a year before, the group of Satoshi Kawata [34] introduced a closely related technique, although much more sensitive. The authors designed hybrid nanoparticles made of a 40 nm gold core nanoparticles functionalized with a 20 nm thick thermosensitive (pNIPAM) polymer layer. Around 32°C, this polymer undergoes a reversible phase transition from an open structure to a collapsed phase. This effect leads to significant increase of the refractive index at the vicinity of the surface of the nanoparticles, and a subsequent redshift of its plasmon resonance. This approach is much more sensitive than counting of the thermal induced modification of the refractive index of the solvent, as proposed by Setoura *et al.* a year after [68].

In 2012, a strategy was proposed to *quantitatively* retrieve a temperature distribution from this local refractive index variation [4]. The idea was no longer to detect scattered light, but to map the wavefront distortion undergone by a planar optical wavefront crossing the system. The local refractive index variation stemming from a local temperature increase acts as a thermal lens (or, equivalently, creates a mirage effects). This thermal induced wavefront distortion can be quantitatively mapped by quadriwave lateral shearing interferometry (QSI) [76]. Albeit the complexity of its appellation, this technique is particularly simple to implement. It is based on the use a special diffraction grating (modified Hartmann grating) placed at a millimetric distance from a regular video camera sensor (see Figure 4.23). Any optical beam impinging on this system creates an interferogram on the camera sensor than can be processed to retrieve the optical wavefront profile, *i.e.*, the phase of the light (along with the intensity). QSI is the only existing wavefront sensing technique that provides high-resolution (400 × 300 pixels in the reported experiments)

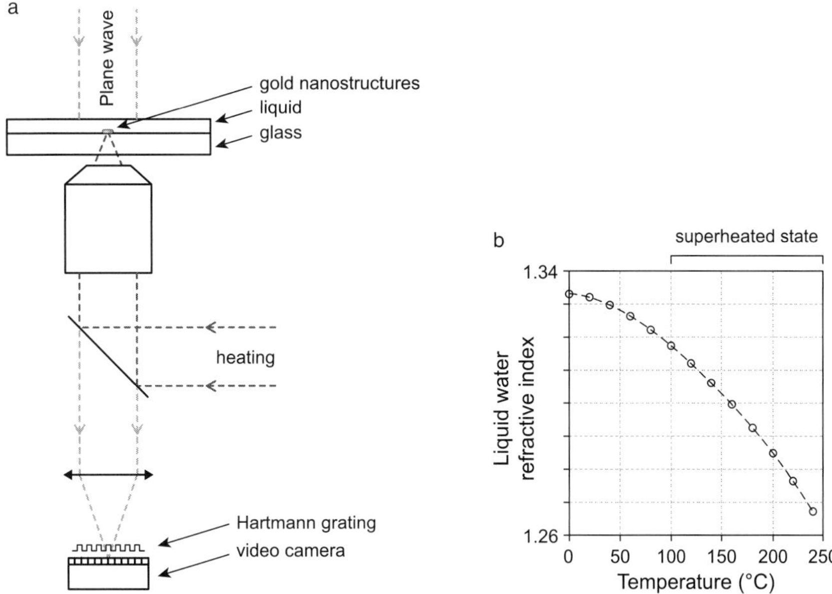

Fig. 4.23 (a) Experimental configuration for temperature imaging using wavefront sensing [4]. (b) Calibration curve used for temperature retrieval taken from Reference [65] (see Appendix B for a fit function).

and *quantitative* measurements at the same time. Subsequently, using an inversion algorithm, one can retrieve the actual three-dimensional temperature field in the sample from the wavefront distortion image with a diffraction limited spatial resolution. The authors named this techniques TIQSI for temperature imaging using QSI.

The advantages of TIQSI are the following: (i) it is fast (standard video rate), (ii) it has a spatial resolution of around 500 nm (diffraction limited), (iii) it has a temperature sensitivity of *ca.* 1 K, (iv) the addressable temperature range is arbitrarily large, a benefit compared to the aforementioned techniques based on the use of fluorescent molecules that usually bleach above 70°C, (v) it is straightforward to implement as one just has to plug the QSI camera into a regular optical microscope, and (vi) unlike other interferometric techniques, QSI is not highly sensitive to mechanical fluctuations and does not require a complex alignment as the camera sensor and the diffraction grating are tightly bound and separated by a millimetric distance.

Another advantage of this technique is that it can also quantitatively retrieve the heat source density (power per unit area), and the actual absorption cross section of nanoparticles [12], no matter their size and composition. Thanks to all these advantages, this technique has been used subsequently to investigate microscale thermal induced processes related to hydrothermal chemistry, hydrodynamics, thermal cooperative effects [5], microscale temperature shaping [7], living cell manipulation [86], phase transitions and superheating [8], where one can hardly use fluorescent molecules.

The drawbacks of this technique are as follows: (i) the surrounding medium has to be liquid, otherwise no measurable refraction index variation occurs; (ii) the source of heat has

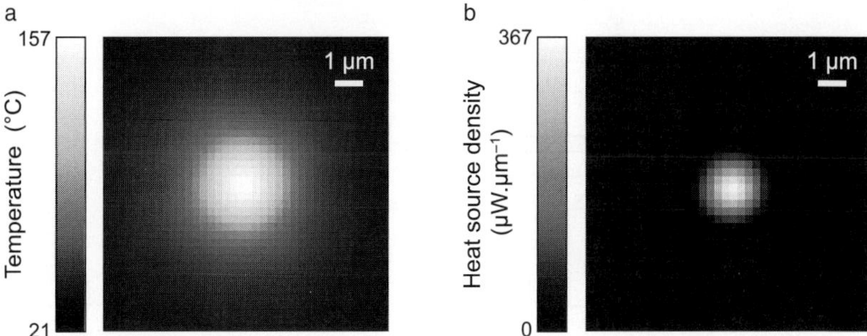

Fig. 4.24 Thermal measurements using TIQSI [4] around a lithographic gold disc, 1 µm in diameter (dimension matching the scale bar of the two images). (a) Temperature distribution. (b) Associated heat source density.

to be two-dimensional, *i.e.*, located, for instance, at the interface between a glass substrate and the liquid (like in Figure 4.1b): the inversion algorithm does not work, for instance, for colloidal nanoparticles dispersed in a liquid as represented in Figure 4.1a; and (iii) one reference image has to be primarily acquired without heating.

Figure 4.24 displays typical thermal images acquired by TIQSI. They represent the temperature and the heat source density associated to a single lithographic gold nanodisc, 40 nm thick and 1 µm in diameter immersed in water and illuminated with a focused laser power of 21 mW at $\lambda = 736$ nm. The spatial resolution of the technique makes the heat source density distribution blurry and the maximum heat source density underestimated. However, by summing all the pixels of this image, one can quantitatively retrieve the total power delivered by the nanoparticle, no matter the lack of resolution (0.93 mW in this measurement).

So far, this technique has been used to investigate the physics of several thermal induced effects in plasmonics, namely fluid superheating [8], bubble generation [8], temperature shaping [7], thermal collective effects [5], hydrothermal chemistry [58], living cell migration [86].

In 2015, Zixuan Chen *et al.* developed a thermal microscopy technique[17] inspired from the Biacore technology[5] (one of the rare practical applications of plasmonics). The authors coined this technique "Plasmonic Thermal Microscopy" (PTM). A thin metal film deposited on a dielectric substrate was illuminated via an objective lens in total internal reflection (see Figure 4.25). In such a configuration, the angle of reflection is dependent on the refractive index of the medium just above the gold layer. This is how Biacore probes the presence of proteins attaching to the gold layer. Here, the refractive index change was not due to the anchoring of molecules, but to the temperature increase. This very sensitive approach is able to detect temperature variations as small as 6 mK with a temporal resolution of 10 µs. However, this technique only works in the presence of a gold film. It is thus not useful for investigate single nanostructures on glass, for instance.

---

[5] www.biacore.com

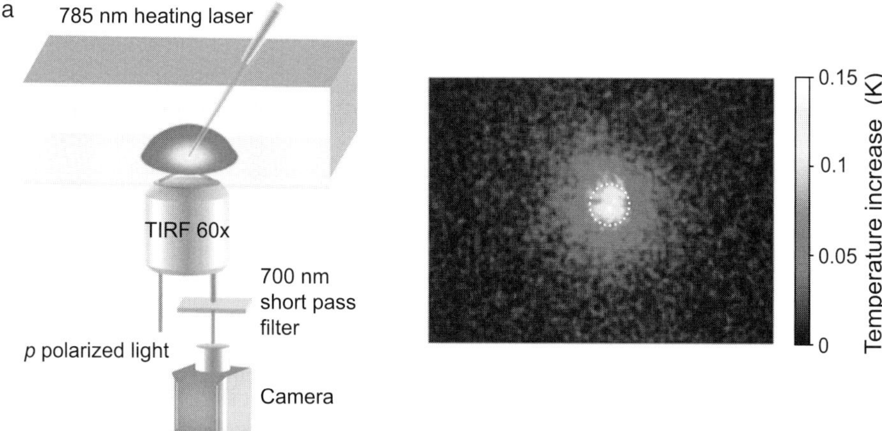

Fig. 4.25 (a) Schematic illustration PTM setup and principle. A 785 nm heating laser is focused on a gold-coated glass slide to locally generate heat and increase the temperature, which leads to a refractive index change of the surrounding medium (water). This refractive index change is imaged by PTM using an inverted optical microscope with a high numerical aperture objective. The plasmonic imaging uses p-polarized light to excite surface plasmons on the gold surface and CCD camera to record the reflected beam. To remove the heating laser on the camera, a short-pass optical filter is placed before the CCD imager. (b) PTM image of temperature distribution generated by 2.8 mW·m^{-2} heating laser. Reproduced with permission from Reference [17]. Copyright 2015, American Chemical Society.

4.5 Raman Scattering Spectroscopy

One speaks about inelastic scattering in optics when photons scattered by a given medium feature an energy different from the incident photons. Raman scattering (RS) is an inelastic scattering process where this energy difference corresponds to the energy of vibrational modes of the scattering molecules. This process is schematized in Figure 4.26a. As in any scattering process, a photon absorption occurs from an eigen molecular state (0) to a non-stationary (also called "virtual") state (2) that instantaneously and coherently relaxes. In RS, the energy of the scattered photon is $h\nu' = h\nu - h\Omega$ where Ω is the vibrational frequency of the excited state (1). There is a second kind of RS process, which is schematized in Figure 4.26b. It consists in the opposite mechanism where a vibrationally excited state (1) is brought down to another eigen state at a lower energy. This second mechanism is called anti-Stokes RS.

RS spectroscopy is intended to study such energy shifts, which constitute molecular signatures and that can be used to identify the nature of molecules in a given sample. In a RS spectrum, the energy locations of corresponding Stokes and anti-Stokes peaks are symmetric around the excitation photon energy. However, their relative intensities are different as schematized in Figure 4.27. This stems from the fact that anti-Stokes RS is much less probable because state (1) is less populated than state (0) in thermodynamic equilibrium. The relative populations of states (1) and (0) simply obeys a Boltzmann distribution: if E_0

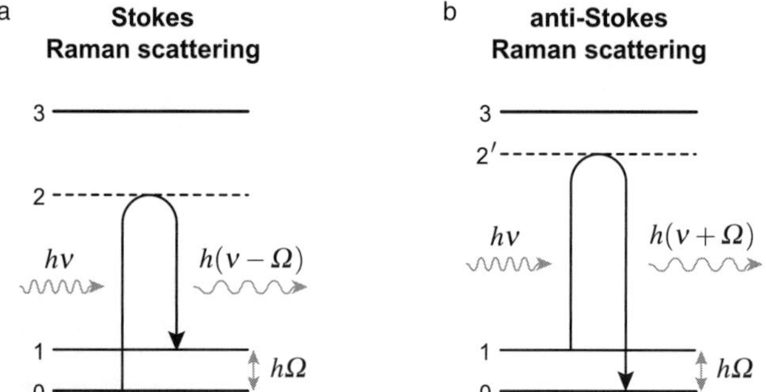

Fig. 4.26 Energy diagram related to stokes and anti-stokes Raman scattering of a photon of energy $h\nu$ by a molecule. (0) and (3) are molecular electronic eigen states, (1) is a vibrationally excited state and (2) and (2′) are non-stationary states, usually called virtual states.

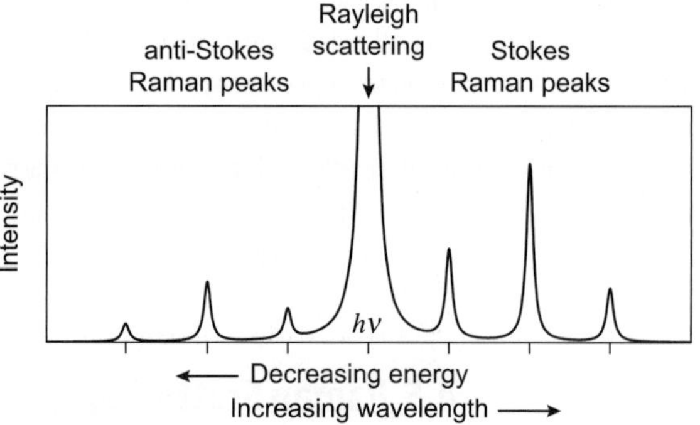

Fig. 4.27 Schematic of the typical line shape of a Raman spectrum.

and E_1 are the energies of states (0) and (1), the average population ratio between these two states reads

$$\frac{N_1}{N_0} = \exp\left(\frac{E_0 - E_1}{k_B T}\right). \tag{4.7}$$

This population ratio dictates the relative intensities of Stokes and anti-Stokes peaks. Interestingly, as this ratio is solely temperature-dependent, RS spectroscopy can be used for temperature measurements.

Importantly, RS can be dramatically enhanced by a factor of *ca.* 10^{11} when the molecules are in the vicinity of a rough metallic surface or of metal nanoparticles [42, 52]. The effect is termed surface-enhanced Raman scattering (SERS) and is the basis of a very active branch of research. Although the actual underlying mechanism of SERS is still a matter of debate in the literature, it is admitted that the large electromagnetic field enhancement at the metal surface plays a dominant role.

In 2009, Rycenga *et al.* [63] proposed for the first time to use SERS measurements to measure a local temperature increase of a plasmonic nanoparticle under illumination. The authors used alkanethiolate molecules chemisorbed on silver nanocubes and considered

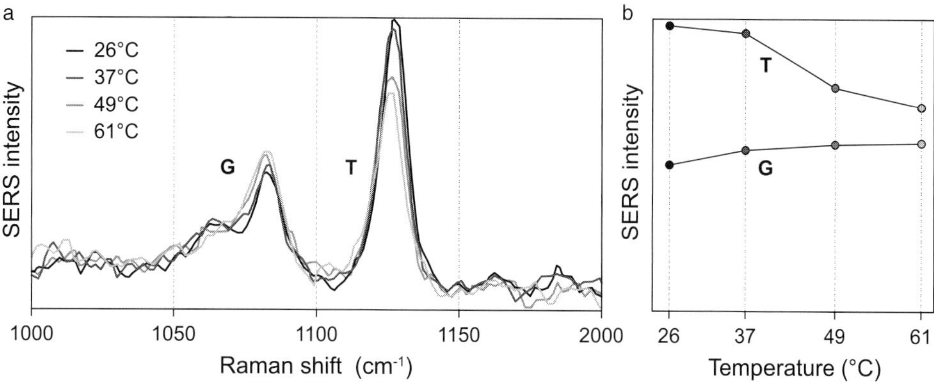

Fig. 4.28 (a) SERS spectra of alkanethiolate-covered Ag nanocubes in water at four solution temperatures with 514 nm Raman excitation [63]. Each spectrum contains the gauche (G, at 1080 cm^{-1}) and trans (T, at 1125 cm^{-1}) carbon–carbon stretch of the molecules. (b) The relationship between temperature and peak intensities of the T and G bands, where an increase in the solution temperature causes the T band to attenuate and the G band to increase. Reproduced from Reference [63]. Copyright 2009, Wiley–VCH Verlag GmbH & Co. KGaA.

the ratio between two Stokes peaks (not Stokes and anti-Stokes peaks) as displayed in Figure 4.28.

In 2011, Yokota *et al.* [84] investigated SERS of molecules at the vicinity of plasmonic nanostructures in order to analyze the theory of electromagnetic field SERS enhancement. This work was the occasion to notice that SERS can be used to retrieve the local temperature because the ratio of Stokes and anti-Stokes intensities is related to the Boltzmann distribution. However, Stokes and anti-Stokes peaks are not equally surface-enhanced. This means that considering Stokes anti-Stokes ratio to retrieve a local temperature increase of a metal nanoparticle must involve a normalization of the SERS intensity that can cancel out the difference in enhancement factors of Stokes and anti-Stokes signals.

In 2015, the group of R. P. Van Duyne reported on the same approach based on the Stokes–anti-Stokes ratio to retrieve a local temperature increase in plasmonics [57]. They conducted SERS measurements on *single* Rhodamine 6G molecules on silver nanoparticle aggregates. They end up with the same observation that the Stokes–anti-Stokes ratio does not only depend on temperature. The remaining factors that may cause variance in the ratio are (i) changes in excited state geometry, (ii) enhancement factor variations between anti-Stokes and Stokes transitions, (iii) local heating, and (iv) vibrational pumping into the $\nu = 1$ state. These results suggest that local heating is observed, but the uncertainty in the wavelength dependence precludes a confident temperature calculation.

In 2015, Yashchenok and coworkers measured the temperature of gold nanoparticles by Raman spectroscopy on carbon nanotubes [83]. The temperature sensitivity was rather poor, around 5 to 40 K depending on the temperature increase.

In 2016, the group of D. G. Cahill used anti-Stokes Raman spectroscopy of the electronic of plasmonic nanoparticles themselves to retrieve the temperature of nanoparticles under illumination [81]. This smart approach does not require the presence on any other molecular probe and can address very large temperature rises, up to 400°C.

4.6 X-ray Absorption Spectroscopy

X-rays correspond to energies ranging from 500 eV to 500 keV, or wavelengths from 25 Å to 0.25 Å. Studying matter using X-rays yields information regarding the atoms' environment in the material. The benefit of X-ray *absorption* spectroscopy compared to X-ray *diffraction* is that it applies for any kind of medium, crystalline or not. A very good review on this subject was written by Newville [51]. When the incident X-ray energy matches the binding energy of an electron of the material, X-ray absorption occurs, causing a drop in the transmitted X-ray intensity. Due to the very high energy of X-rays, X-ray photon absorption is caused by core electrons of atoms (such as the 1s or 2p levels). Spectra are obtained by scanning the incident energy. It is common to convert X-ray photon energies E into the wave numbers k of the associated ejected photo-electrons to plot an X-ray absorption spectrum. As the photo-electrons have an energy of $E - E_0$, one has

$$k = \sqrt{\frac{2m(E - E_0)}{\hbar^2}} \tag{4.8}$$

where m is the electron mass and E_0 the electronic state energy.

In 2011, Van de Broek *et al.* [75] managed for the first time to conduct such X-ray absorption spectroscopy (EXAFS) measurements on plasmonic nanoparticles to determine their temperature under illumination. EXAFS measurements require a synchrotron facility producing high-intensity X-ray beams with tunable energy. The experiments were carried out at the European Synchrotron Radiation Facility (ESRF) in Grenoble (France). Probing variations of the nearest-neighbor interatomic distance of the nanoparticle upon heating would have been the natural approach, but the authors noticed that thermal-induced expansion of a gold nanoparticle crystal structure is very small. The theoretically calculated bond distance changes for a temperature increase of 54°C is 0.0025 Å (*i.e.*, 0.09%), which cannot be detected by EXAFS with sufficient precision. On the contrary, the authors focused on the Debye-Waller factor that tends to damp any signal of an EXAFS spectrum due to atomic disorder, and in particular due to thermal agitation. Measurements to determine the calibration curve are displayed in Figure 4.29.

Thanks to this approach, the authors have been able to evidence a temperature increase of 17 K of gold nanoparticles upon illumination.

Apart from the complexity of the experimental device (a synchrotron), the particularity of this approach is that it measures the inner nanoparticle temperature. In the aforementioned approaches, temperature nanoprobes were dispersed outside the nanoparticles and were measuring in the temperature of the surrounding medium.

4.7 Scanning Thermal Microscopy

Scanning thermal microscopy (SThM) [80, 40, 30] was invented in 1986 and consists in directly measuring a temperature distribution using a sharp tip acting as a nanoscale thermocouple or bolometer. The achieved resolution is on the order of 50 nm.

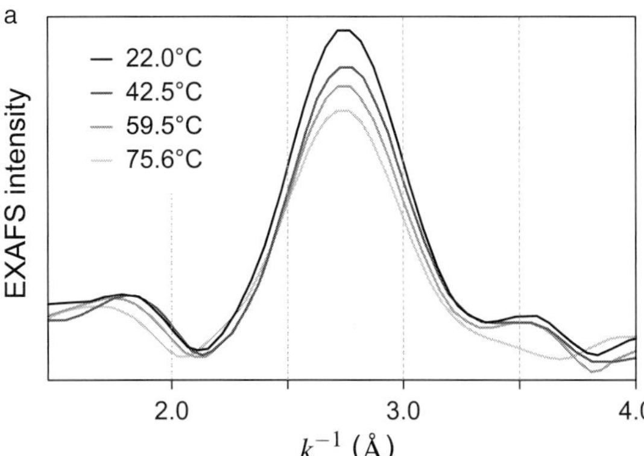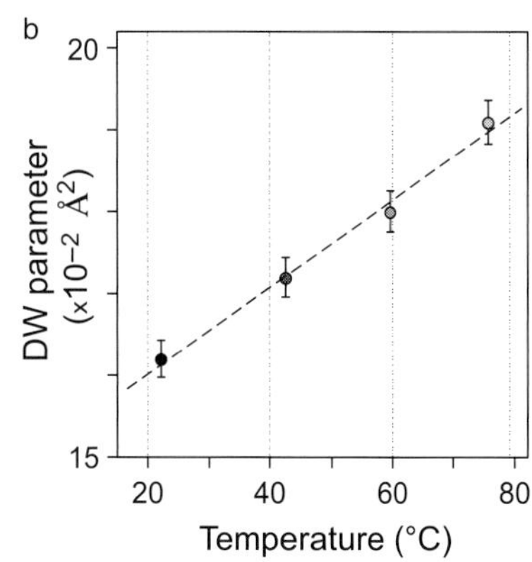

Fig. 4.29 (a) Phase-corrected Fourier transforms of the Au L3-edge k3- weighted EXAFS as a function of k^{-1}, for a branched gold nanoparticles upon calibration with a gas heater gun at different fixed temperatures. (b) Calibration curve of the Debye-Waller term $2\sigma^2$ as a function of temperature for the branched gold nanoparticles. Reproduced from Reference [75]. Copyright 2011, WILEY–VCH Verlag GmbH & Co. KGaA, Weinheim.

In 2014, the group of Uriel Levy [23] proposed for the first time to use SThM to probe a temperature distribution over a plasmonic device. Figure 4.30 presents thermal simulations and measurements using SThM on a silicon plasmonic waveguide excited in the infrared. The authors explain that SThM can circumvent the inherent problem of far-field optical microscopy techniques limited by light diffraction (around 300 nm for typical fluorescence based approaches). However, quite surprisingly, the authors investigated a *micrometric* structure, much larger than the optical diffraction limit, and do not show any high resolution temperature map. Although they call their structure a "nanofocusing device" or a "nanowaveguide," its typical dimensions are rather micrometric, just like the extension of the temperature distribution as evidenced by their numerical simulations represented in Figure 4.30a. One can regret that the authors did not investigate nanometric plasmonic particles in order to evidence the gain in spatial resolution using SThM. We can imagine that the reason why the authors did not investigate smaller systems is that SThM would be too invasive for nano-objects. As mentioned in the introduction of this chapter, a near-field tip cannot act as a noninvasive probe when touching a nanoparticle. On the contrary, it would certainly act as a thermal reservoir. Moreover, SThM is known to achieve quantitative measurement with difficulty for nanoscale devices, as explained in Reference [40].

In 2016, Kinkhabwala *et al.* reported on another near-field technique to map the temperature of plasmonic nanoantennas, although it is not SThM [41]. The experiments were conducted in the context of heat-assisted magnetic recording (HAMR, see Section 6.4). The technique was called "polymer imprint thermal mapping" (PITM) and was intended to break the optical diffraction limit, while fixing the problems related to SThM. The technique is based on the use of a thermosensitive polymer that permanently cross-links upon heating. By depositing a layer of such a polymer on the sample of interest, the local heating of plasmonic structures induced a permanent local change of the polymer thickness that was subsequently mapped using AFM.

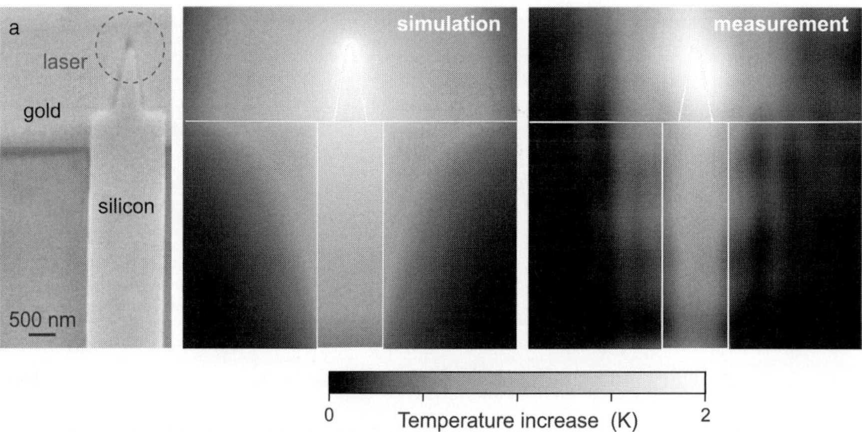

Fig. 4.30 (a) Scanning electron microscope image of a silicon plasmonic waveguide. The tip of the structure is illuminated using at wavelength of 1.55 μm. (b) Numerical simulation of the temperature distribution around the waveguide under illumination. (c) Temperature distribution measured by SThM. Reproduced with permission from Reference [23]. Copyright 2014, American Chemical Society.

4.8 Other Techniques

In this chapter, I reviewed the most general techniques. Some more specific techniques have not been discussed because they don't belong to general families. Let us mention here some valuable techniques that have been skipped.

In 2014 the group of Natelson [32] developed a technique to probe the temperature of a plasmonic nanowire. The technique does not consist in *mapping* the temperature field, but rather in measuring the average temperature of the plasmonic wire by measuring variations of its electrical resistance. This technique is thus restricted to plasmonic structures that are connected to electrodes.

Also in 2014, Schmid *et al.* developed a similar technique in the sense that the plasmonic structure had to be connected to a macroscopic device [67]. The authors investigated the heating of a plasmonic structure (nanorod and nanoslit) placed at the centre of a microstring resonator (see Figure 4.31). The eigen frequency of the resonator was dependent on any local variation of the temperature along the string.

Other techniques have been developed to measure the heat generation in plasmonics, but not on small scales. For instance, in 2014 Pacardo *et al.* developed a device endowed with a resistance temperature detector [53]. The resistance was temperature-dependent, and the temperature of a solution of nanoparticles was measured by following the variation of the resistance.

In 2014 again, Petriashvili introduced a technique to map the temperature of a solution containing gold nanoparticles on the macroscopic scale [54]. The approach was based on the use of thermochromic cholesteric liquid crystals. Micrometric droplets undergoes a gradual color shift from red to blue when the temperature goes from 25°C to 42°C.

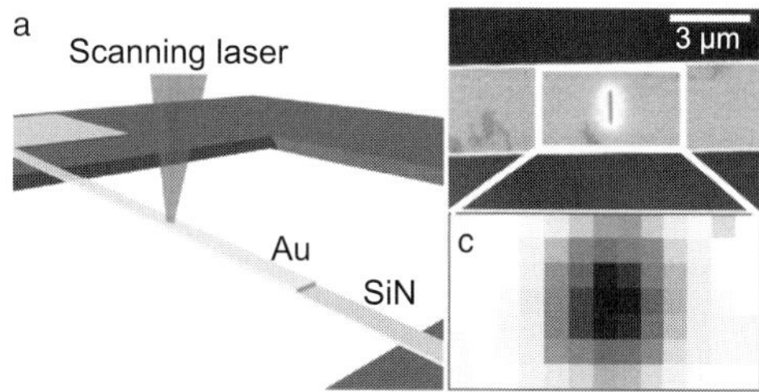

Fig. 4.31 (a) Schematic depiction of the experimental setup. A 900 µm long SiN string resonator is partially coated with a Au layer. A plasmonic nanoslit or nanorod (etched with a focused ion beam) is located in the string centre and probed with a low-power focused laser beam from a laser-Doppler vibrometer. Reproduced with permission from Reference [67]. Copyright 2014, American Chemical Society.

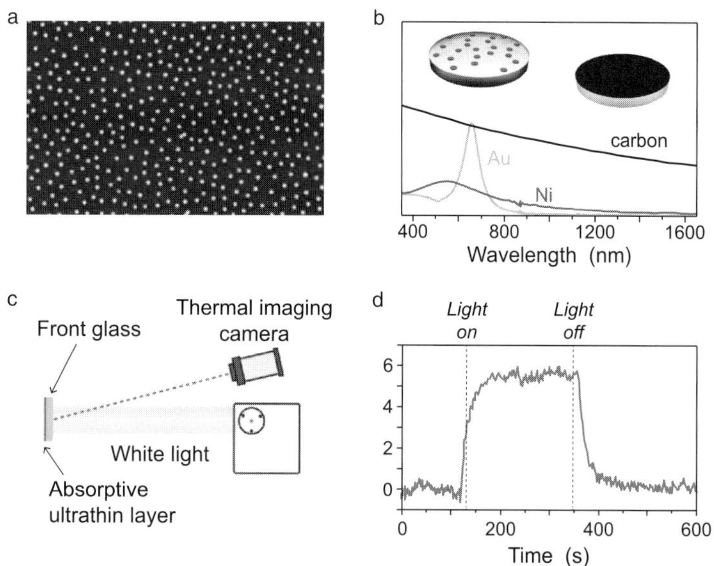

Fig. 4.32 (a) SEM image of Ni nanoparticles. (b) Absorption spectra of Au and Ni nanoparticles and of carbon thin film of the same thickness as the height of the plasmon particles. Insets schematically show both types of absorbing platforms. (c) Experimental setup for measuring temperature and (d) a typical temperature response. Reproduced from Reference [39].

Rarely has simple infrared imaging been used in thermoplasmonics, which is a pity because in many applications in plasmonics, heating is performed over areas much larger than tens of microns, making infrared imaging fully effective. For instance, using infrared imaging could have been an efficient means to evidence thermal effects in nanochemistry experiments [19], where illumination is often performed over large scale areas (typically one-inch large). In 2014, Jonsson *et al.* conveniently used an infrared camera to monitor the heat generation and the temperature increase over arrays of gold nanoparticles, as illustrated by Figure 4.32 [39].

4.9 Conclusion

It seems that after around a decade of investigation, most of the techniques suited to map the temperature in plasmonics have been invented. I am now expected a slowing down of this activity of research, which will rather consist in finding better temperature probes (more fluorescent, more stable molecules, for instance) than in finding new techniques.

References

[1] Acosta, V. M., Bauch, E., Ledbetter, M. P., Waxman, A., Bouchard, L. S., and Budker, D. 2010. Temperature Dependence of the Nitrogen-Vacancy Magnetic Resonance in Diamond. *Phys. Rev. Lett.*, **104**, 070801.

[2] Baffou, G., Kreuzer, M. P., Kulzer, F., and Quidant, R. 2009. Temperature Mapping near Plasmonic Nanostructures using Fluorescence Polarization Anisotropy. *Opt. Express*, **17**, 3291.

[3] Baffou, G., Girard, C., and Quidant, R. 2010. Mapping Heat Origin in Plasmonics Structures. *Phys. Rev. Lett.*, **104**, 136805.

[4] Baffou, G., Bon, P., Savatier, J., Polleux, J., Zhu, M., Merlin, M., Rigneault, H., and Monneret, S. 2012. Thermal Imaging of Nanostructures by Quantitative Optical Phase Analysis. *ACS Nano*, **6**, 2452–2458.

[5] Baffou, G., Berto, P., Bermúdez Ureña, E., Quidant, R., Monneret, S., Polleux, J., and Rigneault, H. 2013. Photoinduced Heating of Nanoparticle Arrays. *ACS Nano*, **7**(8), 6478–6488.

[6] Baffou, G., Rigneault, H., Marguet, D., and Jullien, L. 2014a. A Critque of Methods for Temperature Imaging in Single Cells. *Nature Methods*, **11**, 899–901.

[7] Baffou, G., Bermúdez Ureña, E., Berto, P., Monneret, S., Quidant, R., and Rigneault, H. 2014b. Deterministic Temperature Shaping using Plasmonic Nanoparticle Assemblies. *Nanoscale*, **6**, 8984–8989.

[8] Baffou, G., Polleux, J., Rigneault, H., and Monneret, S. 2014c. Super-Heating and Micro-Bubble Generation around Plasmonic Nanoparticles under cw Illumination. *J. Phys. Chem. C*, **118**, 4890.

[9] Baffou, G., Rigneault, H., Marguet, D., and Jullien, L. 2015. Reply to: "Validating Subcellular Thermal Changes Revealed by Fluorescent Thermosensors" and "The 10^5 Gap Issue Between Calculation and Measurement in Single-Cell Thermometry." *Nature Methods*, **12**, 803.

[10] Baral, S., Johnson, S. C., Alaulamiz, A. A., and Richardson, H. H. 2016. Nanothermometry Using Optically Trapped Erbium Oxide Nanoparticle. *Appl. Phys. A*, **122**, 340.

[11] Bendix, P. M., Reihani, S. N. S., and Oddershede, L. B. 2010. Direct Measurements of Heating by Electromagnetically Trapped Gold Nanoparticles on Supported Lipid Bilayers. *ACS Nano*, **4**(4), 2256.

[12] Berto, P., Bermúdez Ureña, E., Bon, P., Quidant, R., Rigneault, H., and Baffou, G. 2012. Quantitative Absorption Spectroscopy of Nano-Objects. *Phys. Rev. B*, **86**, 165417.

[13] Carlson, M. T., Khan, A., and Richardson, H. H. 2011. Local Temperature Determination of Optically Excited Nanoparticles and Nanodots. *Nano Lett.*, **11**, 1061–1069.

[14] Carlson, M. T., Green, A. J., Khan, A., and Richardson, H. H. 2012a. Optical Measurement of Thermal Conductivity and Absorption Cross-Section of Gold Nanowires. *J. Phys. Chem. C*, **116**, 8798–8803.

[15] Carlson, M. T., Green, A. J., and Richardson, H. H. 2012b. Superheating Water by CW Excitation of Gold Nanodots. *Nano Lett.*, **12**, 1534.

[16] Chen, X. D., Dong, C. H., Sun, F. W., Zou, C. L., Cui, J. M., Han, Z. F., and Guo, G. C. 2011. Temperature Dependent Energy Level Shifts of Nitrogen-Vacancy Centers in Diamond. *Appl. Phys. Lett.*, **99**, 161903.

[17] Chen, Z., Shan, X., Guan, Y., Wang, S., Zhu, J. J., and Tao, N. 2015. Imaging Local Heating and Thermal Diffusion of Nanomaterials with Plasmonic Thermal Microscopy. *ACS Nano*, **9**(12), 11574–11581.

[18] Chiu, M. J., and Chu, L. K. 2015. Quantifying the Photothermal Efficiency of Gold Nanoparticles Using Tryptophan as an in Situ Fluorescent Thermometer. *Phys. Chem. Chem. Phys.*, **17**, 17090–17100.

[19] Christopher, P., Xin, H., and Linic, S. 2011. Visible-Light-Enhanced Catalytic Oxidation Reactions on Plasmonic Silver Nanostructures. *Nature Chem.*, **3**, 467–472.

[20] Clarke, M. L., Grace Chou, S., and Hwang, J. 2010. Monitoring Photothermally Excited Nanoparticles via Multimodal Microscopy. *J. Phys. Chem. Lett.*, **1**, 1743–1748.

[21] Coppens, Z. J., Li, W., Walker, D. G., and Valentine, J. G. 2013. Probing and Controlling Photothermal Heat Generation in Plasmonic Nanostructures. *Nano Lett.*, **13**, 1023–1028.

[22] Debasu, M. L., Ananias, D., Pastoriza-Santos, I., Liz-Marzán, L. M., Rocha, J., and Carlos, L. D. 2013. All-In-One Optical Heater-Thermometer Nanoplatform Operative From 300 to 2000 K Based on Er_3^+ Emission and Blackbody Radiation. *Adv. Mater.*, **25**, 4868–4874.

[23] Desiatov, B., Goykhman, I, and Levy, U. 2014. Direct Temperature Mapping of Nanoscale Plasmonic Devices. *Nano Lett.*, **14**, 648–652.

[24] Doherty, M. W., Manson, N. B., Delaney, P., Jelezko, F., Wrachtrup, J., and Hollenberg, L. C. L. 2013. The Nitrogen-Vacancy Colour Centre in Diamond. *Physics Reports*, **528**, 1–45.

[25] Dolde, F., Fedder, H., Doherty, M. W., Nöbauer, T., Rempp, F., Balasubramanian, G., Wolf, T., Reinhard, F., Hollenberg, L. C. L., Jelezko, F., and Wrachtrup, J. 2011. Electric-Field Sensor Using Single Diamond Spins. *Nature Phys.*, **7**, 459–463.

[26] Donner, J. S., Thompson, S. A., Alonso-Ortega, C., Morales, J., Rico, L. G., Santos, S. I. C. O., and Quidant, R. 2013. Imaging of Plasmonic Heating in a Living Organism. *ACS Nano*, **7**(10), 8666–8672.

[27] Donner, J. S., Thompson, S. A., Kreuzer, M. P., Baffou, G., and Quidant, R. 2012. Mapping Intracellular Temperature Using Green Fluorescent Protein. *Nano Lett.*, **12**, 2107–2111.

[28] Ebrahimi, S., Akhlaghi, Y., Kompany-Zareh, M., and Rinnan, Å. 2014. Nucleic Acid Based Fluorescent Nanothermometers. *ACS Nano*, **8**(10), 10372–10382.

[29] Freddi, S., Sironi, L., D'Antuono, R., Morone, D., Donà, A., Cabrini, E., D'Alfonso, L., Collini, M., Pallavivini, P., Baldi, G., Maggioni, D., and Chirico, G. 2013. A Molecular Thermometer for Nanoparticles for Optical Hyperthermia. *Nano Lett.*, **13**, 2004–2010.

[30] Gomès, S., Assy, A., and Chapuis, P. O. 2015. Scanning Thermal Microscopy: A review. *Phys. Status Solidi A*, **212**(3), 477–494.

[31] Hao Vu, X., Levy, M., Barroca, T., Nhung Tran, H., and Fort, E. 2013. Gold Nanocrescents for Remotely Measuring and Controlling Local Temperature. *Nanotechnology*, **24**, 325501.

[32] Herzog, J. B., Knight, M. W., and Natelson, D. 2014. Thermoplasmonics: Quantifying Plasmonic Heating in Single Nanowires. *Nano Lett.*, **14**(2), 499–503.

[33] Hirsch, L. R., Stafford, R. J., Bankson, J. A., Sershen, S. R., Rivera, B., Price, R. E., Hazle, J. D., Halas, N. J., and West, J. L. 2003. Nanoshell-Mediated Near-Infrared Thermal Therapy of Tumors under Magnetic Resonance Guidance. *Proc. Natl. Acad. Sci. U.S.A.*, **100**(23), 13549–13554.

[34] Honda, M., Saito, Y., Smith, N. I., Fujita, K., and Kawata, S. 2011. Nanoscale Heating of Laser Irradiated Single Gold Nanoparticles in Liquid. *Opt. Express*, **19**(13), 12375–12383.

[35] Hormeño, S., Gregorio-Godoy, P., Pérez-Juste, J., Liz-Marzán, L. M., Juárez, B. H., and Arias-Gonzalez, J. R. 2014. Laser Heating Tunability by Off-Resonant Irradiation of Gold Nanoparticles. *Small*, **10**(2), 376–384.

[36] Ito, S., Sugiyama, T., Toitani, N., Katayama, G., and Miyasaka, H. 2007. Application of Fluorescence Correlation Spectroscopy to the Measurement of Local Temperature in Solutions under Optical Trapping Condition. *J. Phys. Chem. B*, **111**, 2365.

[37] Jaque, D., and Vetrone, F. 2012. Luminescence Thermometry. *Nanoscale*, **4**, 4301–4326.

[38] Jaque, D., Maestro, L. M., Escudero, E., Martín Rodriguez, E., Capobianco, J. A., Vetrone, F., Juarranz de la Fuente, A., Sanz-Rodríguez, F., Iglesias-de la Cruz, M. C., Jacinto, C., Rocha, U., and García Solé, J. 2013. Fluorescent Nano-Particles for Multi-Photon Thermal Sensing. *J. Luminescence*, **133**, 249–253.

[39] Jonsson, G. E., Milijkovic, V., and Dmitriev, A. 2014. Nanoplasmon-Enabled Macroscopic Thermal Management. *Sci. Rep.*, **4**, 5111.

[40] Kim, K., Chung, J., Hwang, G., Kwon, O., and Sik Lee, J. 2011. Quantitative Measurement with Scanning Thermal Microscope by Preventing Thermal Microscope by Preventing the Distortion Due to the Heat Transfer through the Air. *ACS Nano*, **5**, 8700–8709.

[41] Kinkhabwala, A. A., Staffaroni, M., Suze, O., and Burgos, S. 2016. Nanoscale Thermal Mapping of HAMR Heads Using Polymer Imprint Thermal Mapping. *IEEE Trans. Magn.*, **52**(2), 1.

[42] Kneipp, K., Wang, Y., Kneipp, H., Perelman, L. T., Itzkan, I., Dasari, R. R., and Feld, M. S. 1997. Single Molecule Detection Using Surface-Enhanced Raman Scattering (SERS). *Phys. Rev. Lett.*, **78**, 1667–1670.

[43] Kucsko, G., Maurer, P. C., Yao, N. Y., Kubo, M., Noh, H. J., Lo, P. K., Park, H., and Lukin, M. D. 2013. Nanometre-Scale Thermometry in a Living Cell. *Nature*, **500**, 54–59.

[44] Lakovicz, J. R. 2007. *Principles of Fluorescence Spectroscopy*. Springer.

[45] Ma, H., Bendix, P. M., and Oddershede, L. B. 2012. Large-Scale Orientation Dependent Heating from a Single Irradiated Gold Nanorod. *Nano Lett.*, **12**, 3954–3960.

[46] Maestro, L. M., Haro-González, P., Coello, J. G., and Jaque, D. 2012. Absorption Efficiency of Gold Nanorods Determined by Quantum Dot Fluorescence Thermometry. *Appl. Phys. Lett.*, **100**, 201110.

[47] Maestro, L. M., Haro-González, P., del Rosal, B., Ramiro, J., no, Caama Carrasco, E., Juarranz, A., Sanz-Rodríguez, F., García Solé, J., and Jaque, D. 2013. Heating Efficiency of Multi-Walled Carbon Nanotubes in the First and Second Biological Windows. *Nanoscale*, **5**, 7882–7889.

[48] Maestro, L. M., Haro-González, P., Ana, Sánchez-Iglesias, Liz-Marzán, L. M., García Solé, J., and Jaque, D. 2014. Quantum Dot Thermometry Evaluation of Geometry Dependent Heating Efficiency in Gold Nanoparticles. *Langmuir*, **30**, 1650–1658.

[49] Martinez Maestro, L., Martín Roriguez, E., Vetrone, F., Naccache, R., Loro Ramirez, H., Jaque, D., Capobianco, J. A., and García Solé, J. A. 2010. Nanoparticles for Highly Efficient Multiphoton Fluorescence Bioimaging. *Opt. Express*, **18**(23), 23544–23553.

[50] Neuman, P., Jakobi, I., Dolde, F., Burk, C., Reuter, R., Waldherr, G., Honert, J., Wolf, T., Brunner, A., Shim, J. H., Suter, D., Sumiya, H., Isoya, J., and Wrachtrup, J. 2013. High-Precision Nanoscale Temperature Sensing Using Single Defects in Diamond. *Nano Lett.*, **13**(6), 2738–2742.

[51] Newville, Matthew. 2004. *Fundamentals of XAFS*. University of Chicago.

[52] Nie, S., and Emory, S. R. 1997. Probing Single Molecules and Single Nanoparticles by Surface-Enhanced Raman Scattering. *Science*, **275**(5303), 1102–1106.

[53] Pacardo, D. B., Neupane, B., Wang, G., Gu, Z., Walker, G. M., and Ligler, F. S. 2015. A Temperature Microsensor for Measuring Laser-Induced Heating in Gold Nanorods. *Anal. Bioanal. Chem.*, **407**, 719–725.

[54] Petriashvili, G., De Santo, M. P., Chubinidze, K., Hamdi, R., and Barberi, R. 2014. Visual Micro-Thermometers for Nanoparticles Photo-Thermal Conversion. *Opt. Express*, **22**(12), 14705–14711.

[55] Plakhotnik, T., and Gruber, D. 2010. Luminescence of Nitrogen-Vacancy Centers in Nanodiamonds at Temperatures between 300 and 700 K: Perspectives on Nanothermometry. *Phys. Chem. Chem. Phys.*, **12**, 9751–9756.

[56] Plakhotnik, T., Doherty, M. W., Cole, J. H., Chapman, R., and Manson, N. B. 2014. All-Optical Thermometry and Thermal Properties of the Optically Detected Spin Resonances of the NV$^-$ Center in Nanodiamond. *Nano Lett.*, **14**, 4989–4996.

[57] Pozzi, E. A., Zrimsek, A. B., Lethiec, C. M., Schatz, G. C., Hersam, M. C., and Van Duyne, P. 2015. Evaluating Single-Molecule Stokes and Anti-Stokes SERS for Nanoscale Thermometry. *J. Phys. Chem. C*, **119**, 21116–21124.

[58] Robert, H., Kundrat, F., Bermúdez Ureña, E., Rigneault, H., Monneret, S., Quidant, R., Polleux, J., and Baffou, G. 2016. Light-Assisted Solvothermal Chemistry Using Plasmonic Nanoparticles. *ACS Omega*, **1**, 2–8.

[59] Rocha, U., Jacinto da Silva, C., Ferreira Silva, W., Guedes, I., Benayas, A., Martínez Maestro, L., Acosta, Elias, M., Bovero, E., van Veggel, F. C. J. M., García Solé, J. A., and Jaque, D. 2013. Thermal Sensing Based on Neodymium-Doped LaF$_3$ Nanoparticles. *ACS Nano*, **7**(2), 1188–1199.

[60] Rodríguez-Sevilla, P., Zhang, Y., Haro-González, P., Sanz-Rodríguez, F., Jaque, F., García Solé, J., Liu, X., and Jaque, D. 2016. Thermal Scanning at the Cellular Level by an Optically Trapped Upconverting Fluorescent Particle. *Adv. Mater.*, **28**, 2421–2426.

[61] Rohani, S., Quintanilla, M., Tuccio, S., De Angelis, F., Cantelar, E., Govorov, A. O., Razzari, L., and Vetrone, F. 2015. Enhanced Luminescence, Collective Heating, and Nanothermometry in an Ensemble System Composed of Lanthanide-Doped Upconverting Nanoparticles and Gold Nanorods. *Adv. Opt. Mater.*, **3**(11), 1606–1613.

[62] Rondin, L., Tetienne, J. P., Hingant, T., Roch, J. F., Maletinsky, and Jacques, V. 2014. Magnetometry with Nitrogen-Vacancy Defects in Diamond. *Rep. Prog. Phys.*, **77**, 056503.

[63] Rycenga, M., Wang, Z., Gordon, E., Cobley, C. M., Schwartz, A. G., Lo, C. S., and Xia, Y. 2009. Probing the Photothermal Effect of Gold-Based Nanocages with Surface-Enhanced Raman Scattering (SERS). *Angew. Chem. Int. Ed.*, **48**, 9924–9927.

[64] Savchuk, O. A., Carvajal, J. J., Pujol, M. C., Barrera, W., Massons, J., Aguilo, M., and Diaz, F. 2015. Ho,Yb:KLu(WO$_4$)$_2$ Nanoparticles: A Versatile Material for Multiple Thermal Sensing Purposes by Luminescent Thermometry. *J. Phys. Chem. C*, **119**, 18546–18558.

[65] Schiebener, P., Straub, J., Levelt Sengers, J. M. H., and Gallagher, J. S. 1990. Refractive Index of Water and Steam as Function of Wavelength, Temperture and Density. *J. Phys. Chem. Ref. Data*, **19**(3), 677.

[66] Schirhagl, R., Chang, K., Loretz, M., and Degen, C. L. 2014. Nitrogen-Vacancy Centers in Diamond: Nanoscale Sensors for Physics and Biology. *Annu. Rev. Phys. Chem.*, **65**, 83–105.

[67] Schmid, S., Wu, K., Larsen, P. E., Rindzevicius, T., and Boisen, A. 2014. Low-Power Photothermal Probing of Single Plasmonic Nanostructures with Nanomechanical String Resonators. *Nano Lett.*, **14**(5), 2318–2321.

[68] Setoura, K., Werner, D., and Hashimoto, S. 2012. Optical Scattering Spectral Thermometry and Refractometry of a Single Gold Nanoparticle under CW Laser Excitation. *J. Phys. Chem. C*, **116**, 15458–15466.

[69] Setoura, K., Okada, Y., Werner, D., and Hashimoto, S. 2013. Observation of Nanoscale Cooling Effects by Substrates and the Surrounding Media for Single Gold Nanoparticles under CW-Laser Illumination. *ACS Nano*, **7**(9), 7874–7885.

[70] Tetienne, J. P., A., Lombard, Simpson, D. A., Ritchie, C., Lu, J., Mulvaney, P., and Hollenberg, C. L. 2016. Scanning Nanospin Ensemble Microscope for Nanoscale Magnetic and Thermal Imaging. *Nano Lett.*, **16**, 326–333.

[71] Toshimitsu, M., Matsumura, Y., Shoji, T., Kitamura, N., Takase, M., Murakoshi, K., Yamauchi, H., Ito, S., Miyasaka, H., Nobuhiro, A., Mizumoto, Y., Ishihara, H., and Tsuboi, Y. 2012. Metallic-Nanostructure-Enhanced Optical Trapping of Flexible Polymer Chains in Aqueous Solution as Revealed by Confocal Fluorescence Microspectroscopy. *J. Phys. Chem. C*, **116**, 14610–14618.

[72] Tzeng, Y. K., Tsai, P. C., Liu, H. Y., Chen, O. Y., Hsu, H., Yee, F. G., Chang, M. S., and Chang, H. C. 2015. Time-Resolved Luminescence Nanothermometry with Nitrogen-Vacancy Centers in Nanodiamonds. *Nanolett.*, **15**, 3945–3952.

[73] Vaijayanthimala, V., Keun Lee, D., Kim, S. V., Yen, A., Tsai, N, Ho, D., Chang, H. C., and Shenderova, O. 2015. Nanodiamond-Mediated Drug Delivery and Imaging: Challenges and Opportunities. *Expert Opin. Drug Deliv.*, **5**, 735–749.

[74] Valeur, Bernard. 2007. *Molecular Fluorescence: Principles and Applications*. Wiley-VCH.

[75] Van de Broek, B., Grandjean, D., Trekker, J., Ye, J., Verstreken, K., Maes, G., Borghs, G., Nikitenko, S., Lagae, L., Bartic, C., Temst, K., and Van Bael, M.J. 2011. Temperature Determination of Resonantly Excited Plasmonic Branched Gold Nanoparticles by X-Ray Absorption Spectroscopy. *Small*, **7**(17), 2498–2506.

[76] Velghe, S., Primot, J., Guérineau, N., Cohen, M., and Wattellier, B. 2005. Wavefront Reconstruction from Multidirectional Phase Derivatives Generated by Multilateral Shearing Interferometers. *Opt. Lett.*, **30**(3), 245–247.

[77] Vetrone, F., Naccache, R., Zamarrón, A., Juarranz de la Fuente, A., Sanz-Rodríguez, F., Martinez Maestro, L., Martín Rodriguez, E., Jaque, D., García Solé, J., and Capobianco, J. A. 2010. Temperature Sensing Using Fluorescent Nanothermometers. *ACS Nano*, **4**(6), 3254.

[78] Šiler, M., Ježek, J., Jákl, P., Zdeněk, P., and Zemánek, P. 2016. Direct Measurement of the Temperature Profile Close to an Optically Trapped Absorbing Particle. *Opt. Lett.*, **41**(5), 870–873.

[79] Weinert, F. M., and Braun, D. 2009. An Optical Conveyor for Molecules. *Nano Lett.*, **9**(12), 4264–4267.

[80] Williams, C. C., and Wickramasinghe, H. K. 1986. Scanning Thermal Profiler. *Appl. Phys. Lett.*, **49**, 1587–1589.

[81] Xie, X., and Cahill, D. G. 2016. Thermometry of Plasmonic Nanostructures by Anti-Stokes Electronic Raman Scattering. *Appl. Phys. Lett.*, **109**, 183104.

[82] Yamauchi, H., Ito, S., Yoshida, K., Itoh, T., Tsuboi, Y., Kitamura, N., and Miyasaka, H. 2013. Temperature near Gold Nanoprticles under Phtotexcitation: Evaluation Using a Fluorescence Correlation Technique. *J. Phys. Chem. C*, **117**, 8388–8396.

[83] Yashchenok, A., Masic, A., Gorin, D., Inozemtseva, O., Sup Shim, B., Kotov, N., Skirtach, A., and Möhwald, H. 2015. Optical Heating and Temperature Determination of Core–Shell Gold Nanoparticles and Single-Walled Carbon Nanotube Microparticles. *Small*, **11**(11), 1320–1327.

[84] Yokota, Y., Ueno, K., and Misawa, H. 2011. Highly Controlled Surface-Ehnaced Raman Scattering Chips using Nanoengineered Gold Blocks. *Small*, **2**, 252–258.

[85] Zhou, J., Liu, Q., Feng, W., Sun, Y., and Li, F. 2015. Upconversion Luminescent Materials: Advances and Applications. *Chem. Rev.*, **115**, 395–465.

[86] Zhu, M., Baffou, G., Meyerbröker, N., and Polleux, J. 2012. Micropatterning Thermoplasmonics Gold Nanoarrays to Manipulate Cell Adhesion. *ACS Nano*, **6**(8), 7227–7233.

5

Thermal-Induced Processes

A heated nanoparticle is bound to deliver its excess of energy to the surroundings in one way or another. This energy gain experienced by the surrounding medium is likely to be associated with thermal-induced processes. Interestingly, as thermal-induced processes exist in any field of science, a large variety of physical processes are likely to occur in the surrounding medium when photothermal conversion occurs in plasmonics. This is what makes the field of thermoplasmonics so rich and this is why this chapter is the longest in this book.

The purpose of this chapter is to introduce the theoretical backgrounds necessary to understand reported works in the literature, lying at the interface between plasmonics with other fields of science such as chemistry, hydrodynamics, phase transition or solid-state physics. In particular, the basics of the phenomena tackled in this chapter are chemical reactions, Brownian motion, thermophoresis, liquid–gas phase transition, superheating, bubble dynamics under cw and pulsed illumination, shock wave generation, refractive index variations, fluid convection and reshaping of nanoparticles.

5.1 Chemical Reaction

Plasmonic nanoparticles under illumination can be used to enhance the yield of chemical reactions using various identified mechanisms:

- due to the enhanced optical near-field intensity, metal nanoparticles can favor local photochemistry at their vicinity.

- due to the creation of a hot electron subsequent to photon absorption, hot electron injection processes can occur at the nanoparticle–surroundings interface, and enable redox reactions.

- due to the local temperature increase subsequent to light absorption, the reaction rate can be increased according to the Arrhenius law.

The only mechanism that is thermally induced is the third. The two others do not fall into the context of thermoplasmonics and I will therefore not describe them in this book, but invite the reader to refer to the literature [5].

The Arrhenius law consists of a simple equation that describes how the rate constant K of chemical reactions depends on temperature. It reads

$$K = A\,e^{-E_\mathrm{a}/RT} \tag{5.1}$$

where R is the gas constant,[1] T is the temperature, E_a is a constant associated to the reaction, which is termed the activation molar energy and A is another constant that has the same dimension as the rate constant. This equation was first proposed by Van't Hoff in 1884. The contribution of Arrhenius five years later was to provide a physical interpretation of this equation. Most chemical reactions obey the Arrhenius law. Only reactions associated to intricate mechanisms can follow a different behavior, for instance when involving enzymatic activity.

The formulation (5.1) is rather related to chemical systems as it involves molar quantities. Another formulation reads

$$K = A\,e^{-e_\mathrm{a}/k_\mathrm{B}T} \tag{5.2}$$

where e_a is the activation energy (often expressed in eV) and k_B is the Boltzmann constant.[2] The activation energy of most chemical reactions is on the order of $E_\mathrm{a} \sim 10^2$ kJ.mol^{-1}, *i.e.*, $e_\mathrm{a} \sim 1$ eV. A common rule of thumb is that constant rates double for every 10 K.

The Arrhenius law is at the origin of a thermally induced enhancement of reaction rates in plasmonics. Although increasing reaction rates is of obvious interest for many applications, it does not systematically make sense to use plasmonic means. If the aim is to produce large quantities of a material over macroscopic scales, it makes more sense to use a hot plate instead of plasmonic nanoparticles and optical means. This observation is, by the way, very general in thermoplasmonics. Using thermoplasmonic means in chemistry can be beneficial, but only in two situations:

- To achieve fast dynamics. It would not be possible to perform heating with microsecond dynamics with a hot plate, which necessarily features a large thermal inertia.
- To generate products on the nano- or microscales.

These last years, several works reported on the benefit of using thermoplasmonics for chemical synthesis purposes. The literature on this subject is reviewed in Chapter 6.

5.2 Brownian Motion

5.2.1 Mobility

Let us consider a particle in a fluid. When a force $\mathbf{f}$ is applied to this particle, after a possible transient regime, the particle velocity $\mathbf{v}$ may reach a steady state value that is proportional to the magnitude of the applied force. If this happens, one can define the mobility μ of the particle in this fluid by[3]

---

[1] $R = 8.31446$ J.mol^{-1}.K^{-1}.

[2] $k_\mathrm{B} = 8.617$ eV.K^{-1}.

[3] One sometimes speaks about the electrophoretic mobility defined by the relation $\mathbf{v} = \mu_\mathrm{e}\mathbf{E}$ when considering a charged particle and where $\mathbf{E}$ is an applied electric field. The dimension of μ_e is [m^2·s^{-1}·V^{-1}].

$$\mathbf{v} = \mu\mathbf{f}. \tag{5.3}$$

For instance, with a fluid of dynamic viscosity η and a spherical particle of radius a, one ends up with the famous Stokes' law

$$\mathbf{v} = \frac{1}{6\pi\eta a}\mathbf{f}. \tag{5.4}$$

The dimension of μ is [s·kg^{-1}] or [m^2·s^{-1}·J^{-1}].

5.2.2 Particle Diffusivity

The particle diffusivity D [m^2·s^{-1}] is the factor involved in the Fick's law, *i.e.*, the law relating the diffusive flux $\mathbf{J}$ [m^{-2}·s^{-1}] of particles with the particle concentration n [m^{-3}]:

$$\mathbf{J} = -D\nabla n. \tag{5.5}$$

It is also the parameter governing the homogeneous diffusion equation:

$$\partial_t n = -D\nabla^2 n. \tag{5.6}$$

5.2.3 Link between Mobility and Diffusivity: The Einstein Relation

In 1905, Einstein (among others) derived an unexpected relation between mobility and diffusivity that holds his name today [36], and simply reads

$$D = \mu k_{\mathrm{B}}T \tag{5.7}$$

where T is the temperature.

5.2.4 Brownian Motion

Brownian motion is the random motion of a particle resulting from the collisions with the molecules of a surrounding fluid. Although one cannot predict what the trajectory of a Brownian motion will be, there exists a simple relation characterizing a Brownian motion that reads

$$\langle r^2 \rangle = 2dDt \tag{5.8}$$

where $r(t)$ is the distance of the particle from its initial position, d is the dimensionality of the motion and t the elapsed time. This equation means that the particle displacement is, on average, not proportional to the elapsed time, but rather its square root. According to the Einstein relation (5.7), one can also write

$$\langle r^2 \rangle = 2d\mu k_{\mathrm{B}}Tt \tag{5.9}$$

$$\langle r^2 \rangle = \frac{2dk_{\mathrm{B}}T}{6\pi\eta a}t. \tag{5.10}$$

This equation is useful to quantify the amplitude of the fluctuations in position of a small object observed at the micrometric scale as a function of its radius a, the temperature T, the nature of the fluid (via η) and the timescale t under consideration.

5.3 Thermophoresis

5.3.1 Basics of Thermophoresis

When a temperature gradient is applied throughout a fluid, dissolved species tend to move along the gradient.[4] This effect was primarily discovered by Charles Soret in 1879 [117] using saline water. Soret observed that a temperature gradient applied along a tube resulted in a gradient of salt concentration. Although this effect was firstly observed by Carl Ludwig several years before, in 1856 [77], one usually calls this phenomenon the "Soret effect," as Soret investigated in more detail the underlying physics and published a couple of articles, while Ludwig only reported his observation in a one-page long article. The appellations "Ludwig–Soret effect," thermodiffusion or thermophoresis are also employed. All these expressions are commonly considered as equivalent, although some articles make a distinction [54] between thermophoresis (restricted to diluted solutes) and thermodiffusion (for binary mixtures). While thermophoresis in gases is well understood and theoretically described, the multiple origins of thermophoresis in liquids are still a matter of debate.

In this section, we shall focus on thermophoresis in liquid (not gas), with dilute solutions (not binary mixtures) and on the microscale (not on human scale), as it is the situation of interest in plasmonics.

When describing thermophoresis of a solute (particles, ions, molecules, proteins, colloids, DNA strands, etc.) in a solvent, one usually states that the dispersed particles, on top of Brownian motion, display a drift velocity that reads

$$\mathbf{v}_T = -D_T \nabla T. \tag{5.11}$$

Equivalently, it is common to use the particle flux density resulting from the same temperature gradient

$$\mathbf{J}_T = -n D_T \nabla T \tag{5.12}$$

where n [m^{-3}] and T [K] are the concentration of particles and the temperature. In this empirical expression, D_T is often called the *thermal diffusion coefficient*. Dimensionally speaking, however, D_T [m$^2 \cdot$s$^{-1} \cdot$K^{-1}] is not a diffusion coefficient. It should rather be called thermophoretic *mobility*, in analogy with other transport effects such as electrophoresis [97]. D_T is in general not a constant for a given solute/solvent association. In other words, although a temperature gradient is responsible for thermophoresis, the process is not necessarily *proportional* to the temperature gradient. In particular, D_T usually depends on temperature and particle concentration.

In order to determine the resulting distribution of particle concentration (what eventually matters), one also has to consider natural Fick's diffusion of the particles, which tends to smooth any concentration gradient and which counterbalances thermophoresis for this

---

[4] This effect also occurs in gas phase with dust and aerosols.

reason. Considering these two contributions, the expression[5] of the flux density of particles in a solvent at rest (assuming no fluid convection) reads [99]

$$\mathbf{J} = -D\nabla n - n D_T \nabla T. \tag{5.13}$$

$D\,[\mathrm{m^2 \cdot s^{-1}}]$ is the Brownian diffusion coefficient of the particles. The first contribution in the expression (5.13) of $\mathbf{J}$ depicts Fick's law of diffusion, proportional to the concentration gradient. At equilibrium, $\mathbf{J} = \mathbf{0}$ yields

$$-D\nabla n = n D_T \nabla T \tag{5.14}$$

$$-\frac{\nabla n}{n} = S_T \nabla T \tag{5.15}$$

where the Soret coefficient S_T is defined by

$$\boxed{S_T = D_T/D.} \tag{5.16}$$

In general, for aqueous solutions, $|S_T|$ is of the order of 10^{-3} to 10^{-1} $\mathrm{K^{-1}}$. In general S_T (and thus D_T) is positive, which implies that molecules and particles usually tend to move *from hot to cold regions*. Thus, one sometimes speaks about thermophobicity of the solute, but this general rule has exceptions as explained further. Noteworthily, solving Equation (5.15), if one assumes a uniform temperature gradient in one direction (x) and a constant Soret coefficient, the concentration profile is exponential $n(x) = n_0 \exp(-x/\ell_T)$ where the characteristic length ℓ_T is a length defined as

$$\ell_T = (S_T \nabla T)^{-1} = \frac{D}{v_T}. \tag{5.17}$$

5.3.2 Thermophoresis: Motion from Hot to Cold or the Opposite?

As explained by Roberto Piazza in Reference [95], in thermophoresis, most particles in liquids preferentially move from hot to cold (as *always* happens for particle thermophoresis in gases). But in some cases in liquids, some particles show the opposite trend, displaying anomalous thermophilic behavior, characterized by a negative Soret coefficient S_T. This usually occurs at a sufficiently small temperature $T^\star$. For many aqueous suspensions $T^\star$ is about $4°\mathrm{C}$, the temperature of maximum density for water. Although some models have been proposed to explain this coincidence [17], the origin of this behavior is still not clear. But for proteins or DNA, $T^\star$ can be highly system-dependent and can lie around ambient temperature (Figure 5.1).

---

[5] In some circumstances, the expression of $\mathbf{J}$ rather reads $\mathbf{J}' = -D\nabla w - w(1-w)D_T\nabla T$ where w is a dimensionless solute concentration defined as a weight fraction and $\mathbf{J}'$ now has the dimension of a velocity. This expression is rather used when considering a binary mixture where the concentration of the reference component is not necessarily small compared to the other component. Thus, Equation (5.13) appears as a limit case, valid only for small concentration n ($w \ll 1$).

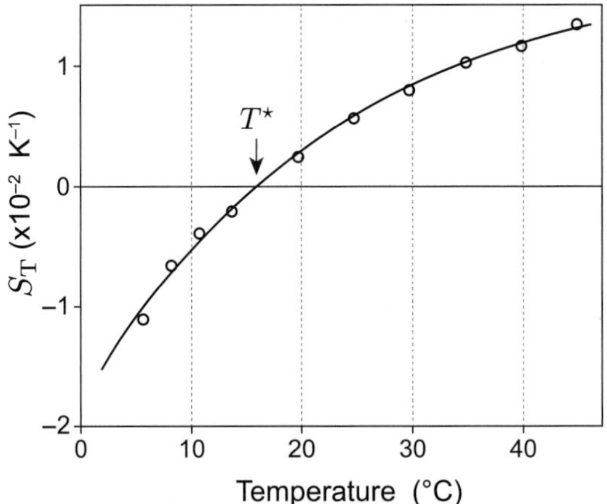

Fig. 5.1   Soret coefficient as a function of temperature for $c = 7\,\mathrm{g\cdot l^{-1}}$ lysozyme solution at pH $= 4.65$, in the presence of 7.5 mM of NaCl. Measurements were performed using the beam-deflection technique. Reproduced from Reference [98]. Copyright © 2004, Royal Society of Chemistry.

5.3.3 Does Thermophoresis Exist with an Ideal Gas?

Although our interest is oriented towards liquids, the case of an ideal gas is always instructive in thermodynamics. Let us consider an ideal gas of volume V, number of particles N, pressure P and temperature T. The intensive quantities P and T are not necessarily uniform. The ideal gas law reads

$$PV = Nk_B T. \tag{5.18}$$

Injecting Equation (5.18) into Equation (5.13) and using the particle concentration $n = N/V$ [m^{-3}] yields

$$\mathbf{J} = -D\nabla\left(\frac{P}{k_B T}\right) - n\,D_T\nabla T \tag{5.19}$$

$$= D\frac{P}{k_B T^2}\nabla T - n\,D_T\nabla T \tag{5.20}$$

$$= \left(\frac{nD}{T} - nD_T\right)\nabla T. \tag{5.21}$$

At equilibrium, $\mathbf{J} = \mathbf{0}$, which yields $nD/T = nD_T$, $i.e.$,

$$S_T = \frac{1}{T}. \tag{5.22}$$

Consequently, an ideal gas submitted to a temperature gradient features a nonuniform density characterized by a weaker density in the hotter region. This could have been anticipated without any calculation just from the ideal gas law (5.18). At constant pressure, if T increases, $n = N/V$ has to decrease. Basically, the origin of this nonuniformity follows this picture: since the particles go slower in the cold region, they spend more time in the cold region, hence the greater particle density in the cold region. A cold region has a sort of trapping effect.

However, this nonuniform particle distribution here only results from a statistical distribution. One cannot say that the nonuniform distribution results from any force applied on the particles (like in electrophoresis) and one usually does not speak about thermophoresis for this reason when the Soret coefficient simply equals $1/T$. Thermophoresis occurs when S_T deviates from its value of $1/T$, *i.e.*, when the particles no longer behave as an ideal gas, *i.e.*, when interaction occurs between the particles themselves, or between the particles and the solvent (which is naturally always the case in liquids).

In liquids, the $1/T$ contribution to the Soret coefficient exists and it is usually called the osmotic pressure. It is not distinguishable from the other mechanisms of thermophoresis experimentally. It is, however, usually completely negligible (on the order of $S_T = 0.0034$ K^{-1} [33]).

5.3.4 Thermophoresis of Dissolved Ions in Water

Let us slightly increase the degree of complexity by discussing thermophoresis of *ions* in liquids. When a temperature gradient is applied in a liquid, ions migrate and not only because of the trapping effect of the cold side as described in the case of an ideal gas. Solvent/ions interactions play a dominant role. In order to clarify the origin of thermophoresis, one can split the ion flux density into two terms:

$$\mathbf{J} = n\hat{\mathbf{v}} - \mu\nabla\Pi \tag{5.23}$$

where $\hat{\mathbf{v}}$ is the contribution to the velocity of the particle due to the interaction of the particle with the solvent. It is assumed to be proportional to the temperature gradient,

$$\hat{\mathbf{v}} = -C\nabla T. \tag{5.24}$$

The second term of Equation (5.23) arises from the gradient of osmotic pressure Π. In analogy with an ideal gas, one can write for the particle pressure $\Pi = nk_B T$. If one considers Π as uniform, an increase of temperature naturally yields a decrease of the particle concentration n. This is what the term $-\mu\nabla\Pi$ means, where μ is the mobility of the solute. By injecting $\Pi = nk_B T$ in Equation (5.23) and comparing the resulting expression to Equation (5.13), one derives the Einstein's relation (see previous Section 5.2)

$$D = \mu k_B T \tag{5.25}$$

and the thermal diffusion coefficient

$$D_T = \mu k_B + C. \tag{5.26}$$

This yields an expression of the Soret coefficient

$$S_T = \frac{1}{T}\left(1 + \frac{C}{\mu k_B}\right). \tag{5.27}$$

Consequently, in the absence of particle–solvent interaction ($C = 0$), the Soret coefficient simply reads $S_T = 1/T$, as stated earlier. Besides, one usually defines the reduced Soret coefficient $\alpha_T = TS_T$. A deviation of α_T from unity means that solvent interactions play a role in the thermophoresis mechanism of the particles (see Table 5.1 on page 152). But if

$\alpha_T = 1$, one comes back to the picture of an ideal gas (see previous Subsection 5.3.3). Yet, in most cases, α_T is different from unity and the solvent–particle interaction current $-C\nabla T$ can largely dominate the osmotic mechanism and even drive the particles in the opposite direction, from cold to hot depending on concentrations and temperature. As stated by Fayolle *et al.* in Reference [40], "these deviations express the failure of the ideal-gas picture for the [particle] and emphasize the importance of particle–solvent interactions."

Let us go one step further in the complexity of the description of thermophoresis of ions in liquids. Ions are always present in water. Even with deionized water, water molecules always break into H^+ and OH^- ions,[6] which subsequently recombine according to the equilibrium reaction:

$$H_2O \overset{K_e}{\rightleftharpoons} H^+ + OH^-.$$

The ionic product of water reads $K_e = [OH^-][H^+]$ and equals 10^{-14} at 25°C. This means that in pure water, concentrations of OH^- and H^+ are of the order of 10^{-7} mol L^{-1}. Other ions are usually present in water, but for the sake of simplicity, only H^+ and OH^- ions are supposed to be present. When a temperature gradient is applied to such a system, the ions migrate along the gradient. The resulting current densities $\mathbf{J}^+$ and $\mathbf{J}^-$ of H^+ and OH^- ions are:

$$\mathbf{J}_t^{\pm} = -\frac{Q_{\pm}^{\star} D_{\pm}}{k_B T^2} n_{\pm} \nabla T. \tag{5.28}$$

$Q_-^{\star}$ and $Q_+^{\star}$ are the heats of transport[7] of both ions, D_+ and D_- their diffusivity, n_+ and n_- their concentration [m^{-3}]. Such an asymmetric (if $Q_+ \neq Q_-$) ion migration leads to the establishment of an electric field in the solvent. Thus, an additional ionic current density, driven by the electric field created by the ions themselves, has to be considered:

$$\mathbf{J}_e^{\pm} = \pm e\mu_{\pm} n_{\pm} \mathbf{E} \tag{5.29}$$

where μ_+ and μ_- are the ionic mobilities of H^+ and OH^- ions. According to the Einstein relation (see Section 5.2 on page 144), it equals $\mu_{\pm} = D_{\pm}/k_B T$. Finally, one also has to consider the ionic current resulting from a non-uniformity of the ions' concentration according to the Fick's law:

$$\mathbf{J}_d^{\pm} = -D_{\pm} \nabla n_{\pm}. \tag{5.30}$$

To summarize, the total ion current densities read

$$\mathbf{J}^+ = \mathbf{J}_t^+ + \mathbf{J}_e^+ + \mathbf{J}_d^+ \tag{5.31}$$

$$\mathbf{J}^- = \mathbf{J}_t^- + \mathbf{J}_e^- + \mathbf{J}_d^-. \tag{5.32}$$

---

[6] I mention H^+ ions for the sake of simplicity, but I should rather write H_3O^+.

[7] Here, $Q_-^{\star}$ and $Q_+^{\star}$ are not molar quantities. using molar quantities is more common in chemistry and in the literature on thermophoresis, but it really complicates the expression, sometimes requiring the use of the Avogadro number. For this reason, I rather consider molecular heats of transports, which have the dimension of an energy, conveniently expressed in eV.

At equilibrium, $\mathbf{J}^+ = \mathbf{0}$ and $\mathbf{J}^- = \mathbf{0}$, which gives

$$\left\{ \begin{aligned} -\frac{\nabla c^+}{c^+} + \frac{e\mathbf{E}}{k_{\mathrm{B}}T} - \frac{Q^\star_+}{k_{\mathrm{B}}T^2}\nabla T &= 0 \\ -\frac{\nabla c^-}{c^-} - \frac{e\mathbf{E}}{k_{\mathrm{B}}T} - \frac{Q^\star_-}{k_{\mathrm{B}}T^2}\nabla T &= 0. \end{aligned} \right. \tag{5.33}$$

By subtracting these two equations, and by discarding the influence of Fick's diffusion, one obtains a relation between the electric field generated by the asymmetric ion migration and the temperature gradient:

$$\mathbf{E} = \frac{Q^\star_+ - Q^\star_-}{2\,e\,T}\nabla T. \tag{5.34}$$

Let us briefly comment on this relation. If $Q^\star_+ = Q^\star_-$, then no electric field is created by the ion migration ($\mathbf{E} = \mathbf{0}$). Indeed, $Q^\star_+ = Q^\star_-$ means that cations and anions migrate at the same speed, in the same direction. Thus, no spatial separation of charges occurs. In analogy with the Seebeck effect[8] encountered in solid-state physics, one can coin $(Q^\star_+ - Q^\star_-)/2eT$ the Seebeck coefficient. One can now inject this expression of $\mathbf{E}$ into Equation (5.29) to obtain an expression of the ionic current densities that only involves the concentration and temperature gradients:

$$\mathbf{J}^\pm = -D_\pm \nabla n_\pm - n_\pm \left(\frac{Q^\star_\pm D_\pm}{k_{\mathrm{B}}T^2} - \frac{D_\pm(Q^\star_+ - Q^\star_-)}{2k_{\mathrm{B}}T^2} \right) \nabla T \tag{5.35}$$

which yields the simpler expression

$$\mathbf{J}^\pm = -D_\pm \nabla n_\pm - n_\pm D_\pm \frac{Q^\star_+ + Q^\star_-}{2k_{\mathrm{B}}T^2}\nabla T. \tag{5.36}$$

Noteworthily, despite the complexity of the process (a temperature gradient that drives migration of ions, which are self-consistently affected by the electric field they produce themselves), the Soret coefficient is rather simple and counterintuitively is the same for both ions:

$$S_{\mathrm{T}}^+ = S_{\mathrm{T}}^- = S_{\mathrm{T}} = \frac{Q^\star_+ + Q^\star_-}{2k_{\mathrm{B}}T^2}. \tag{5.37}$$

We already discussed the case of $Q^\star_+ = Q^\star_-$, but the case of $Q^\star_+ = -Q^\star_-$ is also interesting as it yields $S_{\mathrm{T}} = 0$. Indeed, $Q^\star_+ = -Q^\star_-$ means that the thermal gradient generates equal but opposite fluxes for anions and cations that are stopped by the resulting electric field acting as a restoring force, hence the absence of ionic current.

With such a formalism, S_T contains not only the contribution of the applied thermal gradient, but also of the electric field resulting from the ion migration created by the thermal gradient. Choosing what S_T actually contains is often a matter of taste in the literature, which makes the task of understanding the literature of thermophoresis in liquid rather intricate. For instance, the first term in Equation (5.36) can be also expressed as a function

---

[8] In solid-state physics, the application of a temperature gradient throughout a conducting device may be responsible of the establishment of an electric current. This phenomenon is called the Seebeck effect.

Table 5.1 Reduced Soret coefficients α_T at room temperature and heats of transport $Q^\star$ for common ions, taken from Reference [128].

Ion	H$^+$	Li$^+$	K$^+$	Na$^+$	OH$^-$	Cl$^-$
$Q_i^\star$ (kJ mol^{-1})	13.3	0.53	2.59	3.46	17.2	0.53
$\alpha_{T,i}$	2.7	0.1	0.5	0.7	3.4	0.1

of a thermal gradient because the concentration gradient results from the temperature gradient. If one assumes that the ions behave as an ideal gas regarding their spatial diffusion, one can write

$$-D\nabla n = -D\nabla\frac{P}{k_{\mathrm{B}}T} = D\frac{n}{T}\nabla T. \tag{5.38}$$

By writing the contribution this way, one ends up with another expression of the Soret coefficient that now includes all the contributions:

$$S_{\mathrm{T}} = \frac{1}{T} + \frac{Q_+^\star + Q_-^\star}{2k_{\mathrm{B}}T^2}. \tag{5.39}$$

This is the handier expression of S_{T}. Besides, a normalized Soret coefficient $\alpha_{\mathrm{T}} = TS_{\mathrm{T}}$ is sometimes used:

$$\alpha_{\mathrm{T}} = 1 + \frac{Q_+^\star + Q_-^\star}{2k_{\mathrm{B}}T}. \tag{5.40}$$

We chose to consider enthalpy (heat) of transport $Q^\star$. Sometimes, the entropy is rather used but the physics is the same as its relation with the enthalpy is straightforward and reads $s = Q^\star/T$. However, the concepts of enthalpies (and entropies) of *transport* and *solvation* are usually mixed up in the literature.

The above-developed formalism was restricted to pure water composed only of H$^+$ and OH$^-$ ions. In practice, these results can be extrapolated to a real electrolyte composed of many more ions. For each ion, one can define its Soret contribution by the relation

$$S_{T,i} = \frac{Q_i^\star}{k_{\mathrm{B}}T^2} \tag{5.41}$$

or the reduced Soret coefficient

$$\alpha_{T,i} = \frac{Q_i^\star}{k_{\mathrm{B}}T}. \tag{5.42}$$

To conclude, Table 5.1 displays some values of the reduced Soret coefficients of common ions in solution.

The cases of NaOH and NaCl electrolytes are particularly important. One can see that the sign of $Q_+^\star - Q_-^\star$ is positive for the couple of ions (Na$^+$,Cl$^-$) and negative for (Na$^+$,OH$^-$). According to Equation (5.34), the direct consequence is that the thermally induced electric field is in the opposite direction when comparing NaOH and NaCl solutions. This is why the thermophoresis of larger charged objects such as proteins of colloids depend on the nature of the electrolyte, as we will see in the two forthcoming subsections.

5.3.5 Thermophoresis of Colloids and Macromolecules

This subsection is longer and more detailed compared to the previous one related to thermophoresis of ions, because it has been the subject of an active area of research.

I have long thought, myself, that thermophoresis of particles (molecules, beads, DNA) in liquids was well understood, as suggested by the article of Dieter Braun from 2006 entitled "Why Molecules Move along Thermal Gradients" [33] ... until I came to read the literature more deeply. Actually, even after two decades of investigation, the underlying physics of thermophoresis of particles in liquids is still a matter of active debate, the mechanisms proposed by Braun are not consensual [12, 130] and there remain many open questions.

In general, not only the models are different, but the experimental results as well [130, 12, 129]. For instance, while some groups measured a Soret coefficient S_T proportional to the particle radius [121, 12] (*i.e.*, D_T constant), Duhr and Braun reported on a quadratic dependence of S_T on the radius in 2006 (*i.e.*, D_T proportional to particle radius) [33]. This problem was pointed out by the group of Piazza in 2007 [12], Putnam in 2007 [105] and Würger in 2016 [130], who all questioned the validity of the results of Duhr and Braun. To date, the origin of this inconsistency has never been uncovered.

Many parameters can be adjusted from one experiment to another, in particular the ionic strength[9] of the solvent, which is not always reproducible from one experiment to another. The nature of the particle substantially differs from one experiment to another as well (polymer beads of various sizes and compositions, single and double DNA strands of various lengths, proteins of different natures). The experimental approaches are also very different. While some groups are using fluorescence microscopy means (Dieter Braun), some other groups rather employ a thermal lens effect or a beam deflection method based on refractive index variations (most of the community). Finally, the concentrations of beads are also very different from one experiment to another (1 wt% for the beam deflection technique against 10^{-6} wt% for the fluorescence measurement of Braun [105]). While studies by Wiegand and *et al.* [92] reached the conclusion that particle-particle interactions were unimportant for concentration of up to 1 wt%, Duhr and Braun [33] claimed that their extremely small concentrations are needed to obtain the true single-particle behavior. All these various approaches certainly contribute to the disparity of results and keep the debate going.

In the following, I will successively describe (i) the experimental techniques implemented to measure thermophoresis of particles, (ii) the proposed mechanisms and the associated experimental results, and (iii) the applications.

Experimental Techniques Used For Thermophoresis in Liquids

The major problem hampering studies of thermophoresis is the extremely low value of the thermal diffusivity D_T of particles in liquids, which directly affects the timescale τ for reaching steady state via the relation $\tau = L^2/D$ where L is the typical size of the

---

[9] The ionic strength is defined as $I = \frac{1}{2} \sum_i z_i^2 c_i$, where c_i are the molar concentrations of all the ions in solution and z_i their charge numbers.

heated region. The experimental approaches are thus striving to shrink the dimensions of the device, keeping in mind yet that the smaller the device the poorer the signal to noise ratio.

To date, four main experimental techniques have been developed based on

   (i) Forced Rayleigh scattering
  (ii) Laser beam deflection
 (iii) Thermal lensing
 (iv) Fluorescence microscopy

Let us successively describe these four techniques.

(i) Forced Rayleigh scattering. This technique was introduced by Köhler in 1993 [58, 59] and detailed for instance in Reference [126]. Two laser beams at different incidence angles create an interference pattern at the sample location. The presence of an absorbing layer converts the intensity fringes into a temperature grating, which in turn causes a concentration grating induced by thermophoresis. Both gratings contribute to a combined refractive index grating that is read out by diffraction of a third laser beam. Analyzing the time-dependent diffraction efficiency, three parameters can be retrieved [126]:

- the thermal diffusivity D_{th}, which controls the dynamics of heat dissipation in the heat diffusion equation (named D in Equation (2.14) on page 40). It can be expressed as $D_{th} = \kappa/\rho c_{\mathrm{p}}$ with the thermal conductivity κ, the mass density ρ and the specific heat capacity c_{p}.
- the translational diffusion coefficient D, which describes the diffusion of the solute by Brownian motion. It is defined in Section 5.2.2 on page 145.
- the thermal diffusion coefficient D_T, defined by Equation (5.11), page 146.

The ratio of the thermal diffusion coefficient D_T and the translational diffusion coefficient D also allows the determination of the Soret coefficient $S_T = D_T/D$.

(ii) Beam deflection technique. The principle of this technique, introduced by Giglio and Vendramini in 1975 [47, 48], is represented in Figure 5.2a.

The liquid is located between two parallel plates closely spaced by a distance h of typically 1 mm. The upper plate is set at a higher temperature by resistive heating or using a Peltier element to induce an upward temperature gradient between both plates, within the fluid layer. A laser beam passes along the fluid layer over a distance ℓ. The concentration gradient of the species undergoing thermophoresis creates a vertical gradient of refractive index that induces the deflection of the laser beam. An additional beam deflection created by the temperature gradient itself cannot be ruled out. Indeed, the refractive index of liquids strongly depends on temperature. The timescale of the establishment of the temperature profile is, however, much faster than the thermophoresis of the particles. Thus, both effects can be discriminated by recording the temporal evolution $\delta z(t)$ of the beam deflection. A first rapid beam deflection δz_{th}^0 is observed due to the establishment of the temperature gradient, followed by a much slower variation $\delta z_{\mathrm{d}}(t)$ stemming from the establishment of the gradient of particle

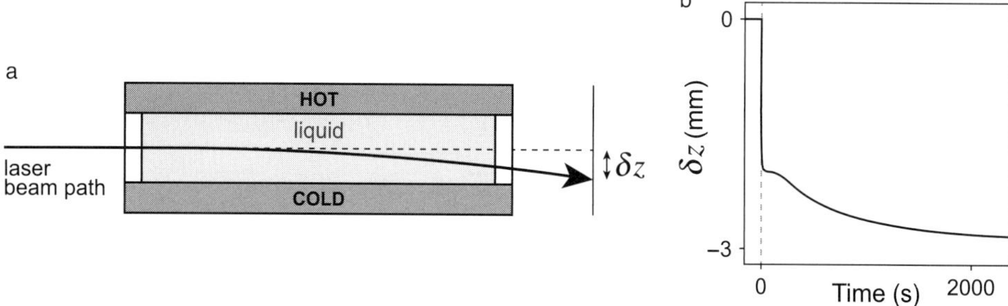

Fig. 5.2 Basic principle of the beam deflection technique to investigate thermophoresis in liquids. (a) A vertical temperature gradient is applied within a 1 mm thick liquid layer. A beam deflection originates both from the temperature gradient and the thermal-induced gradient of particle concentration within the fluid. (b) Time trace of the beam deflection showing two regimes: a fast variation related to the quasi-instantaneous establishment of the temperature gradient and a much slower evolution stemming from the particle thermophoresis.

concentration, eventually leading to a steady state beam deflection $\delta z_{\mathrm{th}}^0 + \delta z_{\mathrm{d}}^0$. A typical time trace of beam deflection is plotted in Figure 5.2b. The evolution of the beam deflection stemming from the concentration gradient is given by [96]

$$\delta z_{\mathrm{d}}(t) = \delta z_{\mathrm{d}}^0 \left[1 + \frac{4}{\pi} \exp\left(-\frac{t}{\tau}\right) \right] \tag{5.43}$$

where $\tau = h^2/\pi D$. A fit of the experimental profile leads to the measurements of the Soret coefficient S_T.

The beam deflection technique depicted above is particularly slow. For instance, for a temperature gradient applied over a distance of $L = 1$ mm, one typically has to wait around 5000 s to reach the steady state [110]. In 2004, Putnam and Cahill introduced a refinement of the technique using a vertical device and a modulated heating [103, 104, 105] aimed at speeding up the acquisition time. This technique is usually called the *ac beam deflection technique*. The idea was to apply a temperature gradient over a much smaller volume using resistively heated microwires (Figure 5.3). This way, the authors could speed up the equilibration times by a factor of 300 compared to previous beam deflection studies. However, this technique still requires large concentrations in order to induce large enough refractive index variations and a sufficient signal-to-noise ratio (a particle concentration larger than 0.3 vol% according to Reference [105]).

Beam deflection techniques, like all the techniques based on a modification of the refractive index, are also restricted to small particles (molecules, proteins, nanoparticles), much smaller than the illumination wavelength, so that they affect the liquid's refractive index and do not behave as light scatterers.

(iii) Thermal lensing. In 1974, Giglio and Vendramini [46] (the same inventive researchers who also pioneered the beam deflection technique in 1975) introduced a thermal lensing technique to investigate thermophoresis in liquids. Thermal lensing is a self-action of a light beam propagating through a partially absorbing medium, generating a local temperature increase associated with a local and radial gradient

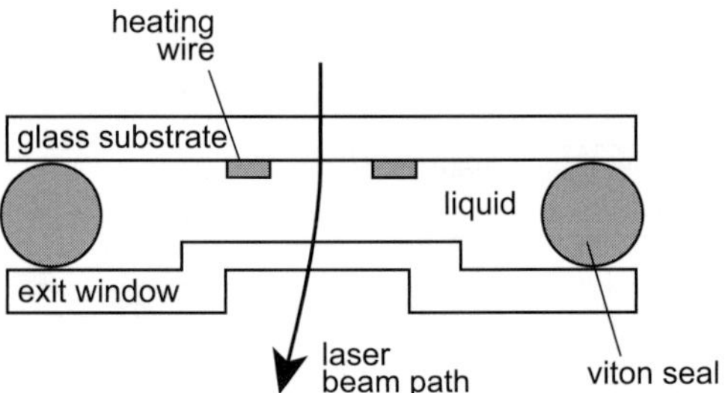

Fig. 5.3   Schematic cross section of the temperature-controlled sample cell for measurements of thermophoresis in liquids using the ac beam deflection technique [103, 105]. The metal heating wires, 250 nm thick, are meant to alternately heat at an angular frequency ω with a high-frequency square-wave current (on the kHz range). The line-heaters are separated by 25 μm and have a width of 5 μm. The chamber height in the sampling region is 300 μm.

of refractive index acting as a lens on the propagating beam itself. When used in a mixture of aniline and cyclohexane, the authors noticed that the effect was noticeably larger than expected. The authors interpreted this observation as a consequence of a temperature-driven concentration gradient, which created an additional gradient of refractive index. This additional "Soret lens" can be divergent or convergent, depending on the preferential direction of motion and density of the different components of the fluid. As a result, spreading of the transmitted beam may further increase, or conversely lessen. The authors suggested that this approach could be exploited to quantitatively investigate thermophoresis in liquids. In this seminal work, heating was due to the absorption of a He–Ne laser beam by the aniline molecules, one of the two components of the mixture.

In 2003, Alve and coworkers [4] proposed a thermal lensing approach to investigate thermophoresis in liquids. A 532 nm laser was used to directly heat an ionic ferrofluid. The high absorbance of the liquid enabled the authors to use an extremely thin layer of liquid of 24 μm. Compared to the seminal work of Giglio and Vendramini [46], they measured a slightly different quantity. They coupled the thermal lensing approach with a single-beam Z-scan technique. Basically, the sample was moved along the laser beam direction and the collected intensity was recorded as a function of the sample position. Using this technique, they could obtain results in good agreement with previous results obtained using forced Rayleigh scattering.

In 2004, in parallel with the ac beam deflection technique developed by Putnam and Cahill, the group of Piazza introduced a thermal lensing technique based on the same concern of shrinking the size of the device [110]. The advance of Piazza compared to the works of Giglio and Alve is that the photothermal approach is based on the optical absorption of the water molecules themselves. Their device is sketched in Figure 5.4. A focused laser beam was sent through a thin layer of the liquid of interest (aqueous micellar solutions or colloidal dispersions were the practical cases in their articles). This laser was meant to locally heat the liquid and create a microscale temperature gradient on the size of the beam waist, and underwent a self thermal lens effect that varies the measured intensity of the detector. The temperature increase

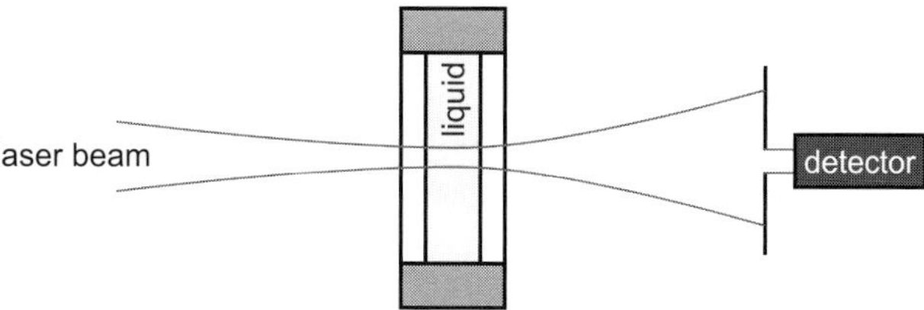

Fig. 5.4   Principle of the thermal lensing technique.

was due to water absorption at the laser wavelength of 980 nm (matching a specific vibrational overtone of the water molecule), corresponding to an absorption coefficient is 500 μm^{-1}. They measure the thermal diffusion constant of 26 nm polystyrene beads.

(iv) Fluorescence microscopy. In 2002, Braun and Libchaber introduced an all-optical technique for investigation of thermophoresis in liquids [13]. A deeper, 10-page study was conducted by Duhr and Braun in 2004 [34]. Their approach was radically different from the above-mentioned approaches. First, they investigated thermophoresis at the micrometric scale through a microscope objective lens. This reduction in size of the system enabled them to speed up the thermophoresis process and achieve much faster readout rates. Second, the measurements were no longer based on a modification of the refractive index of the liquid due to gradients of concentrations. The authors measured all the quantities using fluorescent molecules. They used one type of fluorescent molecules (BCECF or Rhodamine molecules) to map the temperature distribution and they labelled the particles with another fluorescent tag to reveal their nonuniform spatial distribution. This enabled them to possibly achieve much weaker particle concentrations (around six orders of magnitude less) and rule out any contribution of interparticle interaction. Heating was performed by heating the water solvent with an infrared laser beam at 1480 nm.

A problem with Braun's approach is that a substantial thermal-induced fluid convection is usually observed, unless the thickness of the liquid layer is reduced to a few μm. Most of the following reported experiments of Braun actually involved both fluid convection of thermophoresis at the same time. Another problem, addressed later on by the authors [34], is that the temperature-sensitive molecules can also experience thermophoresis, which yields measurement errors (fluorescence intensity variations are supposed to stem from temperature variations, not from molecule nonuniformity). Braun estimated this problem to cause an error of around 8% in the estimation of S_T [34].

Over the last decade, a bunch of experimental studies of thermophoresis in liquids has been conducted using molecules, DNA strands and micrometric beads. The observations have been rather peculiar, as D_T and S_T have been shown to depend on many factors, such as the salinity, pH, surface coating of the beads, size of the particle, molecular weight and even isotopes [30]. D_T and S_T may even be negative is some circumstances, leading to a flux toward hot regions. In particular, low temperatures tend to favor a thermophilic behavior ($D_T < 0$).

Interestingly, although the underlying physics of thermophoresis in *gas phases*[10] is rather simple and well understood [26], its origin in *liquids* is much more complicated. The reason for this higher complexity for liquids is twofold: (i) a fluid involves short-distance molecular interactions (hydrogen bond, Van der Waals interaction, ...) and (ii) there are often dissolved ionic species in liquids (always, in water) that may generate electrostatic forces. Since the nature and the magnitude of the particle–solvent interactions and ionic concentrations may vary from one system to another, particles move along temperature gradients in liquids not always for the same reason. The underlying physics of thermophoresis in liquids is intricate, multiple and still a matter of debate.

To date, thermophoresis in liquids has been investigated using three different kinds of particles that are colloids, DNA strands and molecules. The proposed mechanisms can be divided into two categories:

1. *Dynamical (force) description.* There exist net forces that apply on the particles and make the particles move along thermal gradients. These forces can be either long range forces (electrostatic forces due to the ions in solution for instance) or contact forces (Van der Waals interactions with the surrounding water molecules for instance).
2. *Thermodynamical description.* Thermophoresis is described as a result of a chemical potential of the particle that is temperature-dependent. No net force is involved. Particles explore the medium with Brownian motion and preferentially occupy the location in the fluid where their chemical potential is weaker.

Depending on the nature of the particle, the ionic strength, the nature of the liquid, the involved mechanism can have either a dynamical or a thermodynamical origin. But this question is still a matter of debate. Some groups only focus on dynamical descriptions while other groups mainly consider thermodynamical effects. Sometimes, both descriptions can be considered as equivalent by writing the general expression of forces in thermodynamics from the chemical potential μ:

$$\mathbf{f} = -T\nabla\left(\frac{\mu}{T}\right). \tag{5.44}$$

Here is a list of the more important mechanisms that have been proposed to account for thermophoresis:

(i) Thermoosmosis
(ii) Thermoelectric effect
(iii) London dispersion forces
(iv) Thermal diffusion

Depending on the nature of the system, one mechanism will be dominant compared to the others. Let us briefly describe one by one these four mechanisms. A more detailed review of the dynamical descriptions can be found in Reference [129]. Details on the thermodynamical description can be found in Braun's publications [33, 106].

---

[10] For binary gas mixtures, Chapman theoretically predicted that thermophoresis can occur only if the masses or the sizes of the two components are different, and the component with larger mass or size moves toward the cold. Chapman also evidenced that the significance of thermophoresis in gases also depends on the interaction potential used for modelling collisions.

(i) Thermoosmosis. When a solid particle is exposed to a fluid, charges are likely to appear on the surface of the particle due to surface bond dissociation. For instance, a silica surface becomes negatively charged in contact with water due to the departure of protons and the formation of SiO^- groups on the surface. The occurrence of a surface charge comes along with an accumulation of dissolved counterions in the vicinity of the particle surface, over a typical thickness called the Debye length ℓ_{DH}. Such a picture is called the electric double layer (EDL) model, the first (thin) layer being the charges of the surface and the second (much thicker) layer the ions accumulated over the Debye length in the liquid. The difference of electric potential between the particle surface and the liquid is called the zeta potential ζ. In the mechanism called *thermoosmosis of the EDL*, the density of positive and negative ions in the liquid in the vicinity of the colloid depends on temperature, which creates a force on the bead directed to the cold region, and proportional to ζ^2, as derived by Eli Ruckenstein in 1981 [109]. Thus, according to Ruckenstein's model, thermophoresis of colloids can be considered as a simple Marangoni effect, where a property of the colloid–fluid interface (here the surface tension) is temperature-dependent. This picture, however, only accounts for a thermophobic effect. It cannot explain why colloids sometimes move from the cold to the hot. For this purpose, other mechanisms have to be considered.

(ii) Seebeck effect of dissolved ions. In 2005, Putnam *et al.* [104] depicted another mechanism explaining why colloids could be either thermophobic or thermophilic, depending on the presence of dissolved ions. A temperature gradient in a liquid can drive different ions at different velocities and even in opposite directions, depending on their charge, nature and solvation energy [50] (see previous Subsection 5.3.4). One usually calls this process a Seebeck effect, in analogy with solid-state physics when a temperature gradient generates an electric current. In an open system, this mechanism can generate an electric current, but in a closed system, where the ion currents must vanish in the steady state, charges accumulate at the boundaries of the system, which results in an electric field in the liquid, which acts on the suspended particles. The amplitude and direction of the driving electric force depends on the nature of the dissolved ions and is proportional to the zeta potential of the colloids, or to the charge of the molecules. Thus, this thermoelectric mechanism may dominate the Ruckenstein's mechanism [109] and drive colloids to the hot.

(iii) Dispersion forces. London dispersion forces are Van der Waals molecular interactions that result from induced-dipole/induced-dipole interaction. This type of force is marginally dependent on temperature. It, however, depends on the mean molecular interdistance, *i.e.*, on the fluid density, which is temperature-dependent. For most fluids, density decreases upon heating. This is valid also for water except below 4°C where the density decreases upon cooling. Because of this effect, if a solid–liquid interface is submitted to a temperature gradient along the interface, a given volume element of the fluid at the cold side will be more strongly attracted by the solid surface than at the hot side, resulting in a liquid flow toward the hot side. If this solid surface is now the surface of a colloid, this motion of the fluid to the hot will result in a motion of the colloid to the cold. The significance of this effect is supposed to be independent of the particle size. This effect will naturally reverse if the temperature gradient is established below 4°C. This model was proposed by Brenner *et al.* [10, 17] in order to account for experimental observations of reversed colloid motion (thermophilicity) at low temperatures, below 4°C.

(iv) Entropy contribution. In 2002, the group of Dieter Braun published a now-famous article entitled "Why Molecules Move along Thermal Gradients" [33]. The explanations are based on thermodynamical considerations. The entropies s associated with hydration and with ionic shielding of the particles directly implies the existence of a Soret coefficient according to the relation derived by Braun:

$$S_T = -\frac{s}{k_B T}. \tag{5.45}$$

The experiments conducted by Braun in this article with DNA strands and beads are in good agreement with this formalism, without free parameters. This physics was revisited by the same group in 2014 in an article with a similar title: "Why Charged Molecules Move Across a Temperature Gradient: The Role of the Electric Field." This article is no longer based on entropy considerations, but rather on electrostatic forces. This picture is more consistent with observations related to the effect of salt concentration.

5.3.6 Applications Based on Laser-Assisted Microscale Thermophoresis

In the 2010s, a growing number of biologists have used laser-assisted microscale thermophoresis as a method for quantifying biomolecule binding [55, 113, 127, 71, 112, 114], a technology developed by the company *Nanotemper Technologies GmbH* created in 2008 by Stefan Duhr and Philipp Baaske, previous members of the group of Dieter Braun. Also, central questions of molecular evolution were addressed by thermophoretic traps by the group of D. Braun [84, 107]. However, none of these studies benefited from nanoplasmonics and associated photothermal effects. Heating resulted from water absorption.

The very first study of thermoplasmonic-assisted thermophoresis dates from 2013 and was reported by the group of Cichos [14]. The authors created dynamical temperature gradients to counteract the natural diffusion of 200 nm polystyrene spheres. The position of the nanoparticle was tracked and used in a feedback loop to act on the orientation of the temperature gradient via the position of a laser beam. This way, since the nanoparticles are supposed to move from hot to cold, the natural Brownian motion of the particle could be confined over a specific, few-micron big area (see Figure 5.5).

In 2014, the authors developed a refined procedure for trapping experiments based on thermophoresis. The authors proposed this time a feedback-free method using a steering focused laser beam, mimicking a Paul's trap [15], that created a rotating temperature field. The authors investigated in detail the underlying physics to work out the stability conditions as a function of the steering frequency of the laser. A threshold rotation frequency was derived.

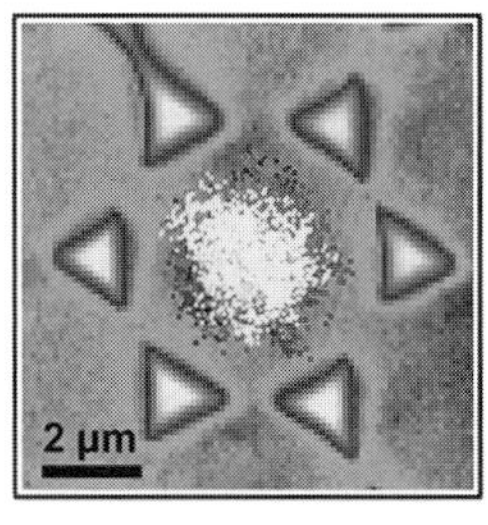

Fig. 5.5 (a) Sketch of the focused heating beam illuminating only one gold pad of an open structure at a time. Trajectory points of a 200 nm polystyrene sphere trapped within an open gold structure at 5 mW heating power. Reproduced with permission from Reference [14]. Copyright 2013, American Chemical Society.

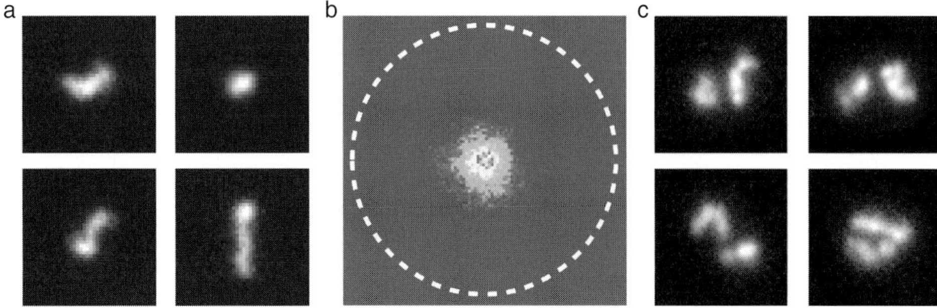

Fig. 5.6 Thermophoretic trapping of λ-DNA molecules. (a) Four snapshots ($4.6 \times 4.6\ \mu m^2$) of a trapped single λ-DNA molecule. (b) Probability density of the trapped molecule with a width $\sigma = 0.75\ \mu m$ at 3.0 mW heating laser power. (c) Four snapshots ($6.9 \times 6.9\ \mu m^2$) of the trapping of two single λ-DNA molecules. Reproduced with permission from Reference [16]. Copyright 2015, American Chemical Society.

In 2015, the group of Cichos extended their technique to the trapping of single DNA strands, as illustrated by Figure 5.6 [16]. This advance nicely introduced an experimental technique to manipulate small objects in liquids using a photothermal approach.

In 2016, following this line of research, Hashimoto's group evidenced reversed thermophoresis (from cold to hot) using a single gold nanoparticle [37]. Molecules such as polyethylene glycol and sodium dodecyl sulfate were observed to move toward heated nanoparticles and attached to their surface, leading to the formation of hybrid nanostructures with tunable optical properties. The reason for the centripetal motion was not clearly established.

In 2016, Lin *et al.* also evidenced a reverse thermophoresis of gold nanoparticles [69]. Gold nanoparticles dispersed in solution themselves underwent a motion from cold to hot due to a positive charge on their surface. This way, the authors could achieve a reversible assembly of nanoparticles triggered by light. The temperature gradient was created by heating a thin gold layer, not the nanoparticles themselves. Using an SLM (spatial light modulator), the authors could even create arbitrarily complex trapping profiles. This work further illustrates how thermophoresis can be used to manipulate nano-objects using small light intensities.

5.4 Liquid Superheating and Liquid–Gas Phase Transition

Phase transition is maybe the most complicated aspect of thermodynamics, just after the concept of *entropy*. Considering phase transition of *water* does not even simplify the problem, on the contrary! Although water is the most abundant compound on Earth, it is definitely not a simple liquid [78]. This makes the physics of liquid–gas phase transition in plasmonics rather intricate.

Superheating characterizes the state of a fluid that remains liquid above its boiling temperature. In theory, such a state can exist above the boiling temperature, up to the so-called spinodal temperature. Such a metastable state is very likely to happen around heated plasmonic nanoparticles. Temperatures as high as $230°C$ have been recently evidenced around gold nanoparticles on a glass substrate [7]. If the water is degassed, this temperature limit can even reach $300°$ [2].

How come superheating can occur in plasmonics? Actually, water is not supposed to boil at $100°C$. There exists an energy barrier, even at $100°C$, that should prevent the molecules of the liquid from forming a gas phase in the liquid bulk. The reason why boiling at $100°C$ is often observed in everyday life is due to the presence of nucleation points, like scratches, dust or microscale roughness. In plasmonics the samples are usually very clean and the substrates free from roughness, as they are usually made of glass. Thus, the absence of bubble formation at $100°C$ in plasmonics is not due to the nanoscale dimension of the nanoparticles that would create a large Laplace pressure if a bubble forms, as is sometimes explained in the literature [86, 80]. It is rather due to the nature of the samples and of the substrates that are naturally free of nucleation points. A well-known example supports this idea: chemists are used to this phenomenon, which is likely to occur when boiling water in glassware using a Bunsen burner. In order to avoid superheating, which is hazardous and can lead to an explosion (called *explosive boiling*), a pumice stone is often introduced in the reaction medium to create nucleation points and force soft boiling at $100°C$. Here again the occurrence of superheating is due to the flatness of the vessel, and not to any nanometric source of heat.

This section recalls some basics on phase transition that explain the underlying physics of superheating.

5.4.1 Origin of Liquid Superheating in Plasmonics

A *fluid* is a general term to indifferently refer to a liquid or a gas. A homogeneous fluid is commonly described by a reduced set of physical quantities that are the pressure P, the temperature T, the volume V, and the number of molecules N, all of them related via the equation of state of the fluid. For instance, for an ideal gas (that consists of a gas composed of point-like, non-interacting particles), the equation of state reads

$$PV = Nk_{\mathrm{B}}T \tag{5.46}$$

where k_{B} is the Boltzmann constant. The molecular volume $v = V/N$ is commonly used instead of the two extensive quantities V and N:

$$Pv = k_{\mathrm{B}}T. \tag{5.47}$$

From this equation, *the occurrence of a phase transition and a spinodal temperature can't be predicted as they precisely originate from molecular interactions.*

Molecular interactions can be divided into attractive interactions at long distances and repulsive interactions at short distances. Without entering the details, the potential energy E as a function of the molecule interdistance r can usually be modeled by the Lennard-Jones expression

$$E(r) = 4A \left[\left(\frac{r_0}{r}\right)^{12} - \left(\frac{r_0}{r}\right)^{6} \right] \tag{5.48}$$

where A measures the strength of the interaction and r_0 is the collision diameter. A and r_0 depend on the nature of the interacting molecules. Figure 5.7 plots such a potential energy.

Deriving a new equation of state of a fluid that incorporates the Lennard-Jones equation would yield overly complex developments. The occurrence of phase transition can be obtained from a simpler model where the fluid obeys the Van der Waals equation of state, which reads

$$\left(P + \frac{a}{v^2}\right)(v - b) = k_{\mathrm{B}}T \tag{5.49}$$

or equivalently

$$P = \frac{k_{\mathrm{B}}T}{v - b} - \frac{a}{v^2}. \tag{5.50}$$

This equation of state (5.49) is similar to the equation of state (5.47) of an ideal gas, with correction elements. The term b models the repulsive interaction at short distances. Indeed, such a correction amounts to considering that the molecules are no longer point-like particles, but endowed with a nonzero volume b. The second correction a/v^2 models the

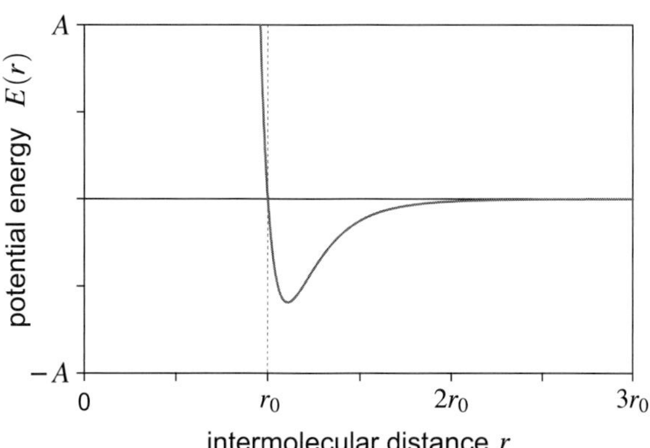

Fig. 5.7 Lennard-Jones potential energy of two molecules separated by a distance r (see Equation (5.48)).

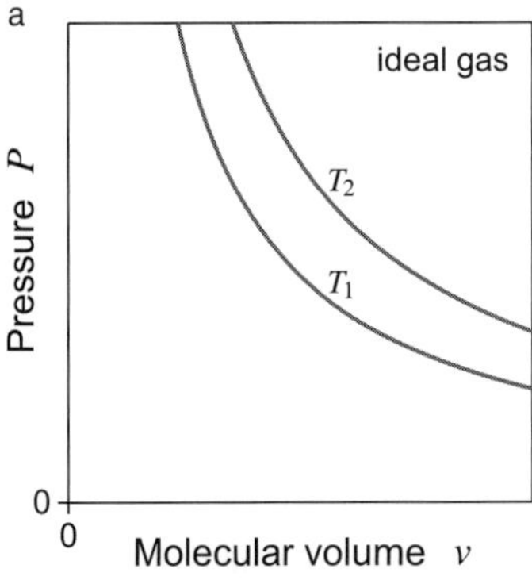
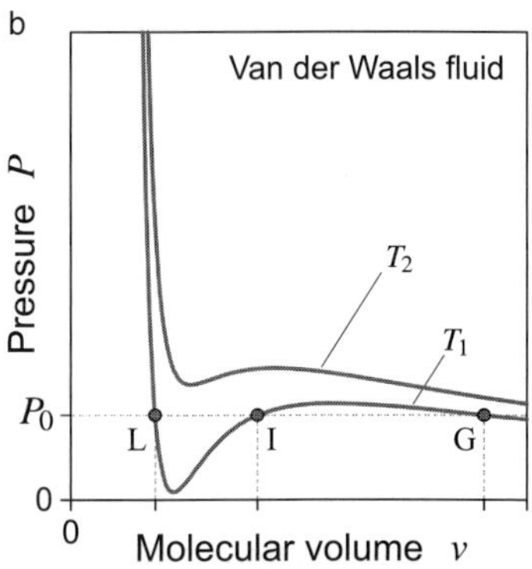

Fig. 5.8 (a) Isotherms in a $P - v$ diagram for an ideal gas corresponding to Equation (5.47), for two different temperatures T_1 and T_2 ($T_2 > T_1$). (b) Isotherms in a $P - v$ diagram for a Van der Waals fluid, corresponding to Equation (5.50), for the same temperatures as case (a). At (P_0, T_1), three states (corresponding to points L, I and G) fulfill the Van der Waals equation of state.

attractive force at longer distances. When a molecule approaches a wall of a container, the molecule experiences a backward attraction from the other molecules that tends to reduce the exerted pressure on the wall. Note that the ideal gas equation of state (5.47) is retrieved from the Van der Waals equation for a diluted fluid, *i.e.*, for a sufficiently large value of molecular volume v.

Figure 5.8 compares the line shapes of $P(v)$ at different temperatures for the ideal gas model governed by Equation (5.47) and for the Van der Waals fluid governed by Equation (5.49). Figure 5.8a corresponds to the case of an ideal gas, governed by the equation of state (5.47). Two isotherms have been represented and as can be expected they feature a monotonic variation: for a given temperature pressure, there exists only one possible molecular volume. This is in contradiction with the occurrence of liquid gas transition where gas and liquid (*i.e.*, two states with different molecular volumes v) can exist at the temperature and a given common pressure.

Taking into account molecular forces (as done with the Van der Waals equation of state) naturally enables the prediction of different phase coexistence. This is what is evidenced in Figure 5.8b where one can see that an isotherm is no longer necessarily monotonic. For instance, three states corresponding to points L, I, G fulfill the Van der Waals equation for the same pressure P_0 and temperature T_1. This observation is at the origin of the phenomena of phase transition and phase coexistence.

Among these three states, the state corresponding to point I is singular. Indeed, it lies on a portion of the isotherm that is increasing, *i.e.*, where one has the condition

$$\left(\frac{\partial P}{\partial V}\right)_T > 0. \tag{5.51}$$

This means that a local dilatation of the system yields a pressure increase. This counterintuitive effect simply means that the system is not stable in this area of the diagram, where

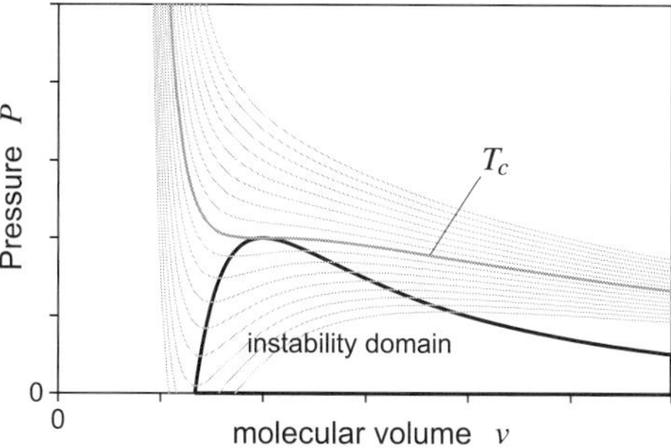

Fig. 5.9 Series of isotherms in a $P - v$ diagram for a Van der Waals fluid. The isotherm corresponding to the critical temperature, along with the instability area are highlighted.

local fluid fluctuations are no longer damped but amplified. In other words, the fluid features a negative elasticity.[11] Consequently, only states L and G can be physically observed. As you may have guessed, state L (which corresponds to the lower molecular volume v) corresponds to a liquid phase while state G corresponds to a gas phase. However, as we will see later, this does not mean that they can coexist (the chemical potential of both states G and L has to be equal as well).

Figure 5.9 plots a series of isotherms on a single $P-T$ diagram. One can notice that for sufficiently high temperatures, the isotherms become monotonic. The consequence is that only one state can exist at a given pressure. The isotherm corresponding to a transition between monotonic and non-monotonic isotherm is highlighted in the figure and corresponds to the so-called critical temperature T_c. This temperature only depends on the nature of the fluid. In the same Figure 5.9 is also represented the domain where the isotherms present a positive slope, which corresponds to an instability domain for the fluid as explained above.

We have now all the elements to understand the occurrence of metastability and superheating. Let us return to Figure 5.8b. The fact that the states corresponding to points L and G are allowed under the same pressure and temperature does not mean that they can coexist. Indeed, where various phases coexist, there is a third condition for the equilibrium, the equality of the chemical potentials $\mu_L = \mu_G$. At a given pressure, this equality occurs for a precise temperature, which is nothing but the boiling temperature T_b. In order to further enter into details, let us focus on Figure 5.10 which plots a series of eight $P - v$ diagrams where the temperature is gradually increased from diagram (a) to diagram (h). The case of Figure 5.8b corresponds to the case of Figure 5.10b. When a molecule passes from state L to state G, it loses an energy that reads

$$U = \int_{L}^{G} P dV \tag{5.52}$$

---

[11] The fluid elasticity is defined as

$$-V \left(\frac{\partial P}{\partial V} \right)_T.$$

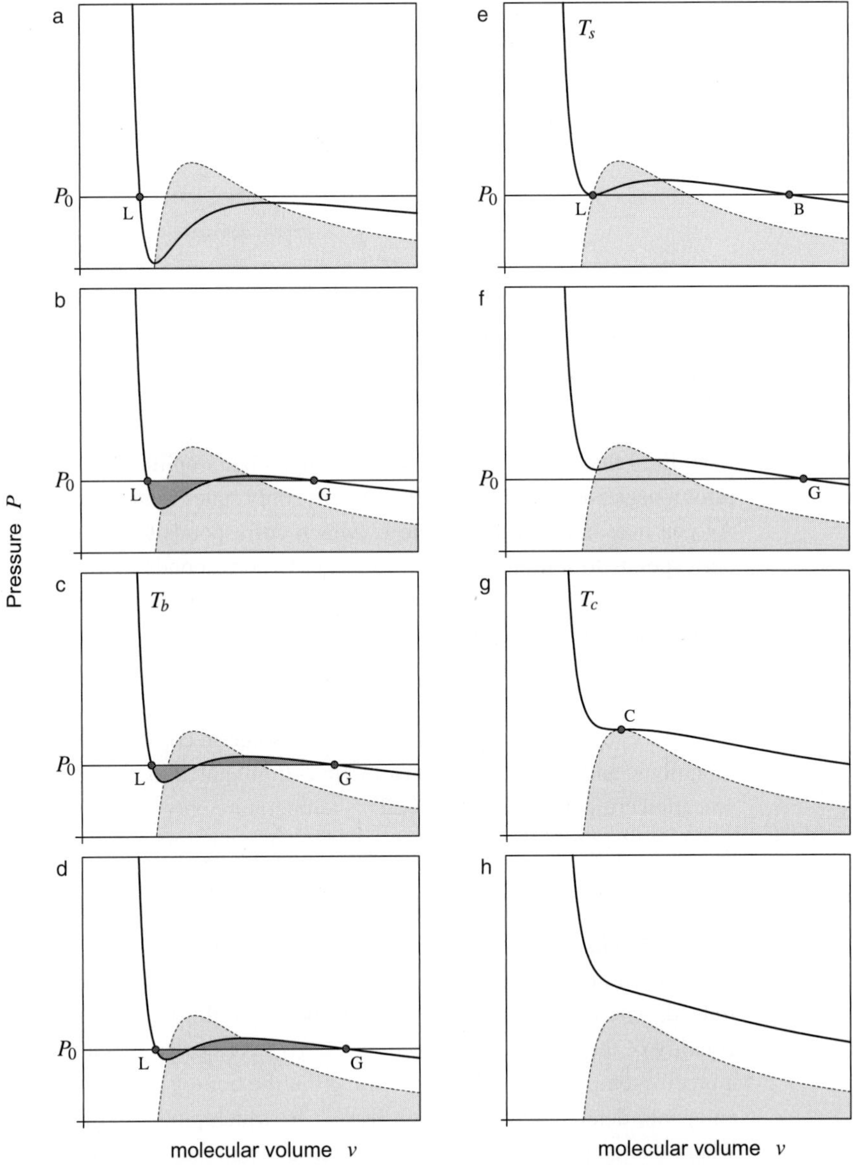

Fig. 5.10　Series of $P - v$ diagrams representing isotherms of increasing temperature from case (a) to case (h). (a) Only a liquid state is possible at the pressure P_0. (b) Two states are permitted, a liquid (L) state and a gas (G) state, but G corresponds to a metastable state (supercooling). (c) Isotherm corresponding to the boiling temperature, gas and liquid can coexist at this temperature. (d) Two states are permitted, a liquid (L) state and a gas (G) state, but L corresponds to a metastable state (superheating). (e) From this temperature, which corresponds to the spinodal temperature, the liquid state is no longer stable. (f) At this temperature only a gas phase can exist. (g) Isotherm corresponding to the critical temperature. (h) Temperature for which no distinction is possible between gas and liquid.

which corresponds to the grey algebraic area in Figure 5.10b. In this example, this area is clearly negative, which means that the molecules gain energy if they go from the liquid to the gas phase, which is not favored for this reason. On the contrary, molecular flux from state G to state L is favorable since the molecular energy decreases during the transfer. Consequently, states L and G in the situation represented in Figure 5.10b cannot be at equilibrium with each other and the molecules in state G are likely to condense into state L.

Consequently, equilibrium between gas and liquid can only occur if the temperature is such that the integral (5.52) equals zero. This condition amounts to saying that the chemical potential in both phases is identical. Such a situation is represented in Figure 5.10c where the upper grey area equals the lower grey area, making the integral (5.52) zero. This isotherm naturally corresponds to the boiling temperature of the fluid at a given temperature. Considering the grey areas represented on the diagram is the basis of the so-called Maxwell construction to determine the boiling isotherm from a $P - v$ diagram.

Let us further increase the temperature to reach the isotherm represented in Figure 5.10d. For the same reason as above, liquid and gas cannot coexist at this temperature as the integral is not zero. This time, the more stable state is the gas phase represented by point G and the liquid is likely to rapidly disappear.

However – and this is where the interesting story begins – in some circumstances the fluid can remain liquid even if the boiling temperature is exceeded. As can be seen in Figure 5.10d, the lower grey area constitutes a small barrier that the molecules have to exceed in order to reach state G. The liquid phase is said to be in a metastable state and the associated phenomenon is called superheating. The reason why superheating of water is rarely observed in everyday life is that most vessels are endowed with *nucleation sites* on the inner walls, like scratches or microscale roughness. For this reason, liquid superheating is more likely to happen in chemistry, where most vessels are made of glass, which is a particularly smooth material even at the microscale. This is also the case of glass coverslips that are commonly used in plasmonics experiments to support plasmonic structures. This is the reason why superheating is likely to happen in nanoplasmonics [7].

Superheating can occur until the limit case represented in Figure 5.10e. This associated temperature is called the spinodal temperature. This situation is characterized by the disappearance of the lower dark area represented in Figures 5.10(b,c,d).

Unlike the critical temperature, the spinodal temperature depends not only on the fluid by also on the pressure. Above this temperature, as represented in Figure 5.10f, only the gas phase is stable and can be observed. Figure 5.10g corresponds to the critical isotherm and Figure 5.10h corresponds to a temperature higher than T_c where no distinction can be made between liquid and gas.

Figure 5.11a presents another kind of diagram that expresses pressure as a function of temperature. The case of water is represented and all the possible states have been represented (gas, liquid but also solid). Domains where liquid water can be in a metastable state have been represented (superheating and supercooling areas). All the important temperature values under atmospheric pressure have been represented, in particular the boiling temperature ($100°C$) and the spinodal temperature (around $280°C$).

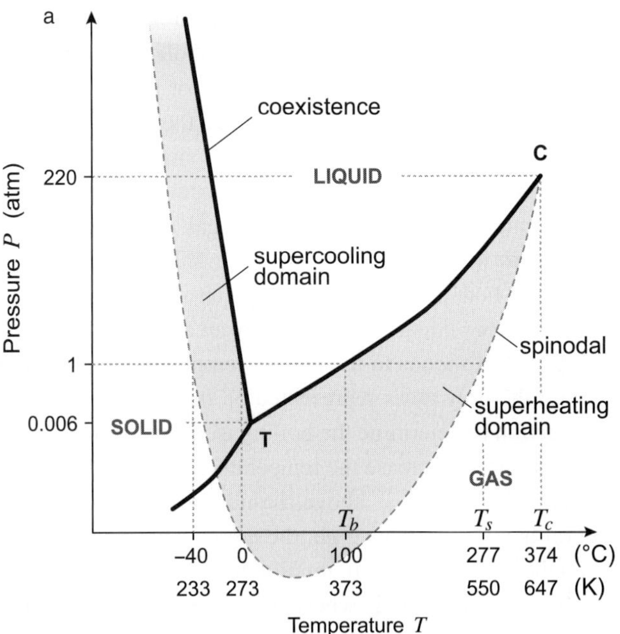

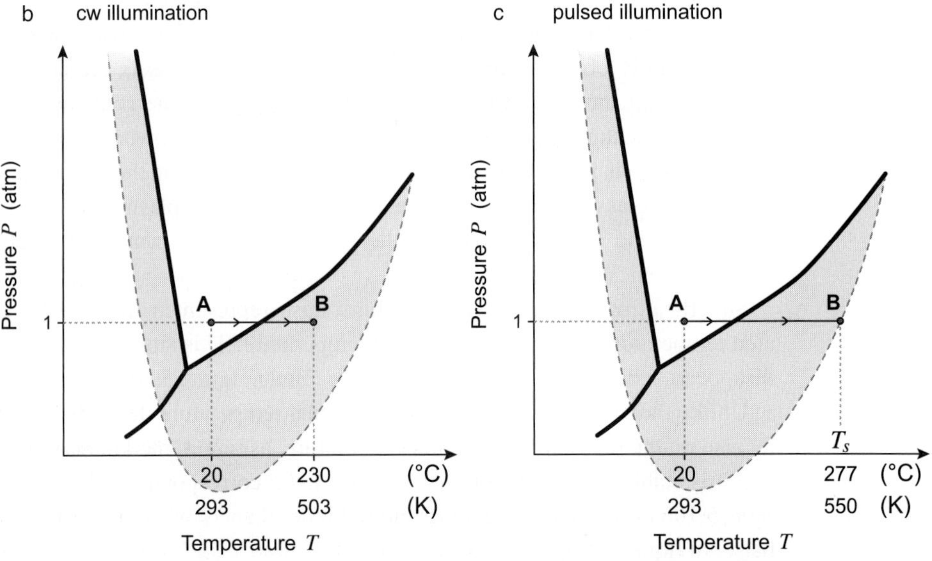

Fig. 5.11 (a) $P - T$ diagram of water showing the three existence domain of solid, liquid and gas separated by coexistence lines. The spinodal line defines the domain of liquid metastability (superheating or supercooling). **T** and **C** stands for the triple point and the critical point. (b) Phase transition in plasmonics under cw illumination. The liquid surrounding the nanoparticles is heated from ambient temperature (point A) to a superheated state (point B) above the boiling temperature usually around 230°C prior to bubble formation [7]. (c) Phase transition in plasmonics under pulsed illumination. The liquid surrounding the nanoparticles is heated from ambient temperature (point A) to a superheated state (point B) at the spinodal temperature of 277°C prior to bubble formation [7].

5.4.2 Difference between CW and Pulsed Illumination for the Liquid–Gas Phase Transition

CW Illumination

Figure 5.11b represents the typical evolution of liquid water surroundings upon heating nanoparticles using cw illumination as recently evidenced in Reference [7]. The nanoparticles originally under ambient temperature and pressure (state A) are heated far above the boiling temperature but the surrounding water remains in a superheated liquid state. Only for a temperature around 230°C occurs bubble formation. The fact that phase transition occurs at a lower temperature than the spinodal temperature $T_s = 277°C$ is expected. 230°C corresponds to a temperature where thermal fluctuations easily help the fluid exceed the energy barrier discussed above. This temperature is sometimes coined the kinetic spinodal temperature. Note that the group of Richardson managed to reach the spinodal temperature and even higher (up to 287°C) without boiling by degassing water [2]. Degassing water is supposed to increase the temperature threshold, but going above the spinodal temperature without phase transition is not physically possible, which casts some doubt about the accuracy of these measurements.

Pulsed Illumination

Under pulsed illumination, the mechanism is slightly different. Boiling has been observed to occur right at the spinodal temperature $T_s = 277°C$, not at the kinetic spinodal temperature around $T_s = 230°C$, although it was never clearly established. The reason is that under femto-to nanosecond-pulsed illumination, the temperature increase is so fast that thermal fluctuations have no time to generate a bubble.

Note that some articles mention that boiling occurs at the critical temperature under pulsed illumination. This is not correct. In any case, when boiling occurs from a super-heated state, the boiling event is violent, and for these reasons is called *explosive boiling*. The subsequent formation of a bubble demonstrates a rich underlying physics and many potential applications. This is the subject of the next section.

5.5 Bubble Dynamics, General Considerations and Physical Laws

As mentioned in the previous section, when plasmonic nanoparticles are immersed in a liquid and optically heated, explosive boiling can occur and the direct consequence is the formation of a small bubble. The physics of bubble dynamics in liquids is particularly rich. This section is devoted to explaining the formation conditions and the dynamics of nanobubbles in plasmonics.

As the mechanisms are radically different under cw and pulsed illuminations, these two cases will be the subjects of two separate subsections, which will be preceded by an introductory subsection on general considerations, drawing an overview of the physical laws involved in the bubble dynamics.

5.5.1 General Considerations on Boiling and Bubbles

Boiling Versus Cavitation

In a liquid, one usually refers to bubble formation as *cavitation* or *boiling*. However, these two terms have different meanings and should not be mixed up. In principle, *boiling* refers to a process where bubble formation results from a temperature increase, while *cavitation* concerns bubble formation that rather results from a local pressure decrease.

There also exists a third kind of spontaneous bubble formation in a liquid, subsequent to a variation of pressure or temperature (like bubbles in glass of champagne). It can happen only when a gas is dissolved in the liquid. But in this latter case, the nature of the bubble is different as it is made of molecules from the dissolved gas, not of steam.

Air Bubbles Versus Steam Bubbles

Two kinds of bubbles in water can be distinguished. *Gas* bubbles (made of molecules that are different from water, for instance molecules from the surrounding atmosphere) and *steam* bubbles. While air bubbles can live for very long periods even at ambient temperature, steam bubbles are not stable at ambient temperature. These two kinds of bubbles can be generated by heating of plasmonic nanoparticles. We will see that cw illumination of plasmonic nanoparticles in water leads to the formation of quasistatic bubbles made of air, while pulsed illumination rather leads to the formation of transient bubbles made of steam, on the nanosecond timescale.

Before going into more detail regarding the physics of these two kinds of bubbles, it is worth mentioning the existence of a peculiar class of bubbles commonly termed "nanobubbles." In 2001, against all odds, Tyrrell *et al.* have shown by atomic force microscopy that any hydrophobic surface immersed in water seems to be endowed with nanometric objects that resemble spherical caps with heights on the order of 10 nm and diameters on the order of 100 nm [120]. These enigmatic bumps have been identified as stable *air nanobubbles*, a supposition that has been further supported by different observations [29, 8]: these presumed nanobubbles can be merged by an AFM tip to form larger bubbles, and their amount depends on gas concentration. They even disappear if the liquid is degassed.

The reason why this observation was the basis of an active area of research is that the presence of such bubbles goes against the common sense. As we will see later, nanobubbles should not be stable and should collapse, but they don't. Note that such nanobubbles have never been investigated in plasmonics. I mention this research area first to highlight that what the plasmonics community calls "nanobubbles" has nothing to do with these surface nanobubbles on a hydrophobic surface, and then to show that physics of small bubbles can hold singular processes.

Let us now introduce the basic physical laws involved in bubble dynamics: the Laplace pressure, Henry's law and the Rayleigh–Plesset equation.

5.5.2 Laplace Pressure

A bubble in a liquid always features an inner overpressure compared to the outer pressure in the surrounding liquid. The pressure difference between the inside and the outside of a bubble in a liquid is termed Laplace pressure and reads

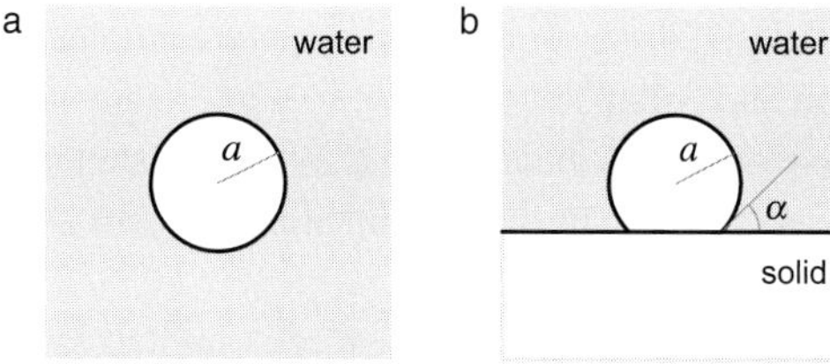

Fig. 5.12 Two common geometries when dealing with bubbles in water. (a) A free bubble in a liquid and (b) a bubble located at solid-water interface. a and α are the curvature radius and the contact angle of the bubble on the surface.

$$\Delta P = \frac{2\gamma}{a} \tag{5.53}$$

where γ is the surface tension (energy per unit area) and a is the curvature radius of the bubble, as represented in Figure 5.12 for two common geometries: a free bubble in water and a bubble located at the interface with a solid surface. In the latter case, one usually defines the contact angle α of the bubble with the solid surface, as represented in Figure 5.12b.

5.5.3 Henry's Law

When considering a system composed of a gas and a liquid *at equilibrium*, Henry's law states that the concentration c_m of dissolved molecules within the liquid is proportional to the partial pressure P_m of these molecules in the gas phase. In mathematical terms, Henry's law reads

$$P_\mathrm{m} = K_\mathrm{H}\, c_\mathrm{m} \tag{5.54}$$

where K_H is the Henry's constant. This law governs for instance the quantity of O_2 and N_2 molecules dissolved in a glass of water.

If this relation is not verified for a given system, there is no equilibrium and a molecular flux occurs either from the liquid to the gas phase if $P_\mathrm{m} < K_\mathrm{H}\, c_\mathrm{m}$, or from the gas phase to the liquid in the other case.

As explained hereinafter, this equation governs the flux of dissolved molecules entering and leaving a bubble in a liquid, especially under cw illumination. Under pulsed illumination, the lifetime of the bubble is so short that molecular transfer across the bubble interface has no time to occur and Henry's law is never satisfied.

5.5.4 Bubble Dynamics: The Rayleigh–Plesset Equation

Once a bubble in a liquid is created, the dynamics of its radius is governed by a subtle interplay between various pressure forces, inertia of the fluid and surface tension. This problem is described by the Rayleigh–Plesset equation. Its simplest form reads [65]

$$\rho a \ddot{a} + \frac{3}{2}\rho \dot{a}^2 = P_\mathrm{i} - P_\mathrm{e} \tag{5.55}$$

where an overdot means differentiation with respect to time. $a(t)$ is the bubble radius, ρ the mass density of the liquid, $P_\mathrm{i}(t)$ the bubble pressure and $P_\mathrm{e}(t)$ the fluid pressure. P_e and P_i can be written as functions of the surface tension, viscosity and bubble radius.

The Rayleigh–Plesset equation describes completely the bubble dynamics from onset to expansion and collapse.

This equation has been used a lot, for instance to demonstrate that bubbles may go through a series of ever-decreasing after-bounces as it returns to its equilibrium size before disappearing [79].

For further information, the reader is invited to refer to References [79] and [65], which nicely depict and use this equation.

5.6 Bubble Formation and Dynamics under Pulsed Illumination

5.6.1 Scenario of Bubble's Life

When illuminating metal nanoparticles with a sufficiently strong pulse energy, bubble formation can occur in the vicinity of the nanoparticles.

Originally, the studies of small bubble generation around light nanoabsorbers under pulsed illumination date from the 90s. The aim was to account for anomalously large photoacoustic effects and nonlinear optical response in experiments involving the pulsed illumination of micro and nanoabsorbers [27, 85, 76]. Madden's group [85, 76] proposed in 1992 that the underlying mechanism could be a vaporization of the surrounding fluid when the spinodal temperature of the fluid was reached (277°C for water, see Section 5.4), a mechanism that still holds today.[12] In the 90s, the involved nanoparticles were rarely metallic. The absorbers mostly used to consist of carbon nanoparticles or organic pigmented nanoparticles. Although Madden was the first to study the bubble generation around gold nanoparticles experimentally [85], the real interest for plasmonic nanoparticles rather dates from 2001 [42].

Another mechanism can be responsible for the formation of a nanobubble in plasmonics under pulsed illumination: the so-called *plasma effect* that consists of an ionization of the liquid in the vicinity of the nanoparticle due to the strong plasmon-enhanced electric near-field of the incoming light pulse.

The occurrence of one mechanism or the other mainly depends on the pulse duration. It is commonly accepted that *plasma*-induced bubble generation requires the use of sub-picosecond illumination (typically using *femtosecond* pulsed lasers), while thermal-induced bubble generation occurs above the *picosecond* pulse range. The plasma mechanism was recently studied by the group of Meunier [11, 62]. The formation of a plasma was not directly evidenced but the authors have shown that the occurrence of bubble generation was dependent on the polarization (linear or circular) of the heating beam, an observation that can be only explained by a plasma-induced mechanism [62]. A plasma mechanism

---

[12] Note that in some articles (not many), the authors sometimes refer to the concept of *critical* temperature instead of *spinodal* temperature, which is, in my opinion, a misuse of language. In some articles, the authors even say "85% of the critical temperature," which is nothing but the spinodal temperature even if the authors do not say it explicitly.

requires large electric near-field enhancements. This is why it is likely to occur with non-spherical nanoparticles (typically nanorods, see Figure 1.10 on page 12), or off resonance to avoid heating.

Once formed, a bubble generated under pulsed illumination is not stable. Once the pulse of light gets absorbed, the associated strong temperature increase rapidly vanishes on the nanosecond scale due to heat diffusion. The freshly generated steam in the vicinity of the particle is therefore bound to re-condense and the bubble to collapse on the nanosecond timescale.

There exists another important timescale governing the fate of a bubble, which is the typical timescale of heat diffusion from the nanoparticle to the surrounding fluid. It simply equals

$$\tau_{\mathrm{d}} = \frac{a^2}{D} \tag{5.56}$$

where a is the nanoparticle radius and D the thermal diffusivity of the surrounding medium. For water, D is on the order of 10^{-7} $\mathrm{W \cdot m^{-1} \cdot K^{-1}}$. In general, τ_{d} is on the nanosecond timescale. Depending on the pulse duration compared to this timescale, two regimes can occur:

- *The short pulse regime.* If the pulse duration is shorter than this timescale, the nanoparticle fully absorbs the laser pulse and thermalizes before exchanging energy with the surrounding (*i.e.*, cooling down). This regime allows for very high nanoparticle temperatures.
- *The long pulse regime.* If the pulse duration is on the order of, or longer than the thermal diffusion timescale, then nanoparticle heating occurs while the nanoparticle is also releasing the heat in the surrounding medium. This situation, which usually occurs using nanosecond pulsed lasers, leads to much less efficient nanoparticle heating, and bubble generation usually requires higher laser fluences.

5.6.2 Detection Techniques of Nano-Bubbles

Over the last decade, the main questions that have been motivating experimental investigation in this field of research are simple: (i) What is the fluence[13] threshold required for bubble formation? (ii) What is the size of the bubble? (iii) What is the bubble lifetime?

Despite the simplicity of the questions, experimental investigation of nanobubbles is complicated for two reasons. First, due to their nanometric size, they cannot be easily visualized. Although they can efficiently scatter light and be detectable for this reason, it is hard to directly measure their diameter using optical microscopy means due to the diffraction limit. Second, their dynamics lies in the picosecond to nanosecond timescale, which is far below the detection speed of many common detectors in optics. This is why, after almost two decades of investigation, some remaining questions are still arousing the interest of an active research community.

---

[13] Let us recall that the fluence is defined as the energy per unit area of an incoming pulse of light.

Table 5.2 Classification of the detection techniques of nanobubbles		
Acquisition: Pump-probe	/	Real time single record
System: Single nanobubble event	/	Ensemble measurements
Technique: X-ray scattering	/	Optical microscopy

All the reported experiments on transient nanobubbles generation in plasmonics can be classified in different binary categories, according to Table 5.2.

In the first category (Acquisition), two kinds of procedures have been reported in the literature to experimentally study transient nanobubbles under pulsed heating:

- *A pump-probe approach* where pulse heating and probing are repeated many times on a single sample (consisting of single or multiple nanoparticles) upon varying delays between heating and probing in order to reconstruct a temporal variation of a signal. The benefit is that the time resolution can be as good as the pulse width of the probe laser. The drawback is that it assumes that the nanoparticles are not damaged between successive pulses. For this reason, the nanoparticle solution is sometimes refreshed after a series of pulses.

- *Real-time measurements* of single nanobubbles are conducted using continuous probe illumination and fast detectors, such as oscilloscopes at 1 GHz [116]. Sub-nanosecond range dynamics was never achieved using such a procedure.

A second classification can be made among the reported studies of transient nanobubbles under pulsed heating: (i) studies of *single nanobubble* events and (ii) studies based on *ensemble measurements*, usually averaged over many nanoparticles in solution over a millimetric area in order to increase the signal to noise ratio and to avoid the implementation of microscopy means. The drawback of averaging many bubble events is that bubble generation is a stochastic process and all the bubbles do not appear at the same time around each nanoparticle in the fluid. Consequently, the evolution of the signal related to the bubble dynamics is necessarily widened and less temporally resolved compared to what really happens around a single bubble. For instance, it is not possible to evidence multiple after-bounces of a bubble.

A third, more fundamental classification can be made among the reported studies: the studies based on the use of *X-ray radiation*, and the studies based on *optical means*. The different configurations related to these two approaches are summarized in Figure 5.13. Here are some more details regarding these two approaches.

X-ray scattering and absorption spectroscopies are powerful approaches to break the diffraction limit inherent in visible light approaches. The counterpart is the necessity to perform the experiments in synchrotron facilities, which are not readily available for daily experiments. This may explain why so few studies have been reported using this approach. Actually, most of the experiments were conducted by the group of Plech [100, 60, 61, 101, 116]. Transient bubbles around gold nanoparticles have been studied by applying pulsed X-ray radiation to probe the response of plasmonic nanoparticle ensembles

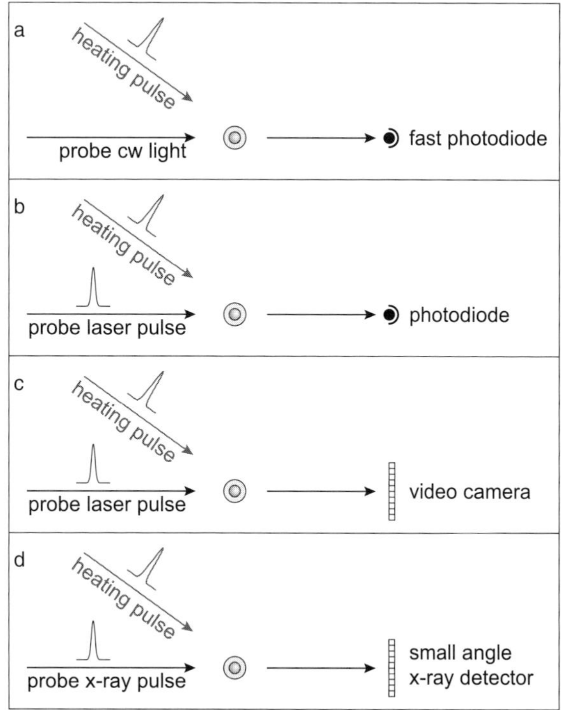

Fig. 5.13 Experimental approaches for studies of bubble dynamics generated by pulsed illumination. (a) Real-time imaging using a cw probe light and a fast photodiode recording scattering or extinction signal. (b) Pump probe approach based on a delayed probe laser pulse. (c) Time-resolved imaging, also called stroboscopic imaging, involving a pump probe approach using a video camera. (d) Pump probe approach based on the use of X-ray pulses and synchrotron facilities.

in solution excited by femtosecond-pulsed laser illumination. The beamline ID09B at the European Synchrotron Radiation Facility (ESRF) is coupled with a femtosecond-pulsed laser, which enables this kind of measurement. The X-ray and laser pulses are synchronized and delayed. The temporal resolution of this approach is given by the X-ray pulse duration, which is on the order of 100 ps. The observable is the scattered X-rays, which are recorded using a charge coupled device (CCD) camera with a phosphor screen (see Figure 5.13d). In the reported experiments, the CCD camera allowed for the detection of a cone of angle θ, corresponding to scattering vectors $Q = 4\pi/\lambda \sin^2(\theta/2)$ ranging from 0.01 to 7.5 Å^{-1}. This range of Q covers different techniques that are commonly termed small angle X-ray scattering (SAXS, for $Q < 0.1$ Å) and wide angle X-ray scattering (WAXS, for $Q > 0.1$ Å). X-rays can be sensitive to molecular correlations and disorder. Depending on the selected X-ray wavelength λ, and the considered Q range, this technique can be either sensitive to the atomic structure of the gold nanoparticles, to molecular correlations of the liquid or to the size of the bubbles. A typical X-ray scattering spectrum is displayed in Figure 5.14.

Small Q values (SAXS range) are related to large length scales and contain information regarding the bubbles. In particular, the frequency shift occurring at $Q < 0.1$ is a measure

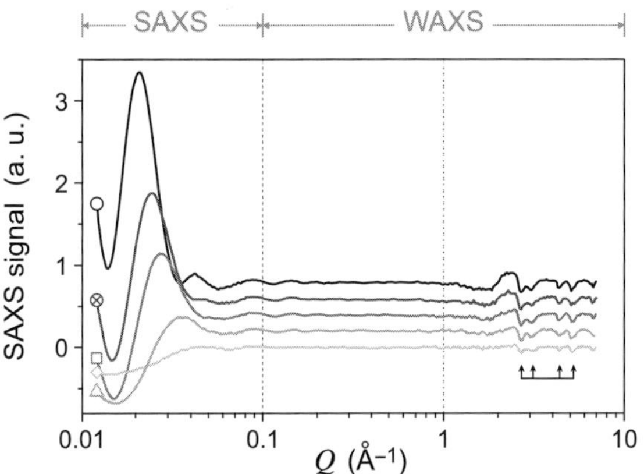

Fig. 5.14 SAXS signal as a function of Q at 300 ps delay after femtosecond pulse illumination of a solution of 9-nm gold nanoparticles. From bottom to top the applied laser fluence is increasing from 300 ($\diamond$), 375 ($\triangle$), 490 ($\square$), 560 ($\otimes$) to 750 J.m^{-2} ($\circ$). The dips marked by the arrows result from Bragg scattering by the atomic lattice of the nanoparticles. Reproduced from Reference [60]. Copyright 2005, AIP Publishing LLC.

of the average size of the bubbles. In parallel, the finer structures at $Q > 1$ (small length scales, WAXS region) contain information regarding the molecular correlation in the liquid and the atomic structure of the gold nanoparticles. In particular, the temperature of the nanoparticles can be retrieved from this part of the spectrum as the lattice dilatation subsequent to the temperature increase produces a frequency shift of Bragg peaks. In X-ray studies, such an observation is usually referred to as X-ray powder scattering (even though there is no powder in the experiment). It allows also for precise determination of nanoparticle morphology, including possible shape modification and melting of the nanoparticles [101]. Note that X-ray powder scattering from the gold nanoparticles and WAXS signal from the liquid molecules overlap in the reported scattering spectra, like in Figure 5.14.

X-ray scattering is thus particularly powerful to probe the system at arbitrarily small length scales. However, only ensemble measurements on many nanoparticles at a time can be done, which prevents from studying peculiar single bubble dynamics, such as multiple after-bounces of a bubble, for instance.

Using optical means is the second approach, which is much more widespread mainly because it does not require the large-scale facilities of X-ray scattering studies. As a nanobubble is a strongly scattering object due to its large refractive index difference compared to the surrounding liquid, nanobubbles can be easily detected by scattering or extinction measurements (see Figure 5.13(a-c)). And although the diffraction limit does not permit the direct assessment of nanobubble diameters, it remains possible to qualitatively evidence the presence of a bubble, and consequently to measure bubble lifetimes and fluence thresholds prior to bubble formation, the two other crucial parameters. Note that it is possible to retrieve nevertheless an estimation of the bubble radius if a calibration curve relating bubble radius and extinction is determined as conducted in Reference [116].

Two general kinds of measurements are usually conducted: (i) Intensity (extinction or scattering) measurements acquired in real time using fast detectors and a cw probe beam (see Figure 5.13a). This approach usually achieves a resolution on the nanosecond scale. (ii) A pump-probe approach where the probe beam consists of a delayed pulse of light and where the detector can be a photodiode, or even a CCD camera (see Figure 5.13b,c).

The Lapotko's group is very active in this field and reported on several studies involving optical pump-probe approaches, and heating in the picosecond-nanosecond range. For instance, in 2009, Lukianova-Hleb and Lapotko [81, 63] investigated the dynamics of plasmon-induced nanobubbles using light scattering imaging. They heated nanoparticles using pulsed laser illumination (532 nm, 0.5 ns) and illuminated the heated region using a pulsed laser (690 nm, 0.5 ns) delayed with respect to the pump pulse by 0.5, 1.5, 2.5 and 9 ns. This approach enabled them to catch instantaneous pictures of single nanoparticles at successive times with a nanosecond dynamics. The authors call this technique *pump probed delayed optical imaging*. The benefit is that it allows the study of the dynamics of *single* bubbles. Typical data using both pump probe approaches in parallel (video camera and photodiode) are displayed in Figure 5.15b.

In Reference [57], Katayama *et al.* used a slightly different configuration than the ones depicted in Figure 5.13. The authors used a white light (continuum) 15 ps pulsed illumination as a probe beam, a polychromator and a photodiode array. This way, they were able to record time-resolved spectra measurements instead of just the intensity (see Figure 5.16). The measured spectra could be fitted by numerical modelling of core–shell nanoparticle bubble to retrieve the evolution of the bubble diameter on the picosecond scale. Such an approach was only applied for ensemble measurements, not for single nanobubble studies, presumably due to the large required signal to obtain a spectrum. Figure 5.17 plots typical

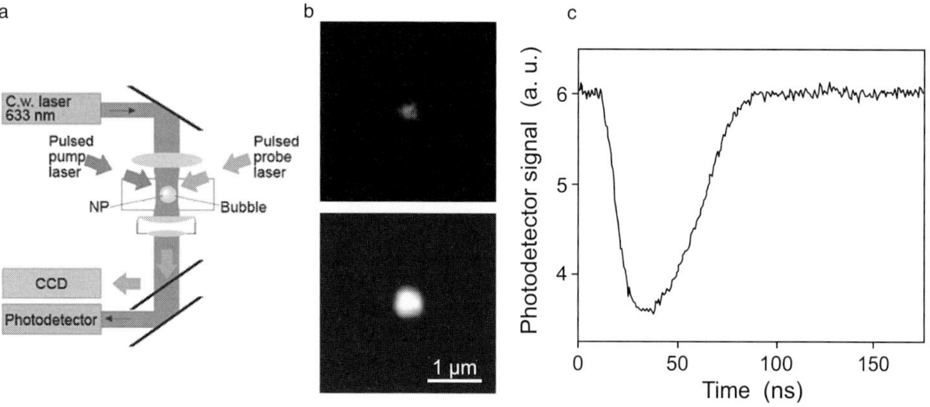

Fig. 5.15 (a) Experimental scheme for nanobubble generation and detection including two different approaches: pump probed delayed optical imaging (using a pulsed probe laser and a CCD camera) and a real-time recording of the optical extinction of nanobubbles induced by laser pulses by probed using cw illumination. (b) Optical scattering image of a gold nanoparticle cluster in water obtained with a pulsed probe laser before heating. (c) Optical scattering time-resolved image of a bubble generated on the same nanoparticle cluster by a single 70 ps pump pulse at a wavelength of 778 nm, probed with a delay of 10 ns. (d) Time response of the same nanobubble as shown in c was obtained with a cw probe laser at 633 nm. Reproduced from Reference [82]. Copyright 2012, American Chemical Society.

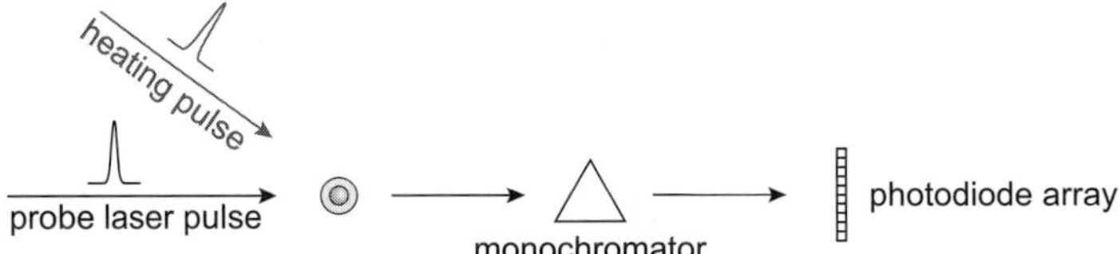

Fig. 5.16 Experimental approach for studies of bubble dynamics generated by pulsed illumination proposed by Katayama *et al.* in Reference [57].

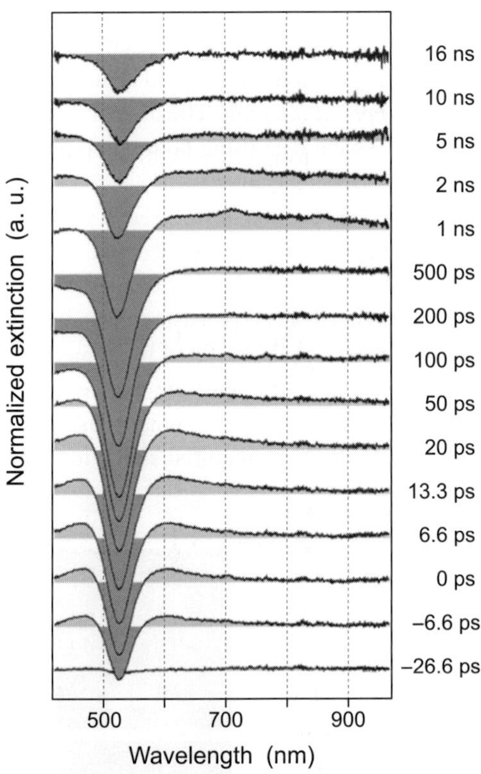

Fig. 5.17 Transient extinction spectra difference of 60 nm gold nanoparticles in water at various delays from the bottom to the top at the fluence of 11 J·m^{-2}. The excitation was provided by a 15 ps laser pulse at a wavelength of 355 nm. The dark areas represent bleaching (negative extinction), and the light areas represent positive extinction. Time delays are given on the right of each graph. The spectra were shifted with time delays from lower to upper. Reproduced with permission from Reference [57]. Copyright 2014, American Chemical Society.

measurements obtained using this approach. The plotted quantity is the normalized extinction $\log_{10}[I_0(\lambda)/I_\mathrm{p}(\lambda)]$ where $I_0(\lambda)$ and $I_\mathrm{p}(\lambda)$ are the extinction spectra without and under heating, respectively.

This was an overview of all the techniques that have been used to study bubble dynamics under pulsed illumination in plasmonics. Let us now present the main experimental results reported in the literature using these techniques. The three important parameters of bubble dynamics will be successively addressed, namely fluence threshold, bubble size and bubble lifetime.

5.6.3 Fluence Threshold for Bubble Formation

The fluence F is a physical quantity of utmost importance in any study related to plasmon-induced bubble experiments under pulsed illumination [87]. The fluence of an incoming laser pulse is defined as its energy per unit area. There are two common units equivalently used in the literature that are $J \cdot m^{-2}$ and $mJ \cdot cm^{-2}$. Apparently, the community did not reach an agreement regarding the use of a unit or another. In the following, all the fluence values will be converted to $J \cdot m^{-2}$.

An important question that is addressed in almost any article is the fluence threshold F_c required for bubble formation. Bubble formation is supposed to occur if the maximum temperature increase in the fluid:

$$T_{max} = \max_{\mathbf{r} \in \text{fluid}, t}[T(\mathbf{r}, t)] \tag{5.57}$$

reaches a certain value. Unlike what one could primarily think, bubble formation around nanoparticles *does not occur* when T_{max} reaches $100°C$. In theory, the fluence threshold F_c should be the fluence that yields a maximum temperature increase T_{max} equal to the spinodal temperature T_s of the surrounding fluid (see Figure 5.11c on page 168). For water, the spinodal temperature is $T_s = 277°C$. This process is often coined *spinodal decomposition* of water.

One can derive a simple estimation of F_c by saying that the whole amount of absorbed energy contributes to a temperature increase of the nanoparticle [6, 87]. The absorbed energy is $U = \sigma_{abs}F$. The temperature increase

$$\delta T_{max} = T_{max} - T_0, \tag{5.58}$$

where T_0 is the ambient temperature, is related to the absorbed energy by the relation $U = C_{th}\delta T_{max}$ where C_{th} is the thermal heat capacity of the nanoparticle.[14] Since $C_{th} = V\rho c_p$ one finally obtains an expression of the fluence threshold

$$\boxed{F_c^0 = (T_s - T_0)\frac{V\rho c_p}{\sigma_{abs}}.} \tag{5.59}$$

In Equation (5.59), the superscript 0 means that this expression involves some approximations (that will be detailed hereinafter). This calculation assumes a temperature continuity at the nanoparticle/fluid interface, so that the nanoparticle temperature equals T_{max}. One can even obtain a simpler expression of F_c by noticing that σ_{abs} is proportional to a^3 (see Chapter 1 on page 9). As V is proportional to a^3 as well, Equation (5.59) can be simplified to obtain an expression of F_c^0 that is independent of the nanoparticle diameter a.

---

[14] This statement assumes that the heat capacity is independent of the temperature. This approximation has to be used with caution. It is not valid in general, especially when δT is larger than a few tens of kelvins (see Appendix B). It can be used mostly to derive orders of magnitude.

$$F_{\text{c}}^{0} = (T_s - T_0)\frac{4\pi\rho c_{\text{p}}}{3\zeta}. \tag{5.60}$$

This expression stands for a universal number in the sense that it only depends on the nature of the nanoparticle and on the nature of the surrounding liquid. It does not depend on the diameter of the particle. For gold NPs in water at $T_0 = 20°C$, the ideal fluence threshold equals $F_{\text{c}}^{0} \approx 6.24 \ \text{J·m}^{-2}$.

Does this really mean that the fluence threshold for bubble generation does not depend on the nanoparticles size? Of course not. This would be in contradiction with experimental observations. The above discussion implied three approximations, which are sometimes valid but rarely all of them fulfilled at the same time [87]. Let us summarize:

Approximation $\mathcal{A}1$: *The timescale of nanoparticle cooling/heating is much longer than the pulse duration.* If the timescale of nanoparticle cooling (via diffusion in the surroundings) is faster (*i.e.*, smaller) than the pulse duration, the nanoparticle is cooling while the pulse is being absorbed, which makes the maximum temperature achieved in the nanoparticle weaker than the ideal temperature $\delta T_{\text{max}}^{0}$ derived in Chapter 2, Equation (2.127) on page 68. The required fluence threshold for bubble formation is therefore larger than F_{c} given by Equation (5.60).

Approximation $\mathcal{A}2$: *The absorption cross section is proportional to a^3.* This assumption is only true for small nanoparticles (typically below a few tens of nanometers). For larger nanoparticles, the absorption cross section is damped compared to the a^3 law (see Figure 1.7 on page 9). Consequently, the fluence threshold becomes larger than the ideal fluence threshold.

Approximation $\mathcal{A}3$: *No Kapitza resistance.* A finite value of thermal interface (Kapitza) conductance (see page 9) of the nanoparticle implies a discontinuity of the temperature at the nanoparticle/surroundings interface. In a steady state regime, a Kapitza resistivity does not affect the temperature of the surrounding fluid, but in the transient regime, the presence of a Kapitza resistivity weakens the fluid temperature at the vicinity of the particle, which, once again, implies a larger fluence threshold.

Figure 5.18 is intended to visually give a physical picture of the question of the fluence threshold F_s for bubble formation in plasmonics. Figure 5.18a plots F_s as a function of the diameter of a gold nanosphere in water. Interestingly, an optimum fluence threshold is observed for nanoparticle diameters around 60 nm. This optimum value results from two effects: (i) an increase of the required fluence for small nanoparticles, due to a fast heat release in the surrounding medium (approximation $\mathcal{A}1$ no longer valid) and (ii) an increase of the required fluence for large nanoparticles due to the fact that the absorption cross section is damped compared to the a^3 law (approximation $\mathcal{A}2$ no longer valid). Then, as a rule of thumb, the fluence threshold is always smaller using short pulses: Figure 5.18 shows that it is always easier to make a bubble using fs-pulsed compared to a ns-pulsed illumination.

Dashed lines in Figure 5.18 represent fluence profiles when some approximations have been done. In Figure 5.18a, the dashed line represent the case where no heat release to the surrounding medium is considered during the pulse absorption (no matter the pulse duration). One can observe that fs-pulsed illumination quasi-systematically achieves this

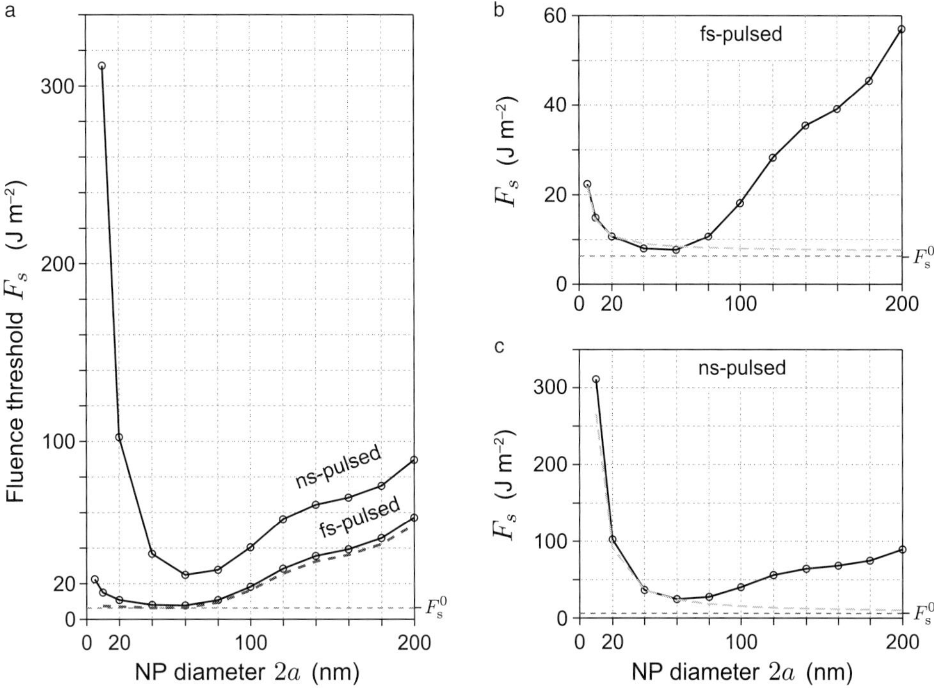

Fig. 5.18 (a) Fluence threshold calculated for a 1 ns pulse and for a femtosecond pulse (solid lines), as a function of the NP diameter. The dashed line represents the fluence threshold in the no-diffusion regime (approximation $\mathcal{A}1$). (b) Fluence threshold in the femtosecond regime (solid line) compared with the fluence threshold calculated assuming a linear relation between the absorption cross section and the NP volume (dashed line) approximation $\mathcal{A}2$. (c) Same as (b) in the case of a 1 ns pulse. In all these graphs F_s^0 values as defined by Equation (5.60) have been indicated using horizontal dotted lines. Reproduced with permission from Reference [87]. Copyright 2015, American Chemical Society.

regime, except for very small nanoparticles, typically smaller than 40 nm. However, under ns-pulsed illumination, this large shift between the dashed line and the nanosecond case means that this regime is never achieved. Dashed lines in Figures 5.18b,c represent the fluence profiles where the absorption cross section is supposed to be proportional to the volume of the nanoparticle (Approximation $\mathcal{A}2$), for any nanoparticle radius. The deviation observed for large nanoparticles evidences that the increase of the fluence threshold for large nanoparticle is due to a nonlinearity of the absorption cross section as a function of the nanoparticle volume.

Let us now look at what happens when Approximation $\mathcal{A}3$ is lifted. So far, no thermal interface resistivity $1/g$ at the metal–liquid interface was considered. Figure 5.19 plots the results of numerical simulations of the fluence thresholds in the fs and ns-pulsed regimes. Two values of surface conductivity have been considered: $g = 150\ \mathrm{MW \cdot m^{-2} \cdot K^{-1}}$, a common value [111, 3, 44], and $g = 50\ \mathrm{MW \cdot m^{-2} \cdot K^{-1}}$, a lower limit according to the literature [44]. The case $g \to \infty$ (no resistivity, as in Figure 5.18) was also represented. The presence of an interface resistivity has a strong influence on the fluence threshold in most

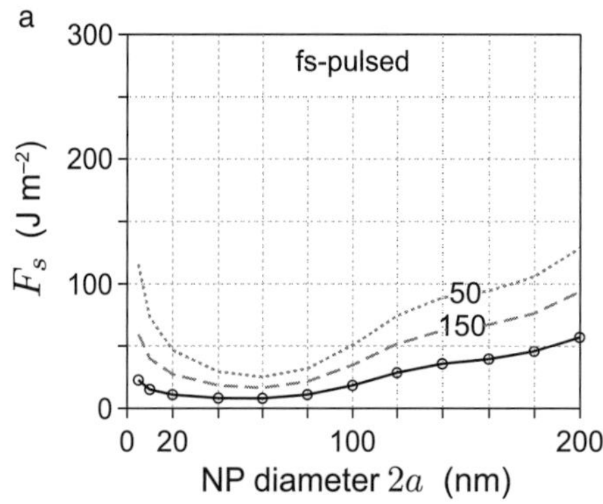

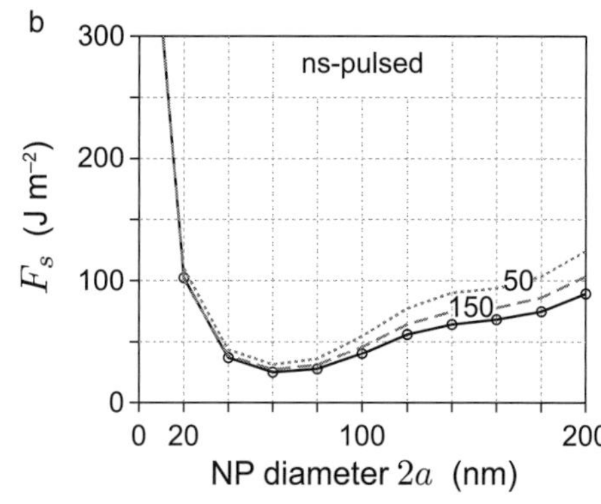

Fig. 5.19 (a) Fluence threshold calculated for a 1 ns pulse and for a femtosecond pulse (solid lines), as a function of the NP diameter. The dashed line represents the fluence threshold when a surface thermal resistivity $1/g$ is considered (Kapitza resistivity). Two cases have been represented: $g = 50$ and $g = 150$ MW·m^{-2}·K^{-1}. Reproduced from Reference [87]. Copyright 2015, American Chemical Society.

cases, except for small nanoparticles in the ns-pulsed regime. This case, characterized by a cooling time that is much shorter than the pulse duration, corresponds to a quasistatic regime where the temperature of the system varies adiabatically. In the static regime, a surface conductivity has indeed no effect on the fluid temperature. It only affects the inner nanoparticle temperature (see Figure 2.6 on page 48).

Because of the presence of a thermal interface resistance, bubble formation is sometimes associated with nanoparticle temperatures than can be particularly high, sometimes higher than the melting point of the nanoparticle itself (around 1000°C for gold). Besides, the nanoparticle is likely to undergo modification of its morphology (reshaping) during nanobubble formation events (see Section 5.11 on page 210).

After this theoretical and numerical description of the relation between fluence and bubble formation, let us discuss the experimental investigation.

In 2006, Kotaidis, Plech and coworkers demonstrated how the fluence threshold for bubble formation could be accurately measured for arbitrary small nanoparticles using X-ray scattering methods [61]. An example of results is reproduced in Figure 5.20.

In 2010, Lukianova-Helb *et al.* [80] experimentally investigated the fluence threshold dependence as a function of the nanoparticle diameter in the ns-pulsed regime ($\tau_p = 0.5$ ns) for a set various nanoparticle diameters, namely 10, 30, 60, 80 and 250 nm. They found a minimum fluence threshold for a NP diameter of 80 nm. Like in several other reported studies, they explained the strong fluence threshold increase for small NPs with surface tension considerations: according to them, overcoming the Laplace pressure would require higher energies for small NPs. This is not in agreement with the interpretation detailed in the previous paragraph, which rather evidences a faster energy release to the surroundings during the pulse duration. Laplace pressure cannot be at the origin of the absence of nucleation because any consideration of a Laplace pressure mechanism already supposes the presence of a bubble, which is paradoxical. Invoking Laplace pressure or surface tension for this process is misleading as there is no creation of liquid–gas interface, strictly speaking.

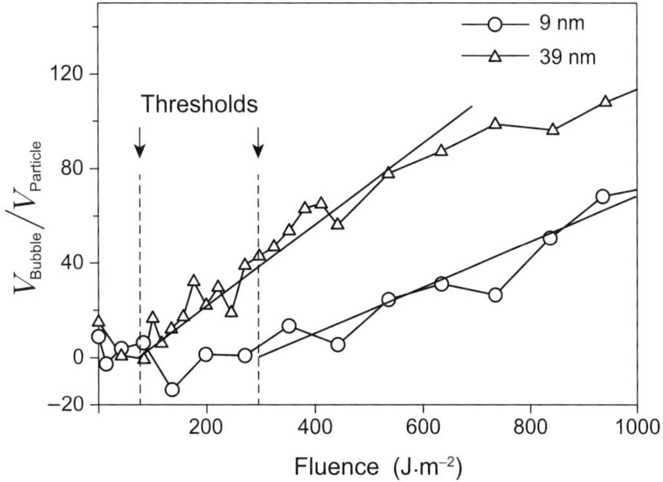

Fig. 5.20 Ratio of bubble volume to particle volume as a function of laser fluence for particle sizes of 39 and 9 nm. The dashed vertical lines indicate the threshold. Reproduced with permission from Reference [61]. Copyright 2006, AIP Publishing LLC.

In 2011, Siems *et al.* [116] investigated the fs-pulsed and ns-pulsed regimes. Supported by numerical simulations, the fs-pulsed regime was investigated experimentally by X-ray scattering and the ns-pulsed regime was investigated experimentally by optical means. In the femtosecond regime, they do not predict an increase of the fluence threshold for nanoparticles larger than 60 nm. Their theoretical fluence threshold profile remains monotonic up to $2a = 100$ nm. This result is in contradiction with the results of Reference [87], presented in Figure 5.18. This may be due to a simplified numerical scheme that considers only analytical expressions. Regarding their experimental measurements, the nanosecond pulse regime was investigated only up to a nanoparticle radius of 60 nm. Regarding the fs-pulsed regime, nanoparticle diameters between 50 and 90 nm have not been investigated and the measurements are rather dispersed. Consequently, the absence of fluence threshold minimum at 60 nm cannot be clearly evidenced by their experimental data.

In 2014, Katayama *et al.* [57] reported on an experimental study of bubble generation using a ps-pulsed illumination of 20–150 nm gold nanoparticles at $\lambda = 355$ nm. They conducted a rich discussion on the underlying physics of bubble formation, and in particular on the fluence threshold. They did observe a minimum fluence threshold at a nanoparticle diameter of 60 nm, as represented in Figure 5.21, in agreement with Figure 5.18 [87]. The occurrence of a non-monotonic shape of F_c with a minimum at 60 nm is also supported by the results of Cavicchi *et al.* [24], who investigated nanoparticle reshaping dependence on laser fluence. In the article, Katayama *et al.* [57] proposed that this non-monotonic dependence of F_c and this dip at 60 nm can be ascribed to the variations of the quantity $(\sigma_{abs}/V)/g_c$, where g_c is the critical thermal interface conductance [20]. This interpretation in not in agreement with the interpretation proposed in the above theoretical part. A thermal resistivity at the NP interface is not supposed to play a role in this non-monotonic shape. In Figure 5.18, a dip at 60 nm is observed even if no interface resistivity is involved.

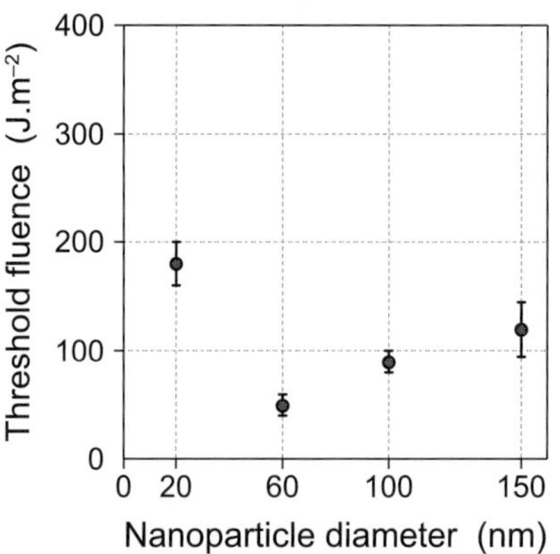

The use of g_c in the context of bubble nucleation in plasmonics is much more complicated than what was described in the literature. I invite the reader to refer to Reference [87] for a more detailed discussion of this issue.

5.6.4 Bubble Size

When illuminating a metal nanoparticle using pulse of light, as soon as the fluence of a light pulse is larger than the fluence threshold F_c (as depicted in the previous subsection), a bubble is bound to appear. The question addressed in this subsection concerns the size of this bubble, which can be an important parameter for some applications, for instance when the bubble is intended to disrupt a cell membrane or to generate a sound wave.

Due to their small size and their transient nature, bubbles formed around nanoparticles are difficult to detect and their sizes are even harder to quantify. Moreover, talking about "the size of the bubble" is ambiguous because a bubble is a dynamic object, characterized by a time-dependent volume and not necessarily spherical. The bubble radius is generally defined as the maximum bubble equivalent radius over time, although the *initial* bubble radius is occasionally discussed. Moreover, a single pulse-absorption event can be associated with a series of successive bubble formations and collapses, called multiple after-bounces, as depicted in Reference [18].

Two experimental approaches have been used to investigate the size of bubbles in plasmonics under pulsed illumination: X-ray scattering measurements and optical extinction measurements. In any case, measurements were averaged over ensembles of nanoparticles and performed using pump-probe approaches. On the one hand, X-ray scattering enables the detection of very small nanobubbles, but requires large-scale facilities, not available on a daily basis. On the other hand, optical techniques are easier to implement, but they cannot directly measure subdiffraction limited nanobubble sizes. Calibration curves have to be established, based on theoretical models of the scattering produced by a nanoparticle surrounded by a gas layer. In the following, I mention the three most important articles

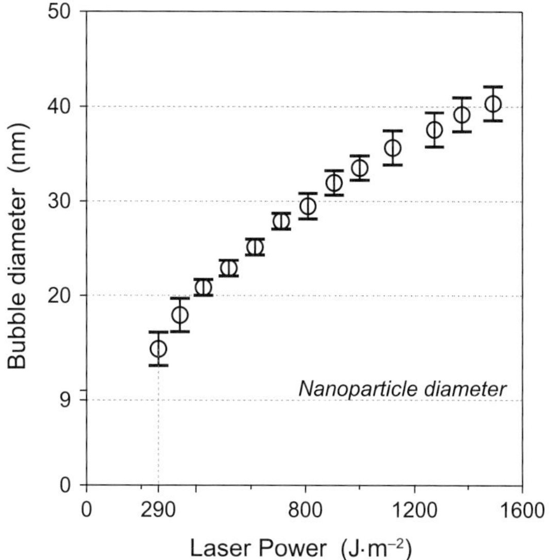

Fig. 5.22 Ensemble measurements of bubble diameters measured by SAXS around 9 nm nanoparticles as a function of the laser fluence. At a fluence of 290 J·m^{-2}, the first signature for bubble formation occurs, no bubbles are detected below that threshold. Reproduced from Reference [60]. Copyright 2005, AIP Publishing LLC.

dealing with the measurement of nanoparticle radii, in order to give a rough idea of how big nanobubbles can be under pulsed illumination, for different nanoparticle sizes and laser fluences.

In 2005, Kotaidis and Plech investigated the dynamics of nanobubbles around gold nanoparticles by X-ray scattering experiments, supported by numerical simulations [60]. Using a pump-probe technique, the authors were able to follow the averaged nanobubble radius as a function of time within a time window of 1 ns. X-ray scattering experiments enabled the measurements of very small radii. Typically, for spherical gold nanoparticles, 9 nm in diameter, a fluence threshold of 290 J·m^{-2} has been evidenced and a fluence of 1600 J·m^{-2} gave rise to nanobubbles with a maximum diameter of 40 nm (Figure 5.22).

In 2011, the same group [116] measured the size of bubbles around gold nanoparticles using time-resolved optical spectroscopy and X-ray scattering. The parallel use of these two techniques enabled the authors to write a comprehensive article addressing both the nanosecond and the femtosecond regimes. The corresponding measurements are reproduced in Figure 5.23. Measuring the size of bubbles using optical means does not benefit from the high spatial resolution of X-ray scattering experiments. Measuring the size of subdiffraction limit bubble is possible provided a calibration curve was determined. But the relation between scattering and bubble size for small bubbles (below 300 nm in diameter) is intricate and non-monotonic as explained by Plech [116].

In 2014, the group of Hashimoto also converted optical extinction signal into bubble diameters using a predetermined calibration curve [57], which proved to be non-monotonic as well. But much larger bubbles were investigated compared to the works of Plech. The average nanoparticle diameter was 60 nm and the bubbles were typically 200 nm big. The evolution of the bubble diameter as a function of time is represented in Figure 5.24.

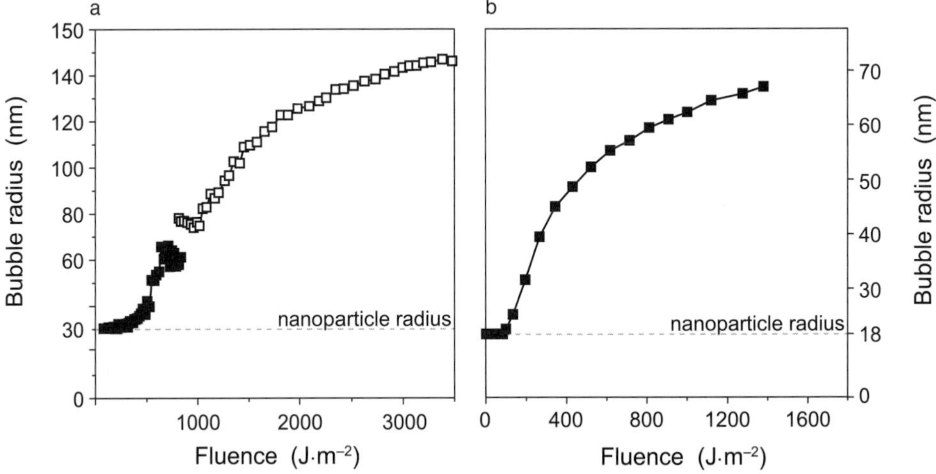

Fig. 5.23 (a) Maximum bubble radius of 60 nm particles excited at 355 nm with 3 ns pulses, determined by optical extinction measurements. (b) Maximum bubble radius of 36 nm nanoparticles excited at 400 nm with 100 fs pulses, determined by X-ray scattering (SAXS). Reproduced with permission from Reference [116]. Copyright 2011, IOP Publishing & Deutsche Physikalische Gesellschaft. CC BY-NC-SA.

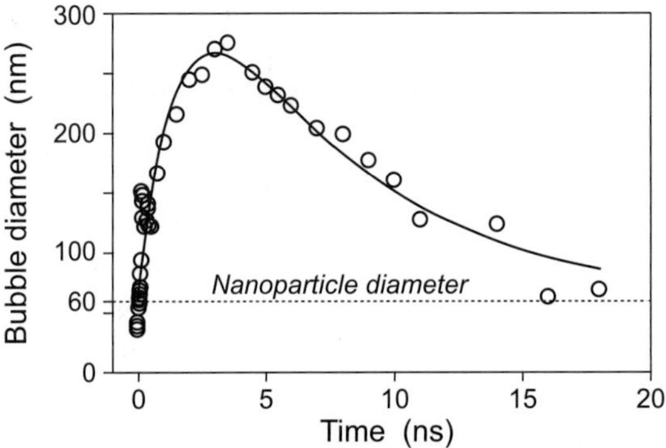

Fig. 5.24 Evolution of bubble diameter as a function of time. In these measurements, the nanoparticle diameter was 60 nm and the fluence was set to 50 J·m^{-2}. Reproduced with permission from Reference [57]. Copyright 2014, American Chemical Society.

All these works evidence that the size of nanobubbles under pulsed illumination is on the order of the size of the nanoparticle. This observation differs from the case of cw illumination, where the bubble can be much larger than the sources of heat, as is detailed further on.

5.6.5 Bubble Lifetime

If a pulse absorbed by a nanoparticle gives rise to a bubble, this bubble is bound to collapse over a timescale called the bubble lifetime. Besides fluence threshold and bubble diameter, bubble lifetime is the third physical quantity of interest in the study of bubble generation in plasmonics under pulsed illumination. Bubble lifetime is much easier to investigate compared with the bubble size (discussed in the previous subsection). No calibration curve is required as the measured signal does not need to be quantitative. For this reason, much more articles have investigated the question of the bubble lifetime.

Bubble lifetime depends mainly on the maximum size of the bubble achieved during its evolution (the larger the bubble, the longer the bubble lives). It was even demonstrated, experimentally and theoretically, that the bubble lifetime is proportional to its maximum diameter (and to the laser fluence) [80]. But this bubble diameter is not a physical quantity that can be directly controlled. It depends on the pulse duration and the absorbed power, and the absorbed power depends on the nature/morphology of the nanoparticle and the laser fluence. At the end, the bubble lifetime depends on many experimental parameters, which explains why very different values are reported in the literature and why they are difficult to compare with each other.

The evolution of the size of a single bubble can be intricate. First, the collapse of a bubble can be followed by multiple after-bounces [79]. Then, the profile of the bubble evolution is different in the fs/ps-pulsed regime and in the ns-pulsed regime: Under fs- and ps-pulsed illuminations, the profile is asymmetric. This was evidenced by Katayama *et al.* [57] (see Figure 5.24) and explained theoretically by Lombard, Biben and Merabia in References [74, 75]: while the expansion is found to be adiabatic, the collapse of a bubble is rather associated with an isothermal evolution. This gives an asymmetrical evolution of the bubble size as a function of time. However, under ns-pulsed illumination, the size of the bubble becomes time-symmetric.

The first report on the measurement of bubble lifetimes in plasmonics dates from 2005 with X-ray scattering measurements performed by Vassilios Kotaidis and Anton Plech [60]. The authors focused on 9 nm gold nanospheres (such small nanoparticles are rarely investigated using optical means, but do not cause any problem using X-ray scattering measurements). 20 nm bubbles have been detected and their lifetime has been estimated to ~ 400 ps under fs-pulsed illumination. The fluence threshold was 290 J·m^{-2}.

These figures markedly differ from the results reported by Lukianova-Helb and Lapotko. These researchers form Rice University have been very active in the field of plasmon-induced bubbles under ps- and ns-pulsed illumination and the use of optical detection means from 2008. In a series of articles [52, 64, 63], they investigated different nanoparticle morphologies using a thermal lens method: nanorods (14×45 nm^2), nanospheres (30 nm in diameter) and nanoshells (170 nm in diameter provided by the group of Halas). They stick to the ns-pulsed regime with pulse durations of 10 ns. The typical bubble lifetimes were around 100 ns (see Figure 5.25), *i.e.*, three orders of magnitude larger than the measurements of Kotaidis and Plech. As Lapotko is using optical detection means (which requires large bubbles to be detected) and longer pulse durations, the measured bubble lifetimes are naturally much longer than the lifetimes measured by the group of Plech, who is using X-ray scattering (which enables the detection of very small bubbles) and femtosecond excitation. As a rule of thumb, bubble lifetimes measured using optical means are usually longer than bubble lifetimes measured using X-ray scattering experiments.

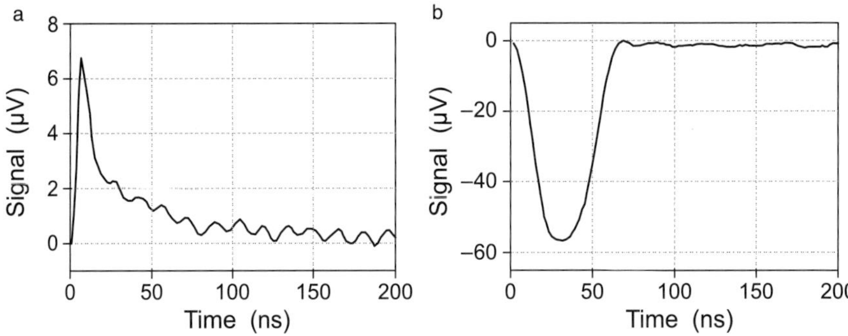

Fig. 5.25 Photothermal integral time responses of the 90 nm NP suspension as registered by a continuous probe laser beam during the application of a single pump pulse (532 nm, 0.5 ns): (a) fast heating and slow cooling of the suspension at a pump pulse fluence of 800 J·m^{-2} and below evaporation threshold; (b) bubble-specific signal of the opposite sign that shows the expansion and the collapse, a pump pulse fluence of 4000 J·m^{-2} is above the evaporation threshold. The ripple patterns were caused by the probe laser noise, and the x coordinate indicates the time from the moment of exposure of NPs to a pump laser pulse. Reproduced with permission from Reference [81]. Copyright 2009, American Chemical Society.

In 2012 and 2014, the group of Meunier investigated the bubble generation under fs-pulsed illumination using optical detection means [11, 62]. Measurements were not performed using microscopy means. Ensemble measurements were performed in a cuvette containing a solution of 100 nm gold nanospheres. Meunier *et al.* measured the variation in transmission due to the formation of bubbles, which scatter a probe beam crossing the cuvette. The typical time-scale of bubble lifetime was a few 100s of nanosecond. Although Meunier used a fs-pulsed illumination, just like Plech, the measured bubble lifetime was much longer for the reason mentioned above: the nanoparticles were much larger (100 nm compared to 9 nm) and larger bubbles need to be formed to create a measurable signal, hence the longer bubble lifetime. Meunier insisted on the asymmetry of the signal: "A very fast expansion in the order of 80 ns followed by a much longer $\sim$ 300 ns recovery phase (see Figure 5.26). This asymmetric growth and collapse phase contrasts with the time-symmetric bubbles observed using nanosecond laser with resonant particles [80]." The authors explained that this asymmetry is stemming from the presence of bubbles of different sizes and lifetimes within the laser focal volume, arising from the Gaussian fluence distribution of the pump laser. Note that Lombard *et al.* numerically evidenced an asymmetric evolution of the bubble size, even around a single nanoparticle. According to the authors, under femtosecond illumination, the expansion of a bubble is found to be adiabatic, while the collapse is best described by an isothermal evolution.

In 2010, the group of Lapotko mentioned that the vapor bubble lifetime is proportional to its maximal diameter.

Other articles have been published until recently. They reported also lifetimes on the order of a few 100s of nanoseconds for the largest nanoparticles and a few 10s of nanoseconds for the smallest [82, 57, 83].

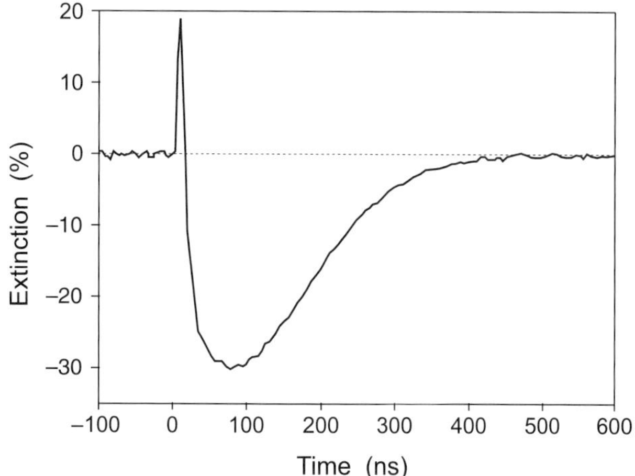

Fig. 5.26   Evolution of the light transmission through a cuvette containing 100 nm gold nanosphere subsequent to a fs-pulsed excitation. The variation is due to the formation of bubbles and enables an estimation of the averaged bubble lifetime. Reproduced with permission from Reference [11]. Copyright 2012, American Chemical Society.

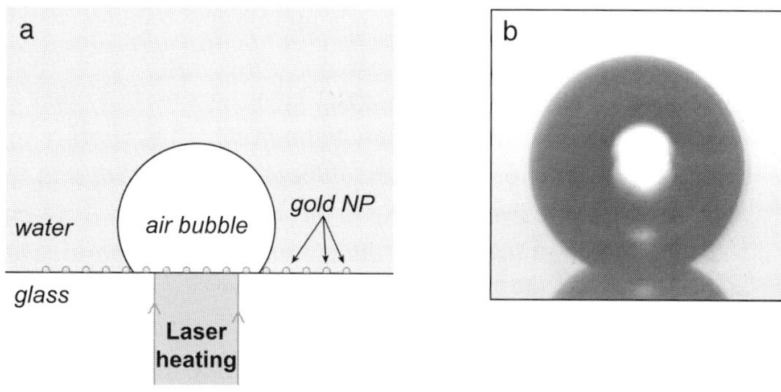

Fig. 5.27   (a) Schematic of the system under study in this section: a bubble formed at the interface between water and a solid substrate (usually glass) where gold nanoparticles have been deposited. (b) Side view image of a bubble. Reproduced from Reference [125]. Copyright 2017, American Chemical Society.

5.7 Bubble Formation and Dynamics under CW Illumination

In this section we consider the formation of a bubble stemming from gold nanoparticles lying on a planar substrate (not nanoparticles dispersed in solution). This is indeed the most general approach when making bubbles from metal nanoparticles using cw illumination. Here is a schematic of the system:

5.7.1 Scenario of Bubble's Life

When illuminating metal nanoparticles with a sufficiently strong laser irradiance (power per unit area), bubble formation can occur in the vicinity of the nanoparticles. As already mentioned in Subsection 5.4.2, the scenario of a bubble evolution under cw illumination is different compared with a pulsed illumination. Under cw illumination and in water, bubble formation is supposed to occur around 230° [7]. As explained in Section 5.4, water can remain in a metastable state above its boiling point until a temperature of 277°C, named the spinodal temperature, where the energy barrier between the liquid and gas states vanishes. Yet, in practice, bubble formation occurs at lower temperature, around 230°C, because at that temperature the energy barrier can be crossed due to thermal fluctuations. This temperature is sometimes coined the *kinetic* spinodal temperature.

Once the bubble is formed, the physics is also very different from the case of pulsed excitation. Under cw illumination in water, the bubble does not disappear. If the nanoparticle is located on a glass substrate (which is the common situation under cw illumination), the remains forms and remains at the glass–water interface. Due to the extended presence of the bubble, a singular mechanism occurs: dissolved gases in the liquid tend to evaporate within the bubble and fill its interior, namely dioxygen and dinitrogen. Molecular diffusion in liquids is quite slow. This filling of the bubble occurs on the second timescale. This is why it never happens under pulsed illumination where the bubble remains exclusively composed of steam. When the heating (the laser) is stopped, a collapse phase begins. The presence of gases (and not only steam) in the bubble makes this collapse quite long, on the order of a few seconds up to a few minutes, depending on the initial bubble size.

This scenario is related to a water environment. For other liquids, the scenario can be different (degassed water, alcanes, alcohol, etc.). The absence of dissolved gasses (although difficult to ensure) can make the collapse much faster, and the use of other liquids can prevent the bubble from sticking to the substrate [53].

The following subsections will depict in detail the physics of the different phases of the bubble dynamics under cw illumination.

5.7.2 Air Bubble Instability

Let us consider a liquid–gas flat interface. The gas phase is characterized by a pressure P and the liquid is characterized by a concentration c of dissolved molecules from the gas phase. At equilibrium, the concentration of dissolved molecules in the liquid phase obeys Henry's law:

$$P = K_{\mathrm{H}}\, c. \tag{5.61}$$

Let us now consider that a bubble of the same gas is present in the liquid phase. In order for the bubble to be at equilibrium, Henry's law would have to be fulfilled at the bubble–liquid interface as well:

$$P + \Delta P = K_{\mathrm{H}}\, c \tag{5.62}$$

where ΔP is the Laplace pressure. However, the two latter equations cannot be valid at the same time because of the Laplace overpressure $\Delta P \neq 0$ in the bubble. The direct consequence is the occurrence of a molecular flux through the interface of the bubble from the bubble to the liquid, creating a shrinkage of the bubble and a disappearance.

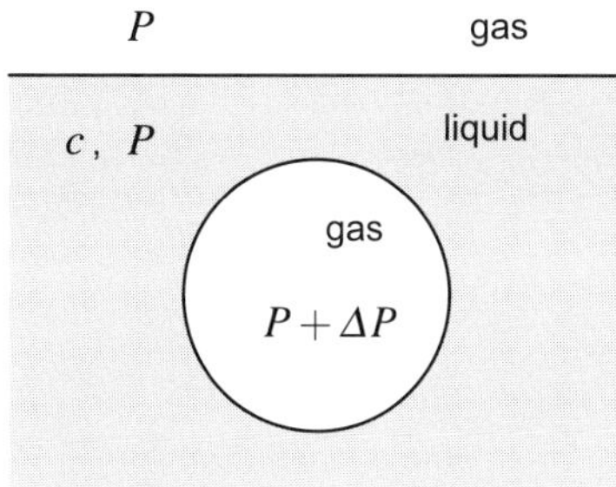

Fig. 5.28 System composed of a gas–liquid flat interface at equilibrium, and a bubble in the liquid phase. Molecules are present in all the phases of the system. They are characterized by a molecular concentration c in the liquid and by partial pressures P and $P + \Delta P$ in the gas phases.

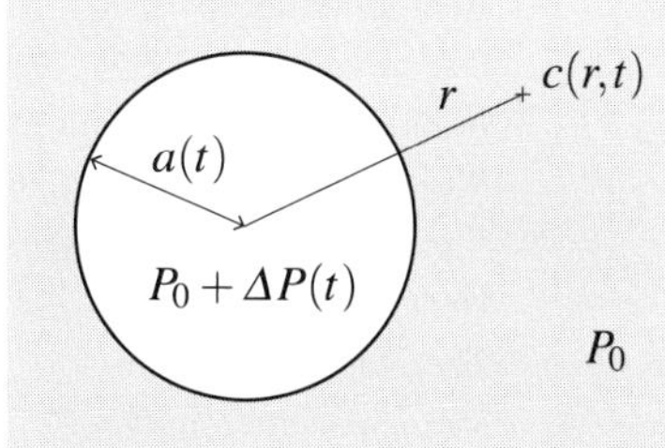

Fig. 5.29 Model with a spherical symmetry used to investigate the lifetime of an air bubble.

This is the reason why the aforementioned presence of *stable* nanobubbles on hydrophobic surfaces goes against the common sense (see page 170).

Note that this discussion assumes the liquid to be at equilibrium with the surrounding air. This happens only if the liquid has remained in contact with the ambient air for a sufficiently long time, and at the same temperature and pressure as during the experimental observation.

5.7.3 Air Bubble Shrinkage and Lifetime

The fact that air bubbles are not stable does not mean that they instantaneously shrink. On the contrary, depending on their size, they can last for seconds, minutes or even hours. The origin of this slow shrinkage is that the molecules escaping from the bubble to the liquid subsequently have to diffuse away in the liquid. As molecules do not diffuse that fast in liquids, an accumulation of molecules occurs in the vicinity of the bubble, which slows down the exhaustion of molecules in the bubble. What determines the bubble lifetime is therefore a subtle interplay between Laplace pressure in the bubble that tends to expel the molecules and molecular diffusion in the liquid.

This process can be analytically described if we assumed a spherical symmetry where a gas bubble stands in a uniform and infinite liquid environment as represented in Figure 5.29. The molecular concentration in the liquid is named $c(r, t)$. If the gas inside the bubble is considered ideal, it follows the ideal gas law

	Table 5.3 Notations used in this section.	
Name	**Description**	**Dimension**
r	Radial coordinate	m
t	Time	s
a	Bubble radius	m
V	Bubble volume	m^3
d	Laser beam diameter	m
P_b	Bubble inner pressure	Pa
P_0	Ambient pressure	Pa
N	Amount of gas molecules in the bubble	mol
T	Temperature	K
Q	Absorbed power by NPs	W
Δt	Heating duration	s
c	Gas molecule concentration in the liquid	$mol \cdot m^{-3}$
c_∞	Initial molecular concentration in the liquid	$mol \cdot m^{-3}$
J	Molecular flux density in the liquid	$mol \cdot m^{-2} \cdot s^{-1}$
γ	Surface tension of the liquid/gas interface	$J \cdot m^{-2}$
K	Henry coefficient	$J \cdot mol^{-1}$
D_T	Thermal diffusivity	$m^2 \cdot s^{-1}$
D	Molecular diffusivity	$m^2 \cdot s^{-1}$
D_s	Soret coefficient	$m^2 \cdot s^{-1} \cdot K^{-1}$
R	Gas constant	$J \cdot mol^{-1} \cdot K^{-1}$

$$P_b V_b = N R T_\infty \tag{5.63}$$

where $V_b = 4\pi a^3/3$ is the bubble volume, N the amount of gas molecules in the bubble (in mole), R is the gas constant and T_∞ is the bubble temperature, which is uniform in the whole system as no heating is performed. With these notation, the Laplace relation (5.53) yields

$$P_b(t) = P_0 + \Delta P(t) = P_0 + \frac{2\gamma}{a(t)}. \tag{5.64}$$

Using these two relations, one gets the constitutive equation relating the gas amount within the bubble and its radius:

$$N(t) = \frac{4\pi \, a^3(t)}{3\,R\,T_\infty} \left(P_0 + \frac{2\gamma}{a(t)} \right). \tag{5.65}$$

The molecule variation rate noted $\dot{N} = dN/dt$ reads thus

$$\dot{N} = \frac{4\pi \, a^2 \, \dot{a}}{R\,T_\infty} \left(P_0 + \frac{4}{3}\frac{\gamma}{a} \right) \tag{5.66}$$

where the (r, t) dependences of N and a have been omitted for the sake of clarity. The spatiotemporal variations of molecular concentration in the liquid $c(r, t)$ are not only governed by a simple diffusion process (Fick's law). They are also affected by the variation of the bubble size, which induces a centripetal overall fluid motion. The equation governing the gas concentration within the liquid is thus [73]

$$\partial_t c = \frac{D}{r^2}\partial_r(r^2\partial_r c) + \frac{a^2\dot{a}}{r^2}\partial_r c \tag{5.67}$$

where D is the molecular diffusivity in the liquid. In the right-hand side of Equation (5.67), the first term represents diffusion and the second term expresses the fact that the bubble pushes the surrounding fluid away upon expanding. In the context of this study, one can consider that the diffusion process is dominant, and neglect the second contribution. This assumption can be justified by dimensional analysis of the equation (see Appendix A) and based on the experimental observations: the order of magnitude of the ratio of diffusive term by the convective term is on the order of $D/(a\,\dot{a}) \gg 1$. This amounts to making a quasistatic approximation where the variation of the bubble size is slow enough in comparison with the timescale of the molecular diffusion processes. Hence, if the convective term is discarded in Equation (5.67), one ends up with a regular diffusion equation and the solution simply reads:

$$c(r,t) = c_\infty + \frac{c(a,t) - c_\infty}{r}\,a(t). \tag{5.68}$$

The boundary condition at the bubble interface is directly related to the inner bubble pressure P via Henry's law:

$$P_{\mathrm{b}}(t) = K_{\mathrm{H}}\,c(a,t). \tag{5.69}$$

Since $P_0 = K_{\mathrm{H}}\,c_\infty$, and using Equations (5.53) and (5.69), one gets

$$c(r,t) = c_\infty + \frac{2\gamma}{K_{\mathrm{H}}\,r}. \tag{5.70}$$

Surprisingly, the molecular concentration in the liquid is not dependent on the size of the bubble, *i.e.*, not dependent on time. This means that, during the whole evolution of the bubble shrinkage, the concentration profile $c(r,t)$ will not be affected. Thus, one can write $c(r,t) = c(r)$. Only its domain of definition $[a(t),\infty]$ will be time-dependent. The concentration profile is plotted in Figure 5.30a.

Let $\mathbf{J}$ be the molecular flux density vector in the liquid. $\mathbf{J}$ is radial and its amplitude reads by definition:

$$J(r) = -D\,\partial_r c(r) \tag{5.71}$$

$$J(r) = \frac{2\gamma D}{K_{\mathrm{H}}r^2}. \tag{5.72}$$

By considerations on mass conservation, the molecular variation rate $\dot{N}$ of the bubble can be expressed as a function of $J(a)$:

$$\dot{N} = -4\pi a^2\,J(a) \tag{5.73}$$

$$\dot{N} = -\frac{8\pi\,\gamma\,D}{K_{\mathrm{H}}}. \tag{5.74}$$

Interestingly, the delivering rate of gas molecules $\dot{N}$ from the bubble to the surrounding liquid turns out to be constant during the bubble shrinkage. This is the origin of the

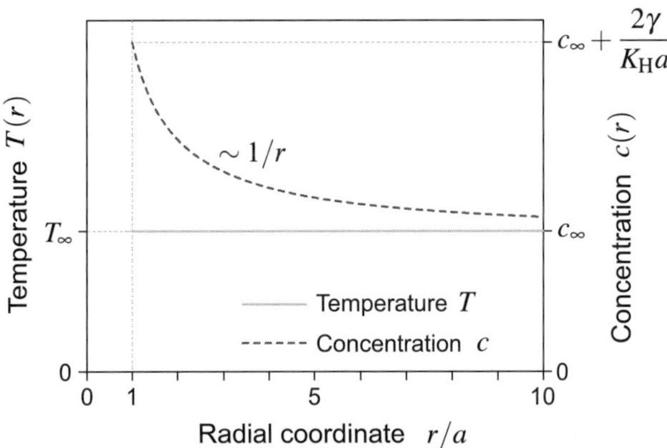

Fig. 5.30   Spatial profiles of the temperature and molecular concentration during bubble shrinkage at ambient temperature. Reproduced with permission from Reference [7]. Copyright 2014, American Chemical Society.

fact that $c(r)$ is constant, since $c(r) = c_\infty + \dot{N}/(4\pi D r)$. From the two expressions of the molecule variation rates obtained in Equations (5.66) and (5.74), we finally obtain the differential equation governing the evolution of the bubble radius:

$$\dot{a} = -\frac{2\gamma\, D R T}{K\, a^2 \left(P_0 + \dfrac{4}{3}\dfrac{\gamma}{a} \right)}. \tag{5.75}$$

Two regimes can be expected depending on which term is dominant between P_0 and γ/a. In water, the transition between these two regimes is obtained for $a \approx 500$ nm. Since the purpose of this work is to investigate long-lived bubbles that are at least a few micrometer big in radius, we can discard the second term. Note that discarding this second term does not mean that the Laplace pressure does not play a role. Indeed, γ is still present in the simplified differential equation. In this approximation, the solution of Equation (5.75) reads:

$$a(t)^3 = -\frac{6 R T D \gamma}{P_0\, K}\, t + \text{const.} \tag{5.76}$$

From this equation, the lifetime of a bubble can be simply estimated:

$$\tau_{\mathrm{MB}} = \frac{P_0\, K}{6 R T D \gamma}\, a^3. \tag{5.77}$$

The bubble lifetime scales thus as a^3, *i.e.*, the volume of the bubble, as observed experimentally (see Figure 5.33c). Note that in this model, the presence of the substrate is not taken into account. Yet, the presence of a surface is supposed to affect Equation (5.73) since the bubble surface will be smaller than $4\pi a^2$, and to affect Equation (5.68) as well, since the surrounding medium no longer features a central symmetry. Taking these corrections into account would add a correction factor to the estimated lifetime, but would not change the power-three dependence.

5.7.4 Air Bubble Steady State upon Heating

Experimentally, bubble shrinkage can be avoided if laser heating is maintained. The purpose of this subsection is to describe this steady state. To simplify the model, we still consider spherical symmetry, and a heat power Q generated within the bubble. Due to this heat generation, the temperature is no longer supposed to be uniform in the medium, which will affect the molecular diffusion in the liquid by thermophoresis. The new molecular flux density vector now reads (see Equation (5.13) on page 147):

$$\mathbf{J} = -D\,\nabla c - D_{\mathrm{T}}\,c\,\nabla T. \tag{5.78}$$

Thermophoresis is a process that induces a diffusion of dissolved species in a liquid due to a temperature gradient (see Section 5.3 on page 146). Most of the time, dissolved species tend to move toward colder regions ($D_{\mathrm{T}} > 0$). As we consider a steady state, $\dot{c} = -\nabla \cdot \mathbf{J} = 0$, which yields

$$\nabla \cdot (-D\,\nabla c) = \nabla \cdot (D_{\mathrm{T}}\,c\,\nabla T) \tag{5.79}$$

$$-D\,\nabla^2 c = D_{\mathrm{T}}\,c\,\nabla^2 T + D_{\mathrm{T}}\,\nabla c \cdot \nabla T. \tag{5.80}$$

As the steady state temperature profile is governed by the Laplace equation, $\nabla^2 T = 0$ and $T(r) = Q/(4\pi\kappa r)$. Using this expression in Equation (5.80) leads to the differential equation governing the molecular concentration around the bubble, under illumination:

$$\frac{1}{r}\partial_r[r\partial_r\, c(r)] = S_T\frac{Q}{4\pi\,\kappa\,r^2}\partial_r c(r). \tag{5.81}$$

The solution of this equation fulfilling the boundary conditions $\lim\limits_{r\to\infty} c(r) = c_\infty$ and $\lim\limits_{r\to\infty} \partial_r c(r) = 0$ is

$$c(r) = c_\infty \exp\left(-\frac{S_T\,Q}{4\pi\,\kappa\,r}\right) \equiv c_\infty\, e^{-r_{\mathrm{d}}/r} \tag{5.82}$$

where $S_T = D_T/D$ is the Soret coefficient and where

$$r_{\mathrm{d}} = \frac{S_T\,Q}{4\pi\,\kappa} \tag{5.83}$$

is the depletion length of the molecular concentration around the bubble induced by thermophoresis.

As a conclusion, during heating, the steady state is characterized by a molecule depletion in the vicinity of the bubble, as represented in Figure 5.31. This figure has to be compared with Figure 5.30 of the previous section that represents the same quantities upon stopping heating (during the bubble shrinkage) and showing the opposite effect: a molecule accumulation is observed around the bubble.

We can now describe and explain the recent experiments in plasmonics based on bubble generation under continuous wave illumination. Plasmon-induced micro-bubbles under cw illumination are usually generated on a solid substrate covered with gold nanoparticles, as represented in Figure 5.32.

Upon shining the sample immersed in water with a focused cw laser, a bubble appears and remains located at the solid–water interface where the nanoparticles are located. The

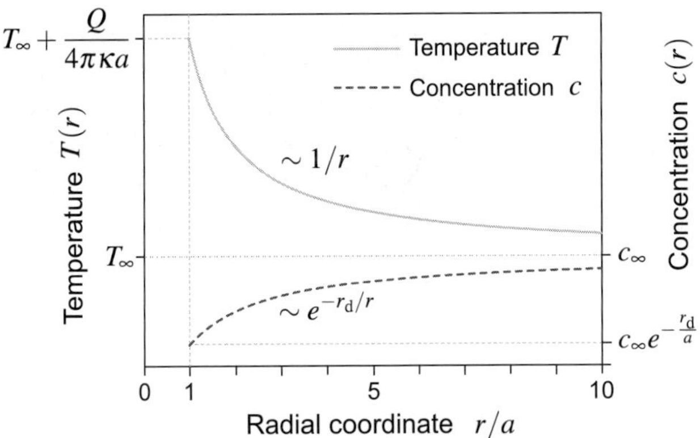

Fig. 5.31   Spatial profiles of the temperature and molecular concentration upon heating, in the steady state. Reproduced with permission from Reference [7]. Copyright 2014, American Chemical Society.

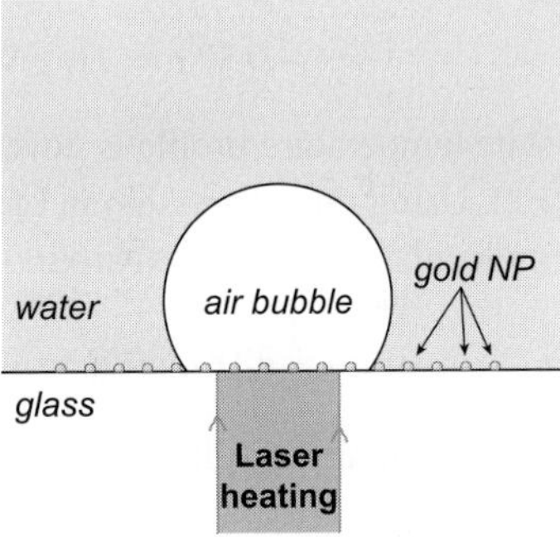

Fig. 5.32   Typical experimental configuration when investigating bubble generation in plasmonics under continuous wave illumination.

bubble nucleation occurs in a superheated liquid and is primarily due to boiling. The primary nature of the bubble is thus steam. If air molecules are present in the surrounding liquid (which is always the case if no particular treatment is conducted), air molecules are bound to diffuse from the liquid to the bubble, which creates a slow swelling of the bubble over a few seconds until a steady state is reached. At this point, the bubble is predominantly composed of air molecules (O_2, N_2, CO_2, etc.).

When the bubble appears, the heated gold nanoparticles are no longer located at a water–glass interface, but a gas–glass interface. The surrounding thermal conductivity of the nanoparticles is thus reduced, which results in a higher local temperature compared to the temperature prior to bubble formation. However, as soon as there is a bubble, superheating is no longer possible and the temperature is bound to remain below 100°C anywhere in the liquid. Only within the bubble can the temperature be higher than 100°C.

When the heating laser is turned off, the temperature increase rapidly disappears. This creates an instantaneous re-condensation of the steam of the bubble along with a slight shrinkage of the bubble. However, for the reason explained above, the air inside the bubble cannot disappear instantaneously and, depending on the size of the bubble, the

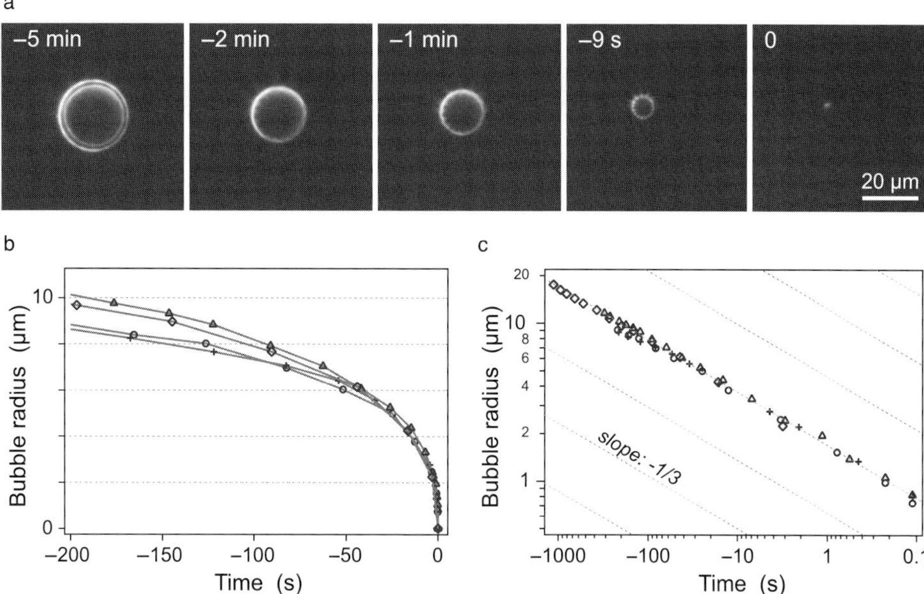

Fig. 5.33 (a) Dark field images of a micro-bubble generated by optical heating of a layer of gold nanoparticles on a glass substrate, at successive times during shrinkage. (b) Evolution of the bubble radius $a(t)$ as a function of time for four different bubbles. The time origin is the bubble disappearance. (c) Same graph with a logarithmic scale. Reproduced with permission from Reference [7]. Copyright 2014, American Chemical Society.

shrinkage can last several minutes to several hours, before the bubble vanishes. Such a plasmon-induced bubble shrinkage has been investigated in Reference [7]. The geometry of the experiments was that represented in Figure 5.32. A power-three scaling of the bubble lifetime as a function of the bubble radius has been evidenced according to Equation (5.76). Measurements are reproduced in Figure 5.33.

5.7.5 Air Bubble Growth under Illumination

In this subsection, we still consider the common case where the bubble is formed from gold nanoparticles deposited on a planar substrate. In this case, the bubble forms at the interface between water and the substrate, and remains there. When heating is performed with the laser at a sufficient power (*i.e.*, when the temperature exceeds $\sim 230°C$), a bubble forms and subsequently grows if the laser is not turned off. This expansion phase is slow and governed by the diffusion of air molecules from the liquid to the inside of the bubble as explained in the previous subsection. This physics of the bubble expansion is described by the group of Detlef Lohse in Reference [125]. Using a fast-imaging side-view microscopy configuration, the research group investigated different parameters, such as the microbubble contact angle, the footprint diameter and the radius of curvature, for two solvents: air-equilibrated water (AEW) and degassed water (DGW).

The main result of this study is reproduced in Figure 5.34. Two cases are compared: the AEW and the DGW cases. At short timescales (below a few 100s of ms), both cases exhibit

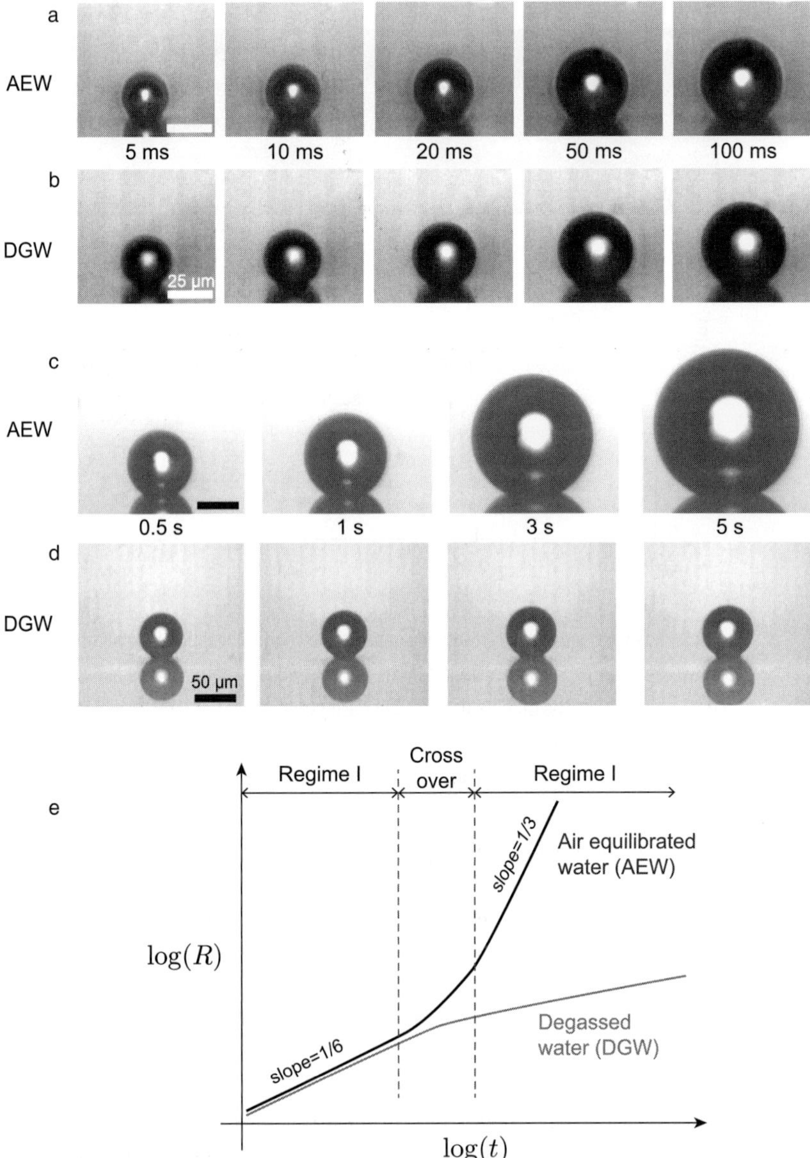

Fig. 5.34 The first two rows of snapshots show the short-term plasmonic bubble dynamics between 5 ms and 100 ms after illumination for a plasmonic bubble in air-equilibrated water ((a), AEW) and in degassed water ((b), DGW). The scale bar is 25 µm. Hardly any difference is seen. The second two rows of snapshots show the long-term plasmonic bubble dynamics between 0.5 s and 5 s after laser illumination for a plasmonic bubble in air-equilibrated water ((c), AEW) and in degassed water ((d), DGW). The scale bar is 50 µm. (e) Illustration of the dynamics of the bubble radius $R(t)$ for air-equilibrated water and degassed water. The microbubble growth process can be divided into two phases, with a crossover in between. Reproduced with permission from Reference [125]. Copyright 2017, American Chemical Society.

a similar expansion phase. However, from around 500 ms, the expansion of the bubble AEW is much faster. This is due to the presence of air molecules (O_2, N_2, etc.) that flow inside the bubble, as mentioned in the previous subsection. Interestingly, the growth of the bubble features the same 1/3 slope coefficient as in the case of the bubble collapse (see Figure 5.33c).

5.7.6 A Recent Research Thematic

The investigation of bubble formation around plasmonic nanoparticles dates from the early 2000s. Surprisingly, the studies focused on pulsed illumination. It was necessary to wait until 2012 to observe the first studies of bubble formation in plasmonics *under cw illumination*. In four years, around 10 articles have been published in this area of research and the main envisioned application is solar steam generation.

In 2012, the group of Richardson reported on the bubble formation around single gold nanoparticles, typically 300 nm big [21]. They couldn't directly measure the temperature threshold for bubble formation because the spatial resolution of their temperature microscopy technique was diffraction limited. The measured temperature increase was only 58°C, much weaker than the boiling point of water or the spinodal temperature. To take into account this lack of resolution, the authors derived that they had to multiply their measured temperature values by a factor of 12.8. Using this procedure, Richardson reached an estimation of the nanoparticle temperature of 320°C, close to the spinodal temperature of 277°C.

In 2013, the group of Halas published a study on the generation of steam from a solution of gold nanoparticles in a cuvette illuminated by solar light focused by a Fresnel lens [90]. Although this work is highly cited (256 times in January 2017), it contains several misleading statements that deserve clarification. For instance, the very first sentence of the abstract reads "Solar illumination of broadly absorbing metal or carbon nanoparticles dispersed in a liquid produces vapor without the requirement of heating the fluid volume." Heating the nanoparticles without heating the fluid is not possible in their configuration using cw illumination. In their experiments, the authors are experiencing collective effects as described in Subsection 2.2.5, but the group was not aware of this effect at that time. Because they did not take into account this effect in their theoretical predictions, they estimated a fluid temperature increase of only 0.04°C, hence their misleading conclusions. Then, what the authors are experiencing is perhaps not boiling, but just evaporation. Evaporation is known to be strong when reaching temperature close to the boiling temperature of water. The loss of water mass in their experiments should stem from that. I personally believe that the physics behind this article is not going further than the regular evaporation of a black solution produced by heating using light. Consideration of nanoscale effects is not relevant in this study.

Still in 2013, the group of Halas performed the same kind of measurements, but with nanoparticles deposited on a glass substrate [38]. The authors were surprised by the "much larger light intensities than the ones used in the nanoparticle-enabled solar steam generation experiment" and gave a wrong interpretation of this observation. Once again, this misunderstanding originates from the fact that the authors were not aware of the physics of thermal collective effects in plasmonics. Heating a black solution is naturally much more efficient compared to heating a coverslip exhibiting only a few percent absorption. One should not think in a "single-nanoparticle" manner when thermal collective effects occur. Moreover, in their reasoning, they consider boiling at $100°C$ while it is supposed to occur around $230°C$ in this configuration (as explained in Subsection 5.4.1 on page 162).

In 2013, the group of Halas pursued their efforts on solar heating of nanoparticles and steam generation for another application [89]. The idea was based on solar heating of gold nanoparticles for sterilization of medical instruments. Once again, the authors insisted on the use of nanoparticles, but the physics of their study did not go beyond the heating of a black solution using light. Any black solution would have led to the same heating process. Using gold nanoparticles as nanosources of heat usually makes sense when a small number of nanoparticles is involved. Otherwise, simply using black paint will certainly do the job. In the study reported by Halas, using *black* medical instruments would certainly be as efficient and much less complicated than using gold nanoparticles.

In conclusion, this series of articles published by the group of Halas in 2013 certainly contributed to the arousal of interest in bubble generation under cw illumination within the plasmonics community [102]. However, many questionable statements are scattered throughout these papers and I consider it important to clarify what can be said and what should not be claimed. Here is an overview of the important, counterintuitive processes that may be misleading. When shining a solution of metal nanoparticles with solar light,

- there is no temperature difference between the nanoparticle and the surrounding fluid due to any reduced thermal conductivity at the metal–liquid interface. The temperature of the metal is the same as the temperature of the fluid and can be considered continuous throughout the fluid. This is due to thermal collective effects (see Section 2.2.5).

- the temperature increase around the nanoparticles is not $0.04°C$. The calculation leading to this value considers that the sunlight is heating only one nanoparticle, while it is heating millions of them. This issue was pointed out by Ni *et al.* in 2015 [91]. Quoting these authors: "models of a single nanoparticle in an infinite medium have been considered [6, 42]. However, this ignores the heating effects of nearby nanoparticles in a real fluid and is only valid for short timescales where the individual heating profile has not reached the neighboring particles [46, 71]."

- Under cw illumination, bubbles in plasmonics are not made of steam, but of air (see Section 5.7.4 on page 195).

- there is no formation of a several-micrometer-thick steam shell around the nanoparticles within a few microseconds. The bubbles are not microscale objects with gold nanoparticles suspended at their centre. This picture is not relevant because the gold nanoparticles are not hot spots, hotter than the liquid, like under pulsed illumination. Under cw illumination, the whole fluid is hot. There is no reason to observe a collection of bubbles forming around isolated nanoparticles. Macroscopic bubbles form within the bulk of the fluid, like in regular boiling without any nano-object.

- when the authors claim they get bubble formation without heating the fluid, they are just experiencing evaporation, just like when water is heated close to $100°C$ in a pan. One can see steam rising (evaporation) without boiling.

The review written by Albert Polman in 2013 about this series of articles is also quite disappointing [102] because none of the problems has been pointed out. Some other groups followed this idea of steam generation in plasmonics using sunlight focusing, like the group of Gang Chen [45, 91] or Demetri Psaltis [132].

In 2013, Chenglong Zhao and coworkers published what can be considered as the first application of a plasmon-induced microbubble on a substrate [131]. The authors used a bubble on a surface made by laser heating of a metal film as a reconfigurable plasmonic lens.

In 2014, the occurrence of metastable water and of the formation of bubbles at 230°C was evidenced experimentally using a label-free microscopy technique [7]. A comprehensive description of the physics of bubble formation and collapse was proposed.

In 2015, the group of Michel Orrit reported on non-conventional bubble dynamics under cw illumination [53]. In their study, the authors did not observe any stable bubble lying at the interface between the coverslip and the liquid. Once formed, the bubble disappears within a few nanoseconds (either because it collapses or detaches from the substrate, the mechanism has not been identified). As a consequence, a new bubble forms, and so on. This results in a periodic formation of bubbles. This picture differs from previous observations of bubbles formation on a substrate under cw illumination [7, 131], where the bubbles were stable. This difference certainly comes from the fact that all the experiments conducted by Orrit and coworkers were performed in pentane, not in water. The nature of the solvent can strongly affect the bubble dynamics. I have myself observed this kind of periodic bubble formation with a mixture of water and ethanol, although the bubbles were not as small as the bubbles observed by the group of Orrit.

In 2017, the group of Detlef Lohse investigated the physics of the bubble growth under cw illumination of gold nanoparticles (see Section 5.7.5). This was the last part of the puzzle describing the dynamics of bubbles in plasmonics. D. Lohse is a specialist in fluid mechanics and, in particular, nanobubbles. But tackling the field of plasmonics was new for him.

5.8 Stress Wave Generation

5.8.1 Definitions

When a light-absorbing medium is heated using a pulse of light, it can result in a very large change in pressure leading to the formation of a pressure wave.

The appellations "pressure wave" or "stress wave" are the most general terms. In the literature, the term "shock wave" is also commonly employed, which is a particular case of a stress wave. Quoting Reference [32], "there is some confusion in the use of the term *shock wave*. Although there are criteria that define a shock wave [88], they are not always applied to the biomedical literature." A *shock* wave is a pressure wave of large amplitude that forms when some molecules constituting the medium happen to move faster than the speed of sound due to some external factor (explosion, rapid motion of an object, collision of solids, etc.). The salient feature of a shock wave is a discontinuity in pressure, density and particle velocity that propagates slightly faster than the sound velocity. Otherwise, the more general term *stress wave* must be used.

The pressure increase δP leading to the formation of a stress wave can be expressed as a function of the absorbed energy $E = \sigma_{abs} F$ by defining the Grüneisen parameter Γ:

$$\delta P = \Gamma E. \tag{5.84}$$

5.8.2 Mechanisms

Laser-induced stress waves (LSW) have been investigated since the 60s, shortly after the invention of the Q-switched laser [22, 23, 32]. Rapidly, LSW have been studied in biological environments in the context of medical applications of pulsed, high-power lasers [41], in particular for ophthalmic applications and ablation procedures [122]. The question was to quantify the deleterious side-effects of stress waves on living cells [39]. The study of the generation of stress waves from the heating of microscale objects came much later, in the 90s, and focused on pigment microparticles (like melanosomes [68], see Figure 5.36). The study of *metal nanoparticles* only dates from the 2000s [93]. Three mechanisms of LSW generation can be involved with micro- and nanoparticles:

- a rapid heating (thermoelastic generation).
- a rapid heating, large enough to yield the formation of a bubble.
- an optical breakdown leading to the formation of a plasma. The intense electric field of the laser pulse ionizes the molecules of the medium and creates a transient bubble. This mechanism occurs over the so-called "optical breakdown threshold" of the medium.

Several reviews describe these physical mechanisms [32, 123], although none of them is specifically dedicated to metal nanoparticles.

5.8.3 History

Figure 5.35 gives *webofknowledge* statistics on the research thematics of stress waves generation using photoacoustic effects of plasmonic nanoparticles, which is becoming a very active area of research.

In 1998, Charles P. Lin and Michael W. Kelly reported on experiments conducted on single melanosome microparticles [68]. Although it does not involve plasmonic nanoparticles, I mention this work because it nicely describes the stress wave generation around microparticles. In particular, nice stroboscopic images of stress wave fronts have been recorded, as shown in Figure 5.36a, and I have never seen such illustrative results with plasmonic nanoparticles. Moreover, the underlying mechanism of stress wave generation around pigment microparticles and metal nanoparticles are similar. Lin and Kelly measured the photoacoustic signal as a function of time and as a function of laser fluence (Figure 5.36b,c). They evidenced a fluence threshold from which the photoacoustic signal is substantially higher (Figure 5.36c). This observation evidenced that bubbles are bound to form

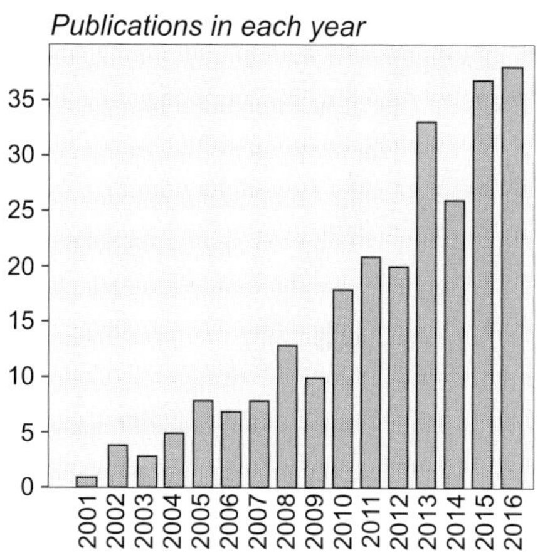
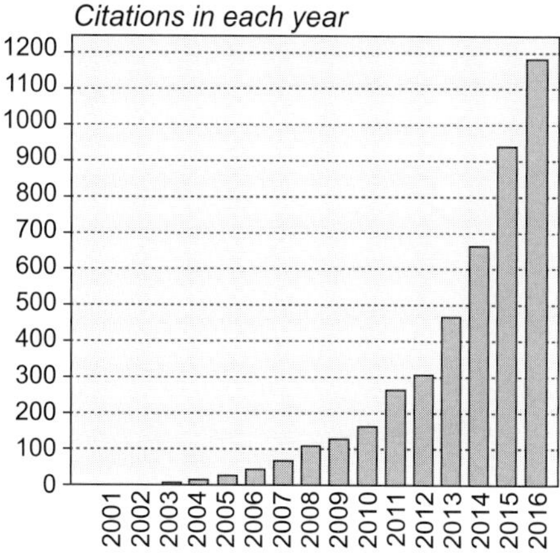

Fig. 5.35   Web of Knowledge statistics [1] (January 2017), corresponding to the search criterion "`TITLE:` `((photoacoustic or optoacoustic or 'stress wave*' or 'shock wave*') and (nanoparticle* or plasmon*) and nano*)`." Results found: 255 / Average citations per item: 17.23 / h-index: 33.

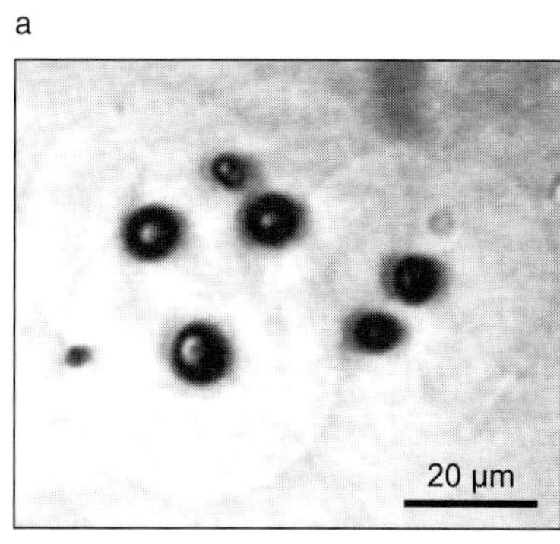

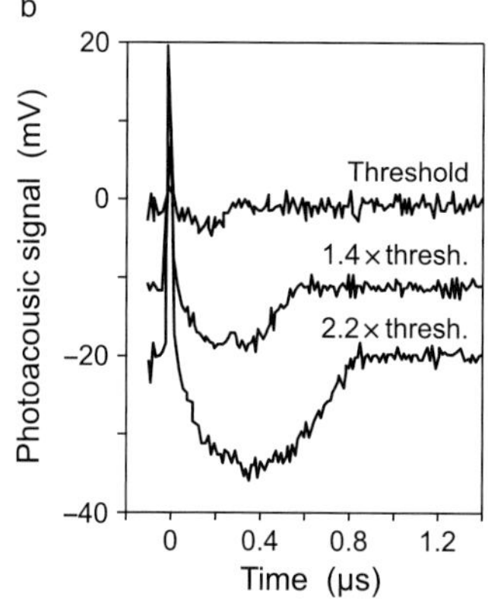

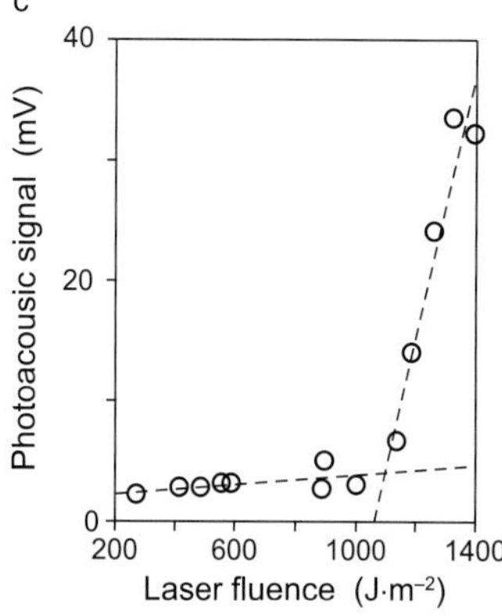

Fig. 5.36   (a) Cavitation and acoustic emission around melanosomes, taken 14 ns after irradiation with a 30 ps laser pulse. (b) Photoacoustic signal around a single melanosome as a function of time. (c) Photoacoustic signal around a single melanosome as a function of the laser fluence. Reproduced with permission from Reference [68]. Copyright 1998, AIP Publishing LLC.

from a given fluence threshold (here 1100 J·m^{-2}), and it highlights a very general rule in this area of research: *photoacoustic signal are weak, unless bubbles form.*

In 2000 and later in 2005, Sun and coworkers published comprehensive numerical simulations on laser absorption of spherical particles (supposed to be melanosomes or metal nanoparticles), and on the subsequent bubble dynamics and shock wave generation [118, 39]. Many aspects of the underlying physics have been addressed in these articles,

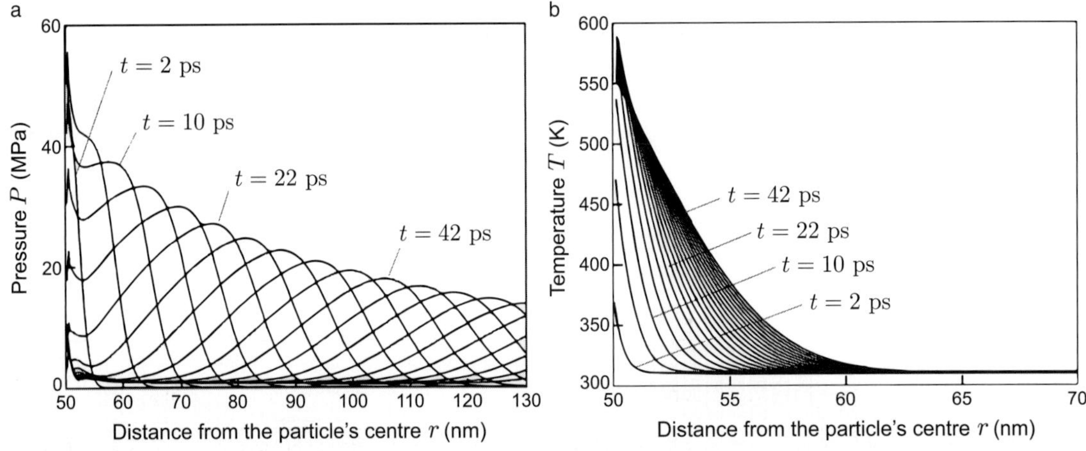

Fig. 5.37 (a) Distributions of pressure P and (b) temperature T in water at every 4 ps after 2 ps. Simulations were performed with a fluence of 60 J·m^{-2}, a pulse duration of 200 fs, and a nanoparticle radius of 50 nm. Reproduced with permission from Reference [124]. Copyright 2007, Elsevier B.V.

namely the shock wave pressure jump, the fluence threshold for bubble formation, the evolution of the bubble size and the dependence on the pulse duration and laser fluence.

In 2001, Oraevsky *et al.* conducted the first experiments on the generation of stress wave from plasmonic nanoparticles [93]. This seminal work is not highly cited (only 40 times between 2001 and 2016), presumably because it was a conference paper.

In 2007, Volkov *et al.* reported on a hydrodynamic numerical study of short pulse laser interaction with gold nanoparticles. The formation of stress waves has been modeled and the main results are presented in Figure 5.37. The smooth spatial profile of the pressure contrasts with the results reported by Sun *et al.* where the emission of a shock wave with a steep front was predicted [118].

In 2009, Egerev *et al.* [35] theoretically derived that there exists a linear relationship between the incident laser fluence and the maximum acoustic pressure for fluence values exceeding the cavitation threshold: $P_{\mathrm{max}} \sim F - F_{\mathrm{c}}$. This theoretical result is consistent with the measurements of Lin *et al.* [68] presented in Figure 5.36c.

In 2010, González and coworkers reported one of the rare articles dedicated to investigating the physics of the photoacoustic wave generation from gold nanoparticles. The authors investigated the influence of many parameters, such as the size of the nanoparticle and the laser fluence. Their major results are reproduced in Figure 5.38. The effect of the fluence threshold is consistent with the measurements of Lin *et al.* performed on melanosomes [68] (see Figure 5.36c). They evidence the presence of a fluence threshold corresponding to a bubble formation. However, the x-scale is logarithmic, the dependence is thus supposed to be nonlinear, in contrast with the prediction of Egerev [35] and the measurements of Lin [68] (see Figure 5.36c).

In 2015, Prost, Poisson and Bossy published a numerical study of the laser-induced stress wave generation by a gold nanosphere. And in 2017, Emil-Alexandru Brujan published a numerical study on the stress wave emission from plasmonic nanobubbles [19]. In his article, Brujan investigated the effects of the three significant parameters in LSW:

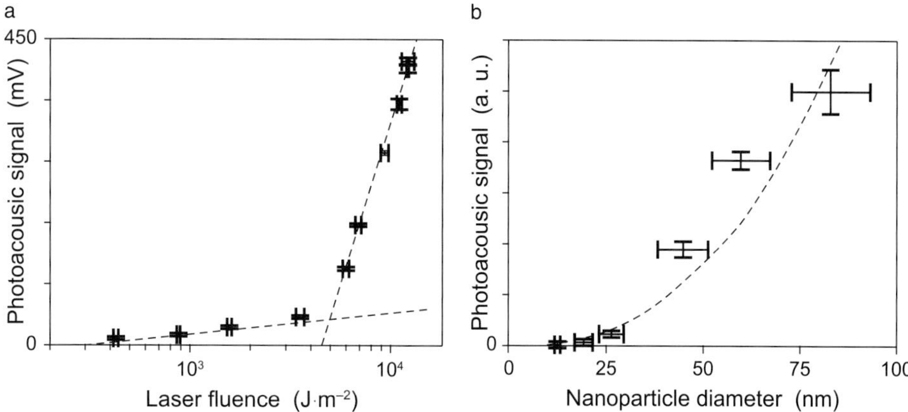

Fig. 5.38 (a) Photoacoustic response as a function of laser fluence for gold nanosphere with a diameter of 12.5 nm. Dashed lines: guide to the eye. (b) Normalized photoacoustic response as a function of the gold nanoparticle diameter. Dashed line: simulations. Reproduced from Reference [49]. Copyright 2010, AIP Publishing LLC.

nanoparticle radius, laser pulse duration, and laser fluence. The fact that such fundamental problems related to simple spheres have been addressed by numerical articles very recently evidences the fact that this research thematics is clearly not as advanced as compared with other fields of research in plasmonics. For instance, no review article exists on the different mechanisms behind the stress wave generation in plasmonics. This conclusion is consistent with the statistics presented in Figure 5.35, which showed a recent gain of interest for this thematics from the 2010s.

5.8.4 Mechanical Effects of Laser-Induced Stress Waves in Biology

Most of the time, research on laser-induced stress wave (LSW) generation has been carried out in the context of medical and biomedical applications. First, LSW can have direct applications for photoacoutic imaging in three dimensions. This application will be reviewed in Section 6.5 on page 242. Second, the deleterious effect of stress waves on cells has been an important concern since the 60s, when pulsed lasers have been used for the first time on living organisms (mostly for applications in laser ablation and ophthalmology). Not only the thermal denaturation or the bubble formation can be deleterious, also the stress wave itself can induce damages. To investigate the biological effects of stress waves in a systematic way, it is necessary to eliminate all other sources of cellular injury [32]. Some more information on the biological effects of LSW can be found in References [32, 66, 122, 39].

5.9 Refractive Index Variation

5.9.1 dn/dT of Common Materials

When a nanoparticle is heated, the temperature increase does not remain confined within the nanoparticle. It naturally expands over the surrounding medium (glass, water, air, ...).

Table 5.4 Optical constants at room temperature and normal pressure, and boiling temperature, for selected materials [43].

Material	n	dn/dT (K^{-1})	Boiling point (°C)
BK7 glass	1.52	$-0.125 \cdot 10^{-4}$	
water	1.33	$-0.90 \cdot 10^{-4}$	100
glycerol	1.473	$-2.7 \cdot 10^{-4}$	290
ethanol	1.36	$-4.4 \cdot 10^{-4}$	78
pentane	1.358	$-6.0 \cdot 10^{-4}$	36
hexane	1.37	$-5.5 \cdot 10^{-4}$	68
decane	1.413	$-6.06 \cdot 10^{-4}$	174
chloroform	1.45	$-6.19 \cdot 10^{-4}$	61
carbon tetrachloride	1.465	$-6.12 \cdot 10^{-4}$	77
carbon disulfide	1.63	$-8.13 \cdot 10^{-4}$	46

This is likely to create a variation δn of the refractive index of the surrounding medium. This variation can be expressed as a function of a temperature variation δT using the coefficient dn/dT:

$$\delta n = \frac{dn}{dT}\delta T. \tag{5.85}$$

This linear relation is only valid for small temperature variations δT. For larger temperature variations (typically larger than 20 K for water), the relation between δn and δT is no longer linear, as discussed hereinafter.

Table 5.4 lists values of dn/dT for current materials at ambient temperature. Only liquids feature appreciable values of dn/dT. Values for gas and solids are usually negligible.

5.9.2 The Case of Water

Water is a very common surrounding medium in plasmonics. Its case is singular because it features a strong nonlinear variation of dn/dT as a function of temperature (see Figure 5.39). n is also dependent on the wavelength. Fortunately, it is tabulated for a large range of temperature, which is far from being the case for many common materials.

The refractive index of liquid water is often given up to 100°C. It is sometimes given at higher temperature but in this case it is measured under high pressure to avoid phase transition. However, dn/dT can be considered as independent of the pressure with a very good approximation [119] (see Figure 5.40a).

n and dn/dT are, however, dependent on the wavelength, as illustrated by Figures 5.39 and 5.40b and Table 5.5.

$n(T)$ can be conveniently estimated using a Taylor development determined using experimental values:

$$n(T) \approx \sum_{j=0}^{M} b_j\, T^j \tag{5.86}$$

Table 5.5 Refractive index of water as a function of temperature T and wavelength λ_0 [9].

T (°C)	λ_0 (nm)					
	266.5 nm	361 nm	404 nm	589 nm	633 nm	1014 nm
0	1.3945	1.34896	1.34415	1.33432	1.33306	1.32612
10	1.39422	1.3487	1.34389	1.33408	1.33282	1.32591
20	1.39336	1.34795	1.34315	1.33336	1.33211	1.32524
30	1.39208	1.34682	1.34205	1.33230	1.33105	1.32424
40	1.39046	1.34540	1.34065	1.33095	1.32972	1.32296
50	1.38854	1.34373	1.33901	1.32937	1.32814	1.32145
60	1.38636	1.34184	1.33714	1.32757	1.32636	1.31974
70	1.38395	1.33974	1.33508	1.32559	1.32438	1.31784
80	1.38132	1.33746	1.33284	1.32342	1.32223	1.31576
90	1.37849	1.33501	1.33042	1.32109	1.31991	1.31353
100	1.37547	1.33239	1.32784	1.31861	1.31744	1.31114

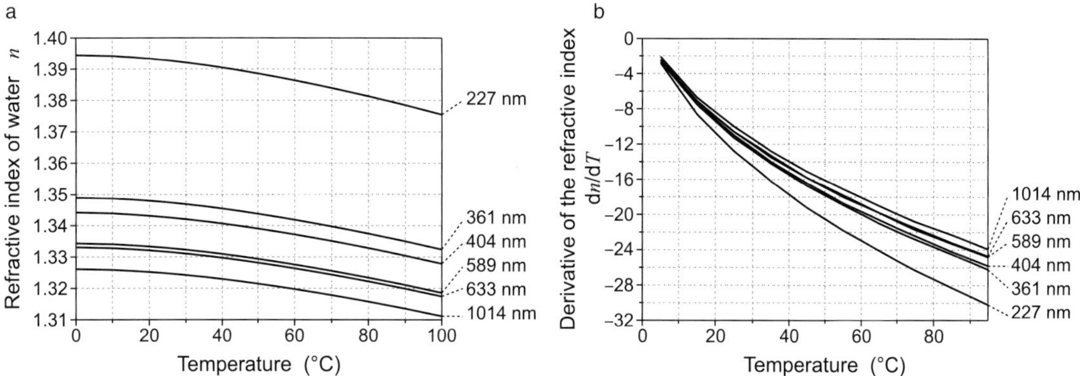

Fig. 5.39 (a) Refractive index of water as a function of temperature, for different wavelengths. (b) Temperature derivative of the refractive index of water as a function of temperature, for different wavelengths. Data collected from [9].

with T in degree Celsius and where the coefficients b_j are empirical parameters. A development at order $M = 4$ is usually sufficient to describe a condensed material over a large temperature range. For water, a fit of the measurements reported in Reference [119] gives at $\lambda = 589$ nm:

$$b_0 = 1.3345$$
$$b_1 = -5.3864e \times 10^{-6}$$
$$b_2 = -2.0985 \times 10^{-6}$$
$$b_3 = 6.8405 \times 10^{-9}$$
$$b_4 = -1.250 \times 10^{-11}.$$

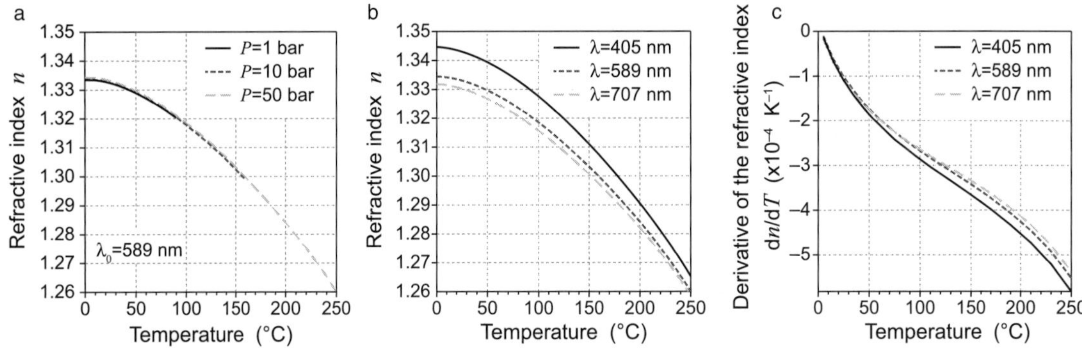

Fig. 5.40   Refractive index of water. (a) $n(T)$ as a function of the pressure, for a wavelength of $\lambda_0 = 589$ nm. (b) Refractive index of water as a function of the temperature for different wavelengths. (c) Derivative of the functions of represented in (b). Data collected from Reference [119].

Figure 5.40 represents the refractive index of liquid water as a function of temperature measured up to 240°C, along with the fitting function.

According to Table 5.4, typical values of dn/dT look small, even for liquids, but they are actually large enough to create measurable effects in plasmonics, namely

- a modification of the plasmon resonance of the nanoparticle, via its dependence on the refractive index of the surrounding medium.
- a thermal lens effect, that can be used in applications such as temperature imaging using wavefront sensing (discussed in Section 4.4 on page 125) or adaptive thermal lensing [31].
- an increase of the scattering of the nanoparticle, with an application in photothermal imaging (introduced in Section 6.7 on page 256) or for the measurements of thermal diffusivity [51].

5.10 Fluid Convection

Consider a nanoparticle deposited on a substrate and immersed in a liquid. If this nanoparticle absorbs light and heats the surrounding fluid, one can expect a Rayleigh–Bénard instability consisting of a centripetal fluid convection around the source of heat and an upward motion of the fluid above, as represented by the fluid velocity lines in this schematic:

This type of convection originates from a decrease of the mass density of the fluid, which undergoes an upward buoyancy force. The natural question is how fast the fluid motion can be. This question was originally raised in the context of plasmonic trapping experiments by the group of Quidant in 2007 [108]. Nanobeads were observed to move toward lithographic nanoparticles under illumination. The underlying physics was assumed to be a near-field optical force, but a substantial fluid convection would have yield the same observation: a centripetal motion of the beads. In this seminal work, the authors

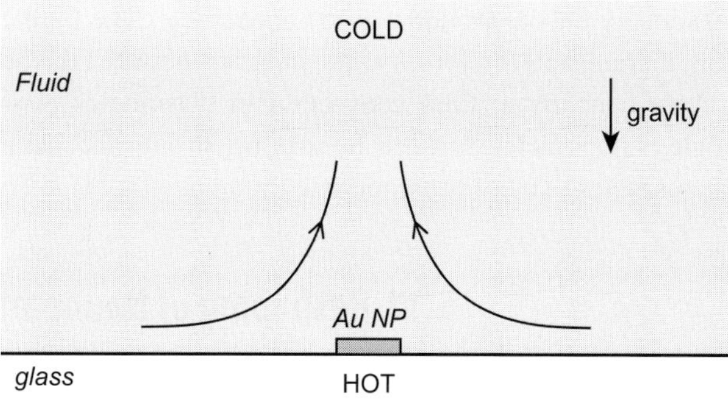

derived a rough estimation of the nanoparticle that was to small (a few K) to generate a substantial fluid motion. In 2007, the same group published a numerical study of the fluid convection around single gold nanodiscs on a glass substrate. The first main result, obtained by dimensional analysis (see Appendix A), was that the characteristic fluid velocity is supposed to be on the order of

$$V = \frac{L^2 \beta g \delta T}{\nu} \tag{5.87}$$

where L is the typical length scale of the heat source, g the gravitational acceleration, β the dilatation coefficient and δT the temperature increase. This equation tells us that the size of the system is a dominant parameter as V scales as L^2. Also, ν is a fast decreasing function of the temperature (one order of magnitude lower at 100°C compared to ambient temperature), which contributes to markedly enhance the fluid convection. For $L \approx 250$ nm, a temperature increase $\delta T = 100°C$, and using $\beta = 10^{-4}$ K, $\nu = 10^{-5}$ Pa·s, and $g = 9.8$ m·s^{-2}, we obtain a characteristic fluid velocity of the thermal-induced convection of $V \sim 100$ nm·s^{-1}. Numerical simulations led to fluid velocities slightly smaller than this rough estimation using Equation (5.87), on the order of 1 nm·s^{-1}.

The main conclusion of this article was that fluid convection is bound to remain insignificant. However, the system was quite restrictive and the conclusion of negligible fluid convection is not a general rule in plasmonics:

- The authors considered a single nanoparticle of 500 nm in diameter. Much investigation is based on the heating of an assembly of nanoparticles undergoing collective thermal effects. In this case, the heated area is much larger and the Rayleigh–Bénard convection is supposed to be stronger. As the amplitude of fluid convection is scaling as L^2, one can expect strong convection when collective effects occur.
- The authors considered a maximum temperature increase of $100^{\circ}irc$C, which was natural at that time because the plasmonics community was not aware that a fluid could be superheated above its boiling point, up to around 230°C without boiling even under cw illumination [7]. What the velocity of a super-heated fluid can be in plasmonics is still an open question, but it is presumably very large because the viscosity is supposed to reach very small values.

Another limitation was lifted in 2017 by Setoura and coworkers, who demonstrated a gigantic fluid convection around plasmonic nanoparticles subsequently to the formation of a stationary bubble around a gold nanoparticle. The fluid convection enhancement results from thermocapillary forces at the interface between the bubble and fluid [115].

As a conclusion, fluid convection in plasmonics is supposed to be negligible with single nanoparticles, but can be a priori strong if collective effects occur (*i.e.*, is many nanoparticles in close proximity are illuminated) or if microbubbles form.

5.11 Reshaping of Nanoparticles

Gold is known to melt at $1064°C$, but this does not mean that gold nanoparticles can be safely heated up to $\sim 1000°C$ without modification. Nanoparticle morphologies are bound to reshape at much lower temperature due to surface effects and because they exhibit important surface-to-volume ratio. Basically, when the temperature is sufficiently high, the nanoparticle becomes spherical. Under cw illumination, this can happen from around $100°C$ to $250°C$, depending on the heating duration. Under pulsed illumination, much higher temperature rises are afforded prior to shape modification, above $700°C$. Nanoparticle reshaping can occur both under pulsed or cw illumination and the conditions for reshaping result from and interplay between the *temperature increase and its duration*. Importantly, the reshaping temperature threshold can be increase by embedding the nanoparticle in a solid matrix (*i.e.*, SiO_2 matrix of SiO_2 coating) (up to round $800°C$ under cw laser heating). Nanoparticle reshaping is becoming a very important concern in plasmonics due to the emergence of new applications involving high temperature under operation (heat-assisted magnetic recording, thermophotovoltaics or photoacoustic imaging, see Chapter 6).

Here is now an overview of the most important articles published in this field that contributed to a better understanding of the underlying physics.

The study of gold nanoparticle reshaping, both theoretically and experimentally, dates from the late 90s [67, 25, 70].

In 1997, the group of J. L. Barrat numerically investigated the reshaping of small nanoparticles (smaller than 3 nm to reduce the computation time) using atomic dynamic simulations (considering a cluster of up to 4000 atoms) [67]. The authors evidenced a temperature melting of the nanoparticle that was highly dependent on the nanoparticle size, and that remained much below $1064°C$, around 400 to $600°C$ as represented in Figure 5.41. The trend observed in the numerical simulations is consistent with a simple theoretical model based on considerations on surface tension energies, which are different for the solid and liquid phases. The method the authors used (the embedded-atom method, EAM) is not accurate in some circumstances. This is why the authors predicted a melting temperature of bulk gold of 1090 K instead of 1338 K. So the values of melting temperatures presented in their work and reproduced in Figure 5.41 should not be considered extremely precise. They just give the good order of magnitude and the good trend (the smaller the nanoparticle, the smaller the reshaping temperature threshold).

In 2006, the group of G. Hartland, in collaboration with L. Liz Marzan and P. Mulvaney, conducted a comprehensive study on gold nanorod stability and compared pulsed

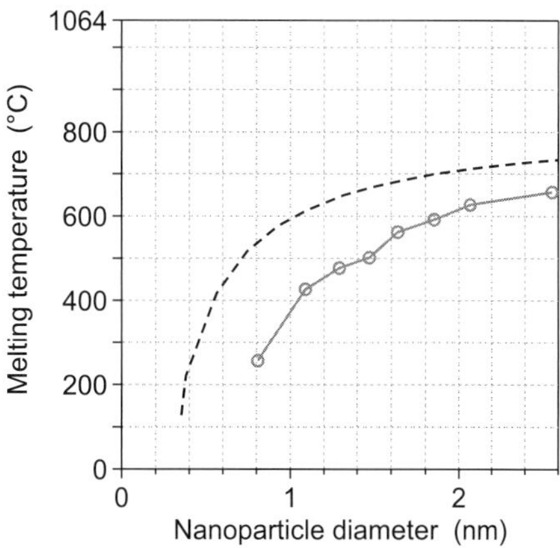

Fig. 5.41 Melting temperature of gold nanoparticles as a function of their diameter (solid line), calculated using atomic dynamics simulations [67]. The melting temperature of gold bulk at 1064°C is indicated. The dashed line represents the predictions of a simple theoretical model.

heating and continuous heating [94]. Nanorods were observed to transform into spherical particles upon melting. Under pulsed laser irradiation they found no significant melting up to nanoparticle temperatures of $700 \pm 50°C$, indicating that the particles maintained their integrity close to the melting temperature of gold bulk. Under continuous heating, however, they observed nanoparticle reshaping at only $150°C$ after typically 1 h, and at $250°C$ in a few minutes. This is significantly below the melting point of bulk gold, which corroborates that the nanoparticle reshaping arises from surface effects. The difference between pulsed illumination and continuous heating evidences the importance of the heating duration. As explained in the article, "the rods do not stay hot for long enough after ultrafast excitation for significant structural transformation to occur. This shows that the timescale for the atomic motion at the surface of the nanorods needed for restructuring must be much longer than the several hundred ps heating time of [the] experiments."

In 2009, Peter Zijlstra, James W. M. Chon and Min Gu studied the reshaping of nanoparticles under 100-fs-pulsed illumination using white light scattering and electron microscopy [133]. They performed the first study on single nanoparticles. Previous studies were restricted to ensemble measurements. An example of the type of observation they achieved is represented in Figure 5.42. They also confirmed that higher aspect ratio particles are thermodynamically less stable. The authors managed to conduct quantitative measurements of the energy absorbed by the nanoparticles, which led to precise estimation of energy threshold before melting.

In 2010, the group of S. Emilianov addressed a problem encountered in photoacoustic imaging [28]. Under intense pulsed heating for the generation of acoustic wave around metal nanoparticles, a reshaping of the nanoparticle is likely to occur, which damps its plasmonic resonance in the infrared and make them inactive as photoacoustic contrast agents, especially for gold nanorods. The authors proposed to coat gold nanorods with a silica layer (SiO_2). Although the temperature was not estimated, they evidenced a clear increase of the thermal stability using this approach, as illustrated by Figure 5.43.

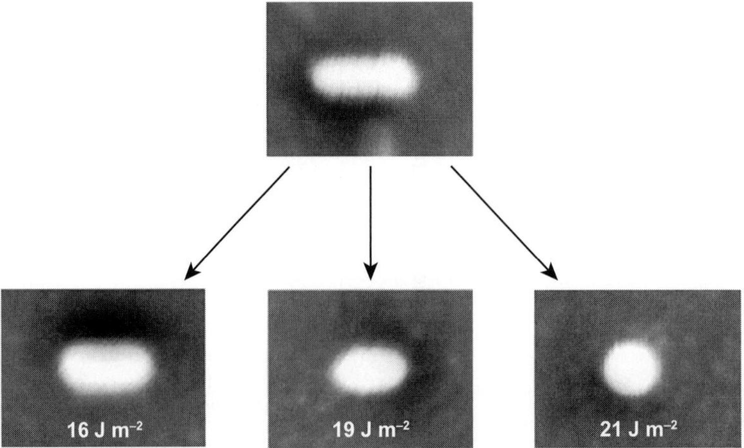

Fig. 5.42 Observed effect of different pulse energy densities on the reshaping of a gold nanorod. The dimensions of all SEM images are 200 nm × 150 nm. Reproduced with permission from Reference [133]. Copyright 2009, Royal Society of Chemistry.

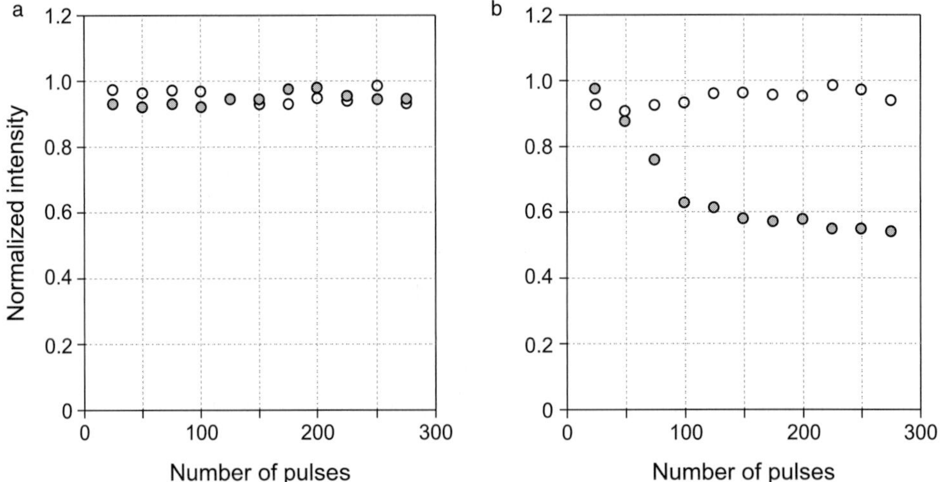

Fig. 5.43 Photoacoustic signal intensity of PEG-coated gold nanorods (hollow circles) and silica-coated gold nanorods (plain circles) versus number of pulses with fluence (a) 40 J·m^{-2} and (b) 18 J·m^{-2}. Reproduced with permission from Reference [28].

5.12 Thermal Radiation

As explained in Section 2.1.4 on page 41, any object with a temperature different from absolute zero radiates light. Planck's law gives the spectral radiance of the light radiated by a body in thermal equilibrium at the temperature T. It represents an electromagnetic power emitted per unit area of the surface of the body, per unit solid angle and per unit wavelength:

$$\mathcal{L}_\lambda^0(\lambda, T) = \frac{2hc^2}{\lambda^5} \frac{1}{\exp\left(\dfrac{hc}{\lambda k_B T}\right) - 1} \tag{5.88}$$

so that the power radiated by a surface area dS over a solid angle $d\Omega$ and over a light frequency range $d\nu$ reads $dP_{\mathrm{rad}} = \mathcal{L}_\lambda^0 dS d\Omega d\lambda$. Figure 2.3 on page 42 plots spectral radiances $\mathcal{L}_\lambda^0(\lambda, T)$ as a function of the wavelength at different temperatures.

Equation 5.88 only holds for a black body. When considering common materials, this expression has to be weighted by the emissivity ϵ_λ:

$$\mathcal{L}_\lambda(\lambda, T) = \epsilon_\lambda(\lambda)\mathcal{L}_\lambda^0(\lambda, T) \tag{5.89}$$

with $\epsilon_\lambda(\lambda) \leq 1$. Interestingly, Kirchoff's law states that the spectral emissivity equals the spectral absorptivity of the surface of the material:

$$\epsilon_\lambda = \alpha_\lambda. \tag{5.90}$$

This means that the more an object absorbs light, the more it radiates light upon heating.

Let us consider now a single plasmonic nanoparticle at the temperature T. If one applies this formalism, the heat power radiated by the nanoparticle should read [56]

$$P_{\mathrm{rad}} = 4\pi \int_0^\infty \epsilon_\lambda(\lambda)\frac{2hc^2}{\lambda^5} \frac{C_{\mathrm{geom}}}{\exp\left(\dfrac{hc}{\lambda k_B T}\right) - 1} d\lambda \tag{5.91}$$

where C_{geom} is the projected area of the nanoparticle (πa^2 in the case of a spherical nanoparticle of radius a). But this equation has to be used with caution. In plasmonics, C_{geom} is typically much smaller than the wavelength. Numerical simulations tend to show that using this expression may yield values of $\epsilon_\lambda(\lambda)$ larger than unity, which is nonsense. A better approach is to consider a formalism based on the use of the absorption cross section, not on any emissivity nor any sub-wavelength projected area. In theory, it suffices to replace $\epsilon_\lambda(\lambda)C_{\mathrm{geom}}$ by the absorption cross section σ_{abs} of the nanoparticle. This way, one ends up with a general and helpful equation suited to compute the total power thermally radiated by a plasmonic nanoparticle:[15]

$$\boxed{P_{\mathrm{rad}} = 4\pi \int_0^\infty \frac{2hc^2}{\lambda^5} \frac{\sigma_{\mathrm{abs}}(\lambda)}{\exp\left(\dfrac{hc}{\lambda k_B T}\right) - 1} d\lambda.} \tag{5.92}$$

However, when assemblies of nanoparticles in close vicinity are illuminated, the standard formalism can be used. In particular, when nanoparticles are deposited on a planar substrate and when heating is performed over an area much larger than the wavelength so that thermal collective effect occur (see Subsection 2.2.5 on page 50), then Equation 5.90 can be used to compute the emissivity of the sample from the absorbance of the sample, and the emitted power can be computed using Expression 5.89.

---

[15] I wish to thank Karl Joulain for letting me know about this equation, which is rarely provided in the literature.

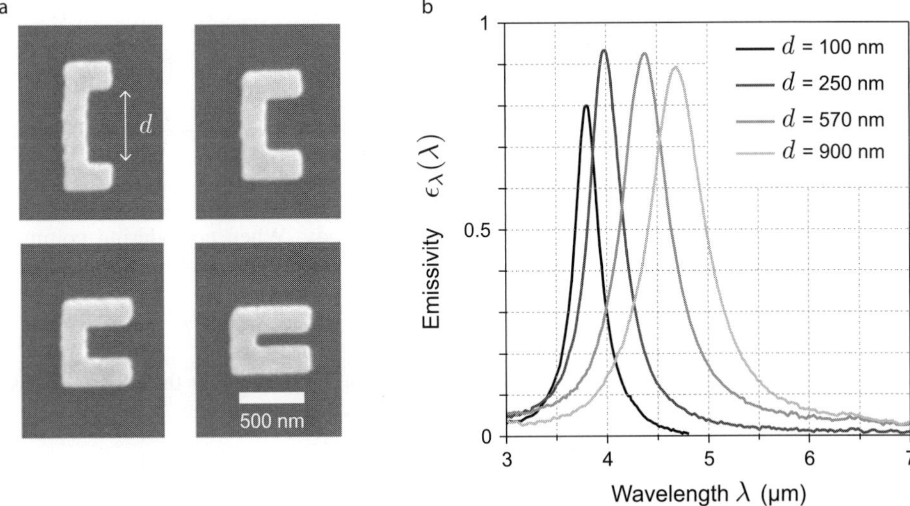

Fig. 5.44   Measured emissivity of arrays of C-shape nanoparticles. (a) SEM images of C-shape nanoparticles with different gap distances d. (b) Experimentally measured y-polarized emissivity spectra of the C-shape resonator arrays with different d. Reproduced with permission from Reference [72]. Copyright 2017, American Chemical Society

Consequently, using gold nanoparticles, the spectrum of the light emitted by thermal radiation matches the plasmonic resonance spectrum. This idea has been used by Liu *et al.* to produce light sources with a "narrow-band" emission [72] (see Figure 5.44).

References

[1] apps.webofknowledge.com, 2016.

[2] Alaulamie, A. A., Baral, S., Johnson, S. C., and Richardson, H. H. 2016. Targeted Nanoparticle Thermometry: A Method to Measure Local Temperature at the Nanoscale Point Where Water Vapor Nucleation Occurs. *Small*, **13**, DOI: 10.1002/smll.201601989.

[3] Alper, J., and Hamad-Schifferli, K. 2010. Effect of Ligands on Thermal Dissipation from Gold Nanorods. *Langmuir*, **26**(6), 3786.

[4] Alvers, S., Bourdon, A., and Figueiredo Neto, A. M. 2003. Generalization of the Thermal Lens Model Formalism to Account for Thermodiffusion in a Single-Beam Z-Scan Experiment: Determination of the Soret coefficient. *J. Opt. Soc. Am. B*, **20**(4), 713–718.

[5] Baffou, G., and Quidant, R. 2014. Nanoplasmonics for Chemistry. *Chem. Soc. Rev.*, **43**, 3898–3907.

[6] Baffou, G., and Rigneault, H. 2011. Femtosecond-Pulsed Optical Heating of Gold Nanoparticles. *Phys. Rev. B*, **84**, 035415.

[7] Baffou, G., Polleux, J., Rigneault, H., and Monneret, S. 2014. Super-Heating and Micro-Bubble Generation around Plasmonic Nanoparticles under cw Illumination. *J. Phys. Chem. C*, **118**, 4890.

[8] Ball, P. 2012. Nanobubbles are not Superficial Matter. *Chem. Phys. Chem.*, **13**, 2173–2177.

[9] Bashkatov, A. N., and Genina, E. A. 2002. Water Refractive Index in Dependence on Temperature and Wavelength: A Simple Approximation. *SPIE*, **5068**, 393–395.

[10] Bielenberg, J. R., and Brenner, H. 2005. A Hydrodynamic/Brownian Motion Model of Thermal Diffusion in Liquids. *Physica A*, **356**, 279–293.

[11] Boulais, E., Lachaine, R., and Meunier, M. 2012. Plasma Mediated off-Resonance Plasmonic Enhanced Ultrafast Laser-Induced Nanocavitation. *Nano Lett.*, **12**, 4763–4769.

[12] Braibanti, M., Vigolo, D., and Piazza, R. 2008. Does Thermophoretic Mobility Depend on Particle Size? *Phys. Rev. Lett.*, **100**, 108303.

[13] Braun, D., and Libchaber, A. 2002. Trapping of DNA Thermophoretic Depletion and Convection. *Phys. Rev. Lett.*, **89**(18), 188103.

[14] Braun, M., and Cichos, F. 2013. Optically Controlled Thermophoretic Trapping of Single Nano-objects. *ACS Nano*, **7**(12), 11200–11208.

[15] Braun, M., Würger, A., and Cichos, F. 2014. Trapping of Single Nano-objects in Dynamic Temperature Fields. *Phys. Chem. Chem. Phys.*, **16**, 15207.

[16] Braun, M., Bregulla, A. P., K., Günther, Mertig, M., and Cichos, F. 2015. Single Molecules Trapped by Dynamic Inhomogeneous Temperature. *Nano Lett.*, **15**, 5499–5505.

[17] Brenner, H. 2006. Elementary Kinematical Model of Thermal Diffusion in Liquids and Gases. *Phys. Rev. E*, **74**, 036306.

[18] Brujan, E. A. 2011. Numerical Investigation on the Dynamics of Cavitation Nanobubbles. *Microfluid Nanofluid*, **11**, 511–517.

[19] Brujan, E. A. 2017. Stress Wave Emission from Plasmonic Nanobubbles. *J. Phys. D: Appl. Phys.*, **50**, 015304.

[20] Carlson, M. T., Khan, A., and Richardson, H. H. 2011. Local Temperature Determination of Optically Excited Nanoparticles and Nanodots. *Nano Lett.*, **11**, 1061–1069.

[21] Carlson, M. T., Green, A. J., and Richardson, H. H. 2012. Superheating Water by CW Excitation of Gold Nanodots. *Nano Lett.*, **12**, 1534.

[22] Carome, E. F., Moeller, C. A., and Clark, N. A. 1964. Generation of Acoustic Signals in Liquids by Ruby Laser-Induced Thermal Stress Gradients. *Appl. Phys. Lett.*, **4**, 95–97.

[23] Carome, E. F., Moeller, C. A., and Clark, N. A. 1966. Intense Ruby-Laser-Induced Acoustic Impulses in Liquids. *J. Acoust. Soc. Am.*, **40**, 1462–1466.

[24] Cavicchi, R. E., Meier, D. C., Presser, C., Prabhu, V. M., and Guha, S. 2013. Single Laser Pulse Effects on Suspended-Au-Nanoparticle Size Distributions and Morphology. *J. Phys. Chem. C*, **117**, 10866–10875.

[25] Chang, S. S., Shih, C. W., Chen, C. D., Lai, W. C., and Wang, C. R. C. 1999. The Shape Transition of Gold Nanorods. *Langmuir*, **15**, 701–709.

[26] Chapman, S. 1912. The Kinetic Theory of a Gas Constituted of Spherically Symmetrical Molecules. *Phil. Trans. R. Soc.*, **211**, 433.

[27] Chen, H., and Diebold, G. 1995. Chemical Generation of Acoustic Waves: A Giant Photoacoustic Effect. *Science*, **270**(5238), 963–966.

[28] Chen, Y. S., Frey, W., Kim, S., Homan, K., Kruizinga, P., Sokolov, K., and Emilianov, S. 2010. Enhanced Thermal Stability of Silica-Coated Gold Nanorods for Photoacoustic Imaging and Image-Guided Therapy. *Opt. Express*, **18**(9), 8867–8877.

[29] Craig, V. S. J. 2011. Very Small Bubbles at Surfaces – The Nanobubble Puzzle. *Soft Matter*, **7**, 40.

[30] Debuschewitz, C., and Köhler, W. 2001. Molecular Origin of Thermal Diffusion in Benzene + Cyclohexane Mixtures. *Phys. Rev. Lett.*, **87**(5), 055901.

[31] Donner, J. S., Morales-Dalmau, J., Alda, I., Marty, R., and Quidant, R. 2015. Fast and Transparent Adaptive Lens Based on Plasmonic heating. *ACS Photonics*, **2**, 355–360.

[32] Doukas, A. G., and Flotte, T. J. 1996. Physical Characteristics and Biological Effects of Laser-induced Stress Waves. *Ultrasound in Med. & Biol.*, **22**(2), 151–164.

[33] Duhr, S., and Braun, D. 2006. Why Molecules Move Along a Temperature Gradient. *Proc. Natl. Acad. Sci. U.S.A.*, **103**, 19678.

[34] Duhr, S., Arduini, S., and Braun, D. 2004. Thermophoresis of DNA Determined by Microfluidic Fluorescence. *Eur. Phys. J. E*, **15**, 277–286.

[35] Egerev, S., Ermilov, S., Ovchinnikov, O., Fokin, A., Guzatov, D., Klimov, V., Kanavin, A., and Oraevsky, A. 2009. Acoustic Signals Generated by Laser-Irradiated Metal Nanoparticles. *Appl. Opt.*, **48**(7), C38–C45.

[36] Einstein, A. 1905. Über die von der Molekularkinetischen Theorie der Wärme geforderte Bewegung von in ruhenden Flüssigkeiten suspendierten Teilchen. *Ann. Phys.*, **332**(8), 549–560.

[37] Enders, M., Mukai, S., Uwada, T., and Hashimoto, S. 2016. Plasmonic Nanofabrication through Optical Heating. *J. Phys. Chem. C*, **120**, 6723–6732.

[38] Fang, Z., Zhen, Y. R., Neumann, O., Polman, A., García de Abajo, F. J., Nordlander, P., and Halas, N. J. 2013. Evolution of Light-Induced Vapor Generation at a Liquid-Immersed Metallic Nanoparticle. *Nano Lett.*, **13**, 1736–1742.

[39] Faraggi, E., Gerstman, B. S., and Sun, J. 2005. Biophysical Effects of Pulsed Lasers in the Retina and other Tissues Containing Strongly Absorbing Particles: Shockwave and Explosive Bubble Generation. *J. Biomed. Opt.*, **10**(6), 064029.

[40] Fayolle, S., Bickel, T., and Würger, A. 2008. Thermophoresis of Charged Colloidal Particles. *Phys. Rev. E*, **77**, 041404.

[41] Fine, S., Klein, E., and Nowak, W. 1965. Interaction of Laser Radiation with Biological Systems. *Proc. Fed. Am. Soc. Exp. Biol.*, **24**, S35–S45.

[42] François, L., Mostafavi, M., Belloni, J., and Delaire, J. A. 2001. Optical Limitation Induced by Gold Clusters: Mechanism and Efficiency. *Phys. Chem. Chem. Phys.*, **3**, 4965–4971.

[43] Gaiduk, A., Yorulmaz, M., Ruijgrok, P. V., and Orrit, M. 2010. Room-Temperature Detection of a Single Molecule's Absorption by Photothermal Contrast. *Science*, **330**, 353–356.

[44] Ge, Z., Cahill, D. G., and Braun, P. V. 2006. Thermal Conductance of Hydrophilic and Hydrophobic Interfaces. *Phys. Rev. Lett.*, **96**, 186101.

[45] Ghasemi, H., Ni, G., Marconnet, A. M., Loomis, J., Yerci, S., Miljkovic, N., and Chen, G. 2014. Solar Steam Generation by Heat Localization. *Nature Commun.*, **5**(4449).

[46] Giglio, M., and Vendramini, A. 1974. Thermal Lens Effect in a Binary Liquid Mixture: A New Effect. *Appl. Phys. Lett.*, **25**(10), 555–558.

[47] Giglio, M., and Vendramini, A. 1975. Thermal-Diffusion Measurements near a Consolute Critical Point. *Phys. Rev. Lett.*, **34**(10), 561–564.

[48] Giglio, M., and Vendramini, A. 1977. Soret-Type Motion of Macromolecules in Solution. *Phys. Rev. Lett.*, **38**(1), 26.

[49] González, M. G., Liu, X., Niessner, R., and Haisch, C. 2010. Strong Size-Dependent Photoacoustic Effect on Gold Nanoparticles by Laser-Induced Nanobubbles. *Appl. Phys. Lett.*, **96**, 174104.

[50] Guthrie, G., Wilson, J. N., and Schomaker, V. 1949. Theory of the Thermal Diffusion of Electrolytes in a Clusius Column. *J. Chem. Phys.*, **17**, 310.

[51] Heber, A., Selmke, M., and Cichos, F. 2015. Thermal Diffusivity Measured using a Single Plasmonic Nanoparticle. *Phys. Chem. Chem. Phys.*, **17**, 20868.

[52] Hleb, E. Y., and Lapotko, D. O. 2008. Photothermal Properties of Gold Nanoparticles under Exposure to High Optical Energies. *Nanotechnology*, **19**, 355702.

[53] Hou, L., Yorulmaz, M., Verhart, N. R., and Orrit, M. 2015. Explosive Formation and Dynamics of Vapor Nanobubbles Around a Continuously Heated Gold Nanosphere. *New. J. Phys.*, **17**, 013050.

[54] Iacopini, S., and Piazza, R. 2003. Thermophoresis in Protein Solutions. *Europhys. Lett.*, **63**(2), 247–253.

[55] Jerabek-Willemsen, M., Wienken, C. J., Braun, D., Baaske, P., and Duhr, S. 2011. Molecular Interaction Studies using Microscale Thermophoresis. *Assay Drug Dev. Technol.*, **9**(4), 342–353.

[56] Kallel, H., Carminati, R., and Joulain, K. 2017. Temperature of a Nanoparticle above a Substrate under Radiative Heating and Cooling. *Phys. Rev. B*, **95**, 115402.

[57] Katayama, T., Setoura, K., Werner, D., Miyasaka, H., and Hashimoto, S. 2014. Picosecond-to-Nanosecond Dynamics of Plasmonic Nanobubbles from Pump Probe Spectral Measurements of Aqueous Colloidal Gold Nanoparticles. *Langmuir*, **30**, 9504–9513.

[58] Köhler, W. 1993. Thermodiffusion in Polymer Solutions as Observed by Forced Rayleigh Scattering. *J. Chem. Phys.*, **98**, 660–668.

[59] Köhler, W., and Rossmanith, P. 1995. Aspects of Thermal Diffusion Forced Rayleigh Scattering: Heterodyne Detection, Active Phase Tracking, and Experimental Constraints. *J. Phys. Chem.*, **99**, 5838.

[60] Kotaidis, V., and Plech, A. 2005. Cavitation on the Nanoscale. *Appl. Phys. Lett.*, **87**, 213102.

[61] Kotaidis, V., Dahmen, C., von Plessen, G., Springer, F., and Plech, A. 2006. Excitation of Nanoscale Vapor Bubbles at the Surface of Gold Nanoparticles in Water. *J. Chem. Phys.*, **124**, 184702.

[62] Lachaine, R., Boulais, E., and Meunier, M. 2014. From Thermo- to Plasma-Mediated Ultrafast Laser-Induced Plasmonic Nanobubbles. *ACS Photonics*, **1**(4), 331–336.

[63] Lapotko, D. 2009a. Optical Excitation and Detection of Vapor Bubbles around Plasmonic Nanoparticles. *Opt. Express*, **17**, 2538.

[64] Lapotko, D. 2009b. Pulsed Photothermal Heating of the Media During Bubble Generation around Gold Nanoparticles. *Int. J. Heat Mass Tran.*, **52**, 1540–1543.

[65] Lauterborn, W., and Kurz, T. 2010. Physics of Bubble Oscillations. *Rep. Prog. Phys.*, **73**, 106501.

[66] Lee, S., Anderson, T., Zhang, H., Flotte, T. J., and Doukas, A. G. 1996. Alteration of Cell Membrane by Stress Waves in Vitro. *Ultrasound in Med. & Biol.*, **22**(9), 1285–1293.

[67] Lewis, L. J. 1997. Melting, Freezing, and Coalescence of Gold Nanoclusters. *Phys. Rev. B*, **56**(4), 2248–2257.

[68] Lin, C. P., and Kelly, M. W. 1998. Cavitation and Acoustic Emission around Laser-Heated Microparticles. *Appl. Phys. Lett.*, **72**(22), 2800–2802.

[69] Lin, L., Peng, X., Wang, M., Scarabelli, L., Mao, Z., Liz-Marzán, L. M., Becker, M. F., and Zheng, Y. 2016. Light-Directed Reversible Assembly of Plasmonic Nanoparticles using Plasmon-Enhanced Thermophoresis. *ACS Nano*, **10**, 9659–9668.

[70] Link, S., Wang, Z. L., and El-Sayed, M. A. 2000. How Does a Gold Nanorod Melt? *J. Phys. Chem. B*, **104**(7867-7870).

[71] Lippok, S., Seidel, S. A. I., Duhr, S., Uhland, K., Hothoff, H. P., Jenne, D., and Braun, D. 2012. Direct Detection of Antibody Concentration and Affinity in Human Serum using Microscale Thermophoresis. *Anal. Chem.*, **84**, 3523–2530.

[72] Liu, B., Gong, W., Yu, B., Li, P., and Shen, S. 2017. Perfect Thermal Emission by Nanoscale Transmission Line Resonators. *Nano Lett.*, **17**(2), 666–672.

[73] Ljunggren, S., and Eriksson, J. C. 1997. The Lifetime of a Colloid-Sized Gas Bubble in Water and the Cause of the Hydrophobic Attraction. *Colloids Surf. A*, **129–130**, 151–155.

[74] Lombard, J., Biben, T., and Merabia, S. 2014. Kinetics of Nanobubble Generation around Overheated Nanoparticles. *Phys. Rev. Lett.*, **112**, 105701.

[75] Lombard, J., Biben, T., and Merabia, S. 2015. Nanobubbles around Plasmonic Nanoparticles: Thermodynamic Analysis. *Phys. Rev. E*, **91**, 043007.

[76] Löwen, H., and Madden, P. A. 1992. A Microscopic Mechanism for Shock-Wave Generation in Pulsed-Laser-Heated Colloidal Suspensions. *J. Chem. Phys.*, **97**(8760).

[77] Ludwig, C. 1856. Diffusion zwischen ungleich erwwärmten orten gleich zusammengestzter lösungen. *Sitz. Ber. Akad. Wiss. Wien Math-Naturw. Kl.*, **20**, 539.

[78] Ludwig, R. 2001. Water: From Clusters to the Bulk. *Angew. Chem. Int. Ed.*, **40**(10), 1808–1827.

[79] Lugli, F., and Zerbetto, F. 2007. An Introduction to Bubble Dynamics. *Phys. Chem. Chem. Phys.*, **9**, 2447–2456.

[80] Lukianova-Hleb, E., Hu, Y. and Latterini, L., Tarpani, L., Lee, S., Drezek, R. A., Hafner, J. H., and Lapotko, D. O. 2010. Plasmonic Nanobubbles as Transient Vapor Nanobubbles Generated around Plasmonic Nanoparticles. *ACS Nano*, **4**, 2109.

[81] Lukianova-Hleb, E. Y., and Lapotko, D. O. 2009. Influence of Transient Environmental Photothermal Effects on Optical Scattering by Gold Nanoparticles. *Nano Lett.*, **9**(5), 2160–2166.

[82] Lukianova-Hleb, E. Y., Sassaroli, E., Jones, A., and Lapotko, D. O. 2012. Transient Photothermal Spectra of Plasmonic Nanobubbles. *Langmuir*, **28**, 4858–4866.

[83] Lukianova-Hleb, E. Y., Volkov, A. N., and Lapotko, D. O. 2015. Laser Pulse Duration is Critical for the Generation of Plasmonic Nanobubbles. *Langmuir*, **30**, 7425–7434.

[84] Mast, C. B., and Braun, D. 2010. Thermal Trap for DNA Replication. *Phys. Rev. Lett.*, **104**, 188102.

[85] McEwan, K. J., and Madden, P. A. 1992. Transient Grating Effects in Absorbing colloidal suspensions. *J. Chem. Phys.*, **97**, 8748.

[86] Merabia, S., Keblinski, P., Joly, L., Lewis, L. J., and Barrat, J. L. 2009. Critical Heat Flux around Strongly Heated Nanoparticles. *Phys. Rev. B*, **79**, 021404.

[87] Metwally, K., Mensah, S., and Baffou, G. 2015. Fluence Threshold for Photothermal Bubble Generation using Plasmonic Nanoparticles. *J. Phys. Chem. C*, **119**, 28586–28596.

[88] Muir, T. G., and Cartensen, E. L. 1980. Prediction of Nonlinear Acoustic Effects at Biomedical Frequencies and Intensities. *Ultrasound in Med. & Biol.*, **6**, 345–357.

[89] Neumann, O., Feronti, C., Neumann, A. D., Dong, A., Schell, K., Lu, B., Kim, E., Quinn, M., Thompson, S., Grady, N., Nordlander, P., Oden, M., and J., Halas N. 2013a. Compact Solar Autoclave Based on Steam Generation using Broadband Light-Harvesting Nanoparticles. *Proc. Natl. Acad. Sci. U.S.A.*, **110**(29), 11677–11681.

[90] Neumann, O., Urban, A. S., Day, J., Lal, S., Nordlander, P., and Halas, N. J. 2013b. Solar Vapor Generation Enabled by Nanoparticles. *ACS Nano*, **7**(1), 42–49.

[91] Ni, G., Miljkovic, N., Ghasemi, H., Huang, X., Boriskina, S. V., Lin, C. T., Wang, J., Xu, Y., Mahfuzur Rahman, M., Zhang, T. J., and Chen, G. 2015. Volumetric Solar Heating of Nanofluids for Direct Vapor Generation. *Nano Energy*, **17**, 290–301.

[92] Ning, H., Buitenhuis, J., Dhont, J. K. G., and Wiegand, S. 2006. Thermal Diffusion Behavior of Hard-Sphere Suspensions. *J. Chem. Phys.*, **125**(20), 204911.

[93] Oraevsky, A. A., Karabutov, A. A., and Savateeva, E. V. 2001. Enhancement of Optoacoustic Tissue Contrast with Absorbing Nanoparticles. *Proc. SPIE*, **4434**, 60–69.

[94] Petrova, H., Perez Juste, J., Pastoriza-Santos, I., Hartland, G. V., Liz-Marzán, L. M., and Mulvaney, P. 2006. On the Temperature Stability of Gold Nanorods: Comparison between Thermal and Ultrafast Laser-Induced Heating. *Phys. Chem. Chem. Phys.*, **8**, 814–821.

[95] Piazza, R. 2008. Thermophoresis: Moving Particles with Thermal Gradients. *Soft Matter*, **4**, 1740–1744.

[96] Piazza, R., and Guarino, A. 2002. Soret Effect in Interacting Micellar Solutions. *Phys. Rev. Lett.*, **88**(20), 208302.

[97] Piazza, R., and Parola, A. 2008. Thermophoresis in Colloidal Suspensions. *J. Phys.: Condens. Matter*, **20**, 153102.

[98] Piazza, R., Iacopini, S., and Triulzi, B. 2004. Thermophoresis as a Probe of Particle-Solvent Interactions: The Case of Protein Solutions. *Phys. Chem. Chem. Phys.*, **6**, 1616–1622.

[99] Platten, J. K. 2006. The Soret Effect: A Review of Recent Experimental Results. *J. Appl. Mech.*, **73**, 5–15.

[100] Plech, A., Kotaidis, V., Grésillon, S., Dahmen, C., and von Plessen, G. 2004. Laser-Induced Heating and Melting of Gold Nanoparticles Studied by Time-Resolved X-Ray Scattering. *Phys. Rev. B*, **70**, 195423.

[101] Plech, A., Kotaidis, V., Lorenc, M., and Boneberg, J. 2006. Femtosecond Laser Near-Field Ablation from Gold Nanoparticles. *Nature Phys.*, **2**, 44–47.

[102] Polman, A. 2013. Solar Steam Nanobubbles. *ACS Nano*, **7**(1), 15–18.

[103] Putnam, S. A., and Cahill, D. G. 2004. Micron-Scale Apparatus for Measurements of Thermodiffusion in Liquids. *Rev. Sci. Instrum.*, **75**(7), 2368–2372.

[104] Putnam, S. A., and Cahill, D. G. 2005. Transport of Nanoscale Latex Beads in a Temperature Gradient. *Langmuir*, **21**, 5317–5323.

[105] Putnam, S. A., Cahill, D. G., and Wong, G. C. L. 2007. Temperature Dependence of Thermodiffusion in Aqueous Suspensions of Charged Nanoparticles. *Langmuir*, **23**, 9221–9228.

[106] Reichl, M., Herzog, M., Götz, A., and Braun, D. 2014. Why Charged Molecules Move Across a Temperature Gradient: The Role of the Electric Field. *Phys. Rev. Lett.*, **112**, 198101.

[107] Reineck, P., Wienken, C. J., and Braun, D. 2010. Thermophoresis in Single Stranded DNA. *Electrophoresis*, **31**, 279–286.

[108] Righini, M., Zelenina, A. S., Girard, C., and Quidant, R. 2007. Parallel and Selective Trapping in a Patterned Plasmonic Landscape. *Nat. Phys.*, **3**, 477.

[109] Ruckenstein, E. 1981. Can Phoretic Motions Be Treated as Interfacial Tension Gradient Driven Phenomena? *J. Coll. Interf. Sci.*, **83**, 77–81.

[110] Rusconi, R., Isa, L., and Piazza, R. 2004. Thermal-Lensing Measurement of Particle Thermophoresis in Aqueous Dispersions. *J. Opt. Soc. Am. B*, **21**(3), 605–616.

[111] Schmidt, A. J., Alper, J. D., Chiesa, M., Chen, G., Das, S. K., and Hamad-Schifferli, K. 2008. Probing the Gold Nanorod–Ligand–Solvent Interface by Plasmonic Absorption and Thermal Decay. *J. Phys. Chem. C*, **112**, 13320.

[112] Seidel, S. A. I., Wienken, C. J., Geissler, S., Jerabek-Willemsen, M., Duhr, S., Reiter, A., Trauner, D., Braun, D., and Baaske, P. 2012. Label-Free Microscale Thermophoresis Discriminates Sites and Affinity of Protein–Ligand Binding. *Angew. Chem. Int. Ed.*, **51**, 10656–10659.

[113] Seidel, S. A. I., Dijkman, P. M., Lea, W. A., van den Bogaart, G., Jerabek-Willemsen, M., Lazic, A., Joseph, J. S., Srinivasan, P., Baaske, P., Simeonov, A., Katritch, I., Melo, F. A., Ladbury, J. E., Schreiber, G., Watts, A., Braun, D., and Duhr, S. 2013. Microscale Thermophoresis Quantifies Biomolecular Interactions under Previously Challenging Conditions. *Methods*, **59**, 301–315.

[114] Seidel, S. A. I., Markwardt, N. A., Lanzmich, S. A., and Braun, D. 2014. Thermophoresis in Nanoliter Droplets to Quantify Aptamer Binding. *Angew. Chem. Int. Ed.*, **53**, 7948–7951.

[115] Setoura, K., Ito, S., and Miyasaka, H. 2017. Stationary Bubble Formation and Marangoni Convection Induced by cw Laser Heating of a Single Gold Nanoparticle. *Nanoscale*, **9**, 719–730.

[116] Siems, A., Weber, S. A. L., Boneberg, J., and Plech, A. 2011. Thermodynamics of Nanosecond Nanobubble Formation at Laser-Excited Metal Nanoparticles. *New J. Phys.*, **13**, 043018.

[117] Soret, C. 1879. Sur l'état d'équilibre que prend au point de vue de sa concentration une dissolution saline primitivement homogène dont deux parties sont portées à des températures différentes. *Arch. Sci. Phys. Nat.*, **2**, 48–61.

[118] Sun, J. M., Gerstman, B. S., and Li, B. 2000. Bubble Dynamics and Shock Waves Generated by Laser Absorption of a Photoacoustic Sphere. *J. Appl. Phys.*, **88**(5), 2352–2362.

[119] Thormählen, I., Straub, J., and Grigull, U. 1985. Refractive Index of Water and its Dependence on Wavelength, Temperature and Density. *J. Phys. Chem. Ref. Data*, **14**(4), 933–945.

[120] Tyrrell, J. W. G., and Attard, P. 2001. Images of Nanobubbles on Hydrophobic Surfaces and their Interactions. *Phys. Rev. Lett.*, **87**(17), 176104.

[121] Vigolo, D., Brambilla, G., and Piazza, R. 2007. Thermophoresis of Microemulsion Droplets: Size Dependence of the Soret Effect. *Phys. Rev. E*, **75**, 040401.

[122] Vogel, A., and Venugopalan, V. 2003. Mechanisms of Pulsed Laser Ablation of Biological Tissues. *Chem. Rev.*, **103**, 577.

[123] Vogel, A., Noack, J., Hüttman, G., and Paltauf, G. 2005. Mechanisms of Femtosecond Laser Nanosurgery of Cells and Tissues. *Appl. Phys. B*, **81**, 1015.

[124] Volkov, A. N., Sevilla, C., and Zhigilei, L. V. 2007. Numerical Modeling of Short Pulse Laser Interaction with Au Nanoparticle Surrounded by Water. *Appl. Surf. Sci.*, **253**, 6394.

[125] Wang, Y., Zaytsev, M. E., Le The, H., ECornelis Titus Eikel, J., Zandvliet, H. J. W., Zhang, X., and Lohse, D. 2017. Vapor and Gas Bubble Growth Dynamics around Laser-Irradiated Water-Immersed Plasmonic Nanoparticles. *ACS Nano*, **11**(2), 2045–2051.

[126] Wiegand, S. 2004. Thermal Diffusion in Liquid Mixtures and Polymer Solutions. *J. Phys.: Condens. Matter*.

[127] Wienken, C. J., Baaske, P., Rothbauer, U., Braun, D., and Duhr, S. 2010. Protein-Binding Assays in Biological Liquids using Microscale Thermophoresis. *Nature Commun.*, **1**, 100.

[128] Würger, A. 2008. Transport of Charged Colloids Driven by Thermoelectricity. *Phys. Rev. Lett.*, **101**, 108302.

[129] Würger, A. 2010. Thermal Non-Equilibrium Transport in Colloids. *Rep. Prog. Phys.*, **73**(12), 126601.

[130] Würger, A. 2016. Hydrodynamic Boundary Effects on Thermophoresis of Confined Colloids. *Phys. Rev. Lett.*, **116**, 138302.

[131] Zhao, C., Liu, Y., Zhao, Y., Fang, N., and Huang, T. J. 2013. A Reconfigurable Plasmofluidic Lens. *Nature Commun.*, **4**, 2305.

[132] Zielinski, M. S., Choi, J. W., La Grange, T., Modestino, M., Hashemi, S. M. H., Pu, Y., Birkhold, S., Hubbell, J. A., and Psaltis, D. 2016. Hollow Mesoporous Plasmonic Nanoshells for Enhanced Solar Vapor Generation. *Nano Lett.*, **16**, 2159–2167.

[133] Zijlstra, P., Chon, J. W. M., and Gu, M. 2009. White Light Scattering Spectroscopy and Electron Microscopy of Laser Induced Melting in Single Gold Nanorods. *Phys. Chem. Chem. Phys.*, **11**, 5915–5921.

This chapter is intended to present all the applications based on the use of metal nanoparticles as nanosources of heat, namely protein denaturation, photothermal cancer therapy, drug and gene delivery, heat-assisted magnetic recording, photoacoustic imaging, plasmonic-induced nanochemistry, photothermal imaging, solar steam generation generation and single living cell experiments. For each application, particular attention will be paid to (i) the pioneering works and how the thematics were born, (ii) the subsequent pivotal works that introduced the variants and new concepts and (iii) the current state of the art and remaining challenges. For each application, I shall also adopt a chronological, story-like description, and occasionally propose various reading templates (chronological, experimental, theoretical, ...), even if it yields redundancy.

6.1 Protein Denaturation: The Very First Application of Thermoplasmonics (1999)

Although one usually presents photothermal cancer therapy (2003) and photothermal imaging (2002) as the two very first applications of thermoplasmonics, there exists a pioneer article published in 1999 by Hüttmann and Birngruber [87]. This work benefited from the photothermal effects of gold nanoparticles and I believe this work can be considered as the very first article reporting on an application of thermoplasmonics. This is the reason why I dedicate a full section to this article, although protein denaturation cannot be considered as an important application of thermoplasmonics today. The last section of this chapter, dedicated to the application in biology, will be the occasion to recall this work.

The idea of this article was to study the thermal-induced denaturation of proteins using a pulsed laser to heat gold nanoparticles. At that time, the authors already understood that the temporal and spatial confinement achieved when heating nanoparticles with a sub-nanosecond laser could help achieve temperature as high as 470 K without boiling. The authors evidenced the denaturation of chymotrypsin proteins within 300 ps at temperatures below 380 K. This work was carried out in the context of photothermal treatment of vessels or pigmented cells.

At that time, I do not think the authors realized that most of the problems they raised in this article were about to be the subject of a large number of forthcoming articles. The authors also made correct predictions, for instance when saying: "It is expected that only solid-state absorbing particles (*e.g.,* metal spheres, melanin, graphite, or iron oxide particles) can be used as such an energy acceptor for thermal microeffects. Dye molecules do probably not have the required photostability and will, therefore, rather produce photochemical damage than photothermal effects."

At that time, two other researchers, Charles P. Lin and Michael W. Kelly, were using photothermal effects of microparticles to induce cancer cell death [117, 118], but they were not using *metal* nanoparticles yet. This work on protein denaturation using gold nanoparticles has certainly something to do with the subsequent pioneering use of gold nanoparticles by Charles P. Lin in the treatment of cancer in 2002 (see Section 6.2), because "stimulating discussions" with Lin were acknowledged by Hüttmann and Birngruber in their article.

6.2 Plasmonic Photothermal Therapy (PPTT)

Conventional cancer therapies, such as surgery, chemotherapy or radiation, lack cancer cell specificity. Moreover, chemotherapy and radiation have deleterious side effects, are harmful and adversely impact a patient's overall well-being [44].

To circumvent these major problems, it was proposed to use gold nanoparticles as photothermal agents. The idea of using gold nanoparticles as nanosources of heat for photothermal cancer therapy is one of the most ancient and the most promoted application of thermoplasmonics. After a general introduction of the thematic, I will proceed with an historical presentation of the most important seminal works, from in vitro investigation to clinical trials. A final section will be dedicated to a critical discussion of the main remaining unsolved challenges that explain why this therapy is still not at its final stage of development.

6.2.1 Hyperthermia for Cancer Therapy

Killing cells by heating them above a certain temperature threshold has long been considered a means to cure cancer, since as early as the late 1800s [66, 154, 139], sometimes applied as an adjunctive therapy with various established cancer treatments such as radiotherapy and chemotherapy [105]. A temperature rise at around 41–48°C is in principle sufficient to induce cell death. This process is called *hyperthermia*. The application of even higher temperatures (48–60°C) is termed *ablation*. In any case, an efficient photothermal treatment relies on a subtle interplay between temperature and exposure time [154, 139]. The golden rule is that a temperature increase of 1K reduces by a factor of 2 the required time to induce a given dose of damage [141]. This rule is valid from 43°C.

Here is a summary of the variety of effects caused by thermal treatments, according to Reference [95]:

- 37–41°C: DIATHERMIA
  - Cellular homeostasis can be maintained
  - Increment in diffusion rates across membranes
  - Increment in blood flow

- 41–48°C: HYPERTHERMIA

  - Unfolding and aggregation of proteins
  - Increased susceptibility to radiation and chemotherapy Irreversible
  - damage for long exposures (> 60 min)

- 48–60°C: HYPERTHERMIA

  - Severe and irreversible denaturation of proteins
  - DNA damage and denaturation
  - Irreversible damage for short exposures (4–6 min)

- > 60°C: HYPERTHERMIA
  Near instantaneous protein coagulation

Various approaches have been developed to induce hyperthermia in tumors. Heating has been applied using different types of incoming energy, namely microwaves, radiowaves, ultrasound waves, visible laser light and oscillating magnetic fields. Irrespectively of the approach, the risk of irreversible damage to surrounding healthy tissues is what prevented hyperthermia from becoming a widely used treatment in cancer therapy [116].

6.2.2 Hyperthermia Using Plasmonic Nanoparticles

Plasmonic nanoparticles can be advantageously used to artificially enhance the optical absorption contrast between cancerous and healthy cells and to use moderate laser intensities. This way cancer cells can be heated and destroyed using a (laser) light illumination at the tumor location, at least in theory. For an efficient cancer treatment following this approach, several requirements have to be fulfilled.

Active and passive targeting. First, gold nanoparticles have to be specifically located in cancer cells and not elsewhere. For this purpose, two approaches are usually considered to achieve specific targeting of the nanoparticles [74, 145]: passive targeting and active targeting. In *passive targeting*, the nanoparticles are injected intravenously and the specific localization of the nanoparticles inside the tumor after a few hours relies on the natural presence of vasculatures (up to 2 µm in size) that facilitate nanoparticle uptake by the cancer cells. Additionally, the lymphatic drainage of tumors is reduced compared with healthy tissues, making it harder for nanoparticles to leave the tumor once they get into it. This aspect is often referred to as the enhanced permeability and retention (EPR) effect [18]. A consequence of the EPR effect is that macromolecules or nanoparticles can accumulate in tumors at concentrations five to ten times higher than in normal tissue. In *active targeting*, the nanoparticles are also injected intravenously, but the nanoparticle uptake by cancer cells is further favored by coating the nanoparticle surface with cell surface receptors (*e.g.*, epidermal growth factor receptors, EGFRs), peptides or antibodies that have a specific binding affinity with receptors overexpressed at the membrane of cancer cells. Nanoparticle internalization can then occur by receptor-mediated endocytosis. In any case, the size of the nanoparticles has to remain smaller than 100 nm to allow for long circulating times in the bloodstream and to enable effective nanoparticle incorporation into cells [95].

Human transparency window. Second, using light instead of microwaves, ultrasound or magnetic fields causes a major problem: human bodies are not especially transparent.

In order to minimize this issue, one has to use a wavelength that matches the so-called biological window of human tissue. The biological window is composed of two separate wavelength ranges [149]. The first window spans from around 700 nm to 900 nm. In this wavelength range, the penetration depth inside tissues can reach a couple of centimeters [8]. The absorption is minimized but a strong scattering still limits the light penetration. The second biological window [166] spans from 1000 to 1400 nm. In this range, absorption does not vanish but scattering is much reduced. This is the reason why most of the investigation in photothermal cancer therapy using metal nanoparticles were based on the use of gold nanoparticles with a non-spherical shape to redshift the resonance, such as nanorods [83, 171, 55], introduced by the group of El-Sayed, core–shell dielectric-gold nanoparticles [53, 73, 105, 120], rather promoted by the group of Halas or nanocages [38]. To some extent, the use of spherical gold nanoparticles can also be efficient in the infrared, although gold nanosphere resonances are in the green region of the spectrum, due to agglomeration of nanoparticles that tends to red-shift the nanoparticle absorption spectrum [188, 126, 134].

Typical preclinical trial procedure. First experiments on plasmonic photothermal therapy (PPTT) of cancer were made in living cells in culture, but rather rapidly, preclinical trials have been made on mice and most of them are based on the same procedure (see Figure 6.1(a)). Subcutaneous tumors were grown in mice up to a certain size, typically one centimeter big. Half the mice population subsequently received gold nanoparticles via in situ deposition or via tail injection, while the remaining mice only received an injection of a saline solution, as a reference. After a few hours, most of the nanoparticles were supposed to have reached the tumor. Laser illumination was thus performed right at the tumor location for a few minutes, at a given laser intensity, sometimes upon controlling the temperature (see Figure 6.1(b)). This process was repeated several days and at the end of the treatment, comparison was made between the mice with and without nanoparticle injection. Figure 6.1(c) shows a mouse before and after effective treatment.

Three photothermal mechanisms leading to cell death. One can identify three different mechanisms leading to cell death. (i) As mentioned above, a simple temperature increase up to 45°C. However, reaching a uniform temperature increase of 45°C in all cancer cells and no deleterious temperature increase in neighboring healthy cells may seem unrealistic. First, the precise value of 45°C is difficult to control. Then, the temperature increase cannot be restricted to the tumor volume because of heat diffusion. (ii) A more promising method consists in using a (nanosecond- to femtosecond-) pulsed laser illumination [147]. The sudden temperature bursts following each pulse of light remain confined at the vicinity of each nanoparticles and can reach huge values, close to 280°C, with no bubble formation (see Section 2.5 on page 64). The direct consequence is the local perforation of cell membrane and destruction of organelles, leading to cell death. (iii) Still under pulsed illumination, a further increase of the laser power can lead to the formation of transient nanobubbles. The sudden formation and collapse of a bubble generates a shock wave that propagates through the medium and can disrupt cell membranes and lysosomes, leading to cell death [107, 123].

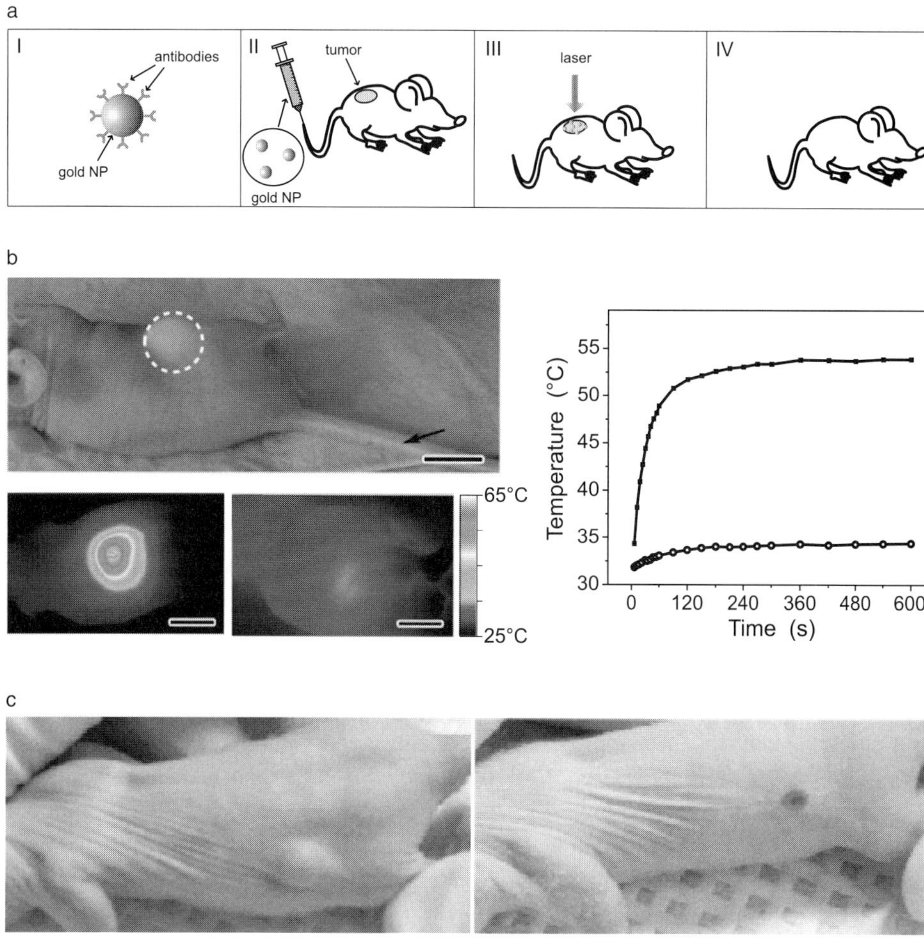

Fig. 6.1 (a) Schematic illustrating the usual approach in plasmonic photothermal therapy (PPTT). First, gold nanoparticles are functionalized with small molecules or antibodies that specifically target cancer cells. Then, a nanoparticle solution is directly injected into the tumor location or via tail vein injection. After a given period of incubation, the tumor is illuminated to heat the nanoparticles and generate hyperthermia. This procedure is repeated until healing is complete. Reproduced with permission from Reference [11]. Copyright 2012, WILEY–VCH Verlag GmbH & Co. KGaA, Weinheim (b) (Top-left) Photograph of a tumor-bearing mouse. The arrow indicates the location of injection of the nanocage or saline solutions. The dash circle indicates the size of the laser beam. (Bottom left) Thermographic images of nanocage-injected and saline-injected tumor-bearing mice. (Bottom right) Control. (Right) Plots of average temperature within the tumors (dashed circle) as a function of irradiation time. All scale bars are 1 cm. Reproduced with permission from Reference [38]. Copyright 2010, Wiley–VCH Verlag GmbH & Co. KGaA, Weinheim. (c) Photothermal tumor ablation: (left) Mouse before treatment. (right) Mouse after treatment. Reproduced with permission from Reference [168]. Copyright 2008, American Urological Association.

6.2.3 History of Plasmonic Photothermal Therapy

The first mention of the principle of plasmonic photothermal therapy (PPTT) dates from 2001. Dou *et al.* [58] chose to conduct numerical simulations of bubble formation on *gold* nanoparticles to match preliminary unpublished experiments and private discussions with the group of Charles P. Lin, who was working on PPTT at the cellular level. In the article, Dou *et al.* mentioned for the first time the possibility to use plasmonic nanoparticles as photothermal agents for selective cell damaging, stating that cell lethality could be due to bubble nucleation that results in loss of membrane integrity. The authors underlined that a benefit of using metal nanoparticles compared to absorbing organic compounds is that they remain stable and inert in cells for extended periods.

In June 2003, two years later, the group of Charles P. Lin published his experimental demonstration [147]. This group was already experienced in heating of micro-absorbers under pulsed laser illumination for photothermal treatment of cancer [117, 118]. However, they had only focused on organic microparticles such as melanosomes. This article introduced for the first time the use of metal nanoparticles. The proposed mechanism was based on the cell membrane permeabilization due to the transient generation of microbubbles.

In September 2003, a few months later, the groups of Jennifer L. West and Naomi J. Halas *et al.* [79] published a work based on the same idea of cell damaging using gold nanoparticles. However, the approach was based on a cw laser illumination and a regular heating of the whole cell (not on microbubbles induced by pulsed illumination). The specificity of their approach was also the use of silica-gold core–shell nanoparticles (called nanoshells), developed by the group of Halas. Nanoshells have been designed to have a 55 nm core radius and a 10-nm-thick shell to exhibit a peak absorption around 800 nm. In this work, human breast carcinoma cells were incubated with gold nanoshells in vitro. The cells were found to have undergone photothermally induced morbidity upon exposure to NIR light (820 nm, 35 W·cm^{-2}). Conversely, cells without nanoshells displayed no loss in viability using the same NIR illumination conditions. In vivo experiments on mice were also conducted in this pioneer work. The gold nanoparticle solution was directly injected interstitially into the tumor volume (control tumor sites received a saline injection). During the light exposure (820 nm, 4 W·cm^{-2}, 5 mm spot diameter, $<$ 6 min), temperature was monitored using MRI (magnetic resonance imaging). Analysis revealed that nanoshell-treated tumors resulted in an average temperature increase of 37.4±6.6°C on light exposure for around 5 min. Conversely, non-treated tumors saw average temperature increases of 9.1 ± 4.7°C. Statistics have been collected for 12 mice.

Shortly after, in 2004, the same group conducted similar in vivo experiments with, this time, the implementation of an *intravenous* nanoshell injection (instead of a direct interstitially injection). This was the first demonstration of (passive) targeting of nanoparticles in PPTT. The tumor temperature was no longer measured using MRI. Only the skin surface temperature was measured using an infrared thermometer. The treatment consisted of one laser-mediated heating performed during 3 min, 6 h after injection in order to let the delivered nanoshells accumulate in the tumors. Ten days after nanoshell treatment, complete resorption of the tumor was observed with the eight treated mice, and they all remain healthy and free of tumors even after 90 days. Meanwhile, the tumors of the nanoparticle-free mice kept on growing despite the laser treatment. This work convincingly illustrates the capability of a PPTT approach to benefit cancer therapy, at least in the treatment of superficial tumors.

In 2005, the teams of Halas, West and Drezek joined their effort to make another substantial contribution to the field by introducing the concept of *active* targeting. In vitro experiments were conducted on breast cancer cells incubated with either anti-HER2-antibody functionalized nanoshells or nanoshells functionalized with nonspecific antibodies. Under near-infrared illumination, a much higher morbidity was observed in the population of cells incubated with the anti-HER2 functionalized nanoparticles.

In 2006, the group of Mostafa A El-Sayed, another very active group in PPTT, also conducted pioneer experiments on in vitro *active* targeting by publishing two similar articles [63, 84]. The authors compared in vitro experiments between malignant carcinoma cells (cancer cells) and benign epithelial cell line (healthy cells). 40 nm gold nanospheres were functionalized with anti-EGFR (anti-epithelial growth factor receptor), a clinically related cancer biomarker. Heating was performed with a laser illumination at 514 nm. It was found that the cancer cells suffered irreversible photothermal injury at a much lower power density (19–25 $W{\cdot}cm^{-2}$) compared to healthy cells (57 $W{\cdot}cm^{-2}$).

In 2006, the group of El-Sayed introduced the use of gold nanorods [83] as photothermal agents, illustrated with in vitro experiments. Just like nanoshells, the resonance of gold nanorods can also be tuned in the near infrared range. The use of nanorods was concomitantly proposed by the group of Alexander Wei [171]. In vivo experiments using gold nanorods were conducted by El-Sayed in 2008 [55].

In 2007, Chen and coworkers [37] introduced the use of gold nanocages ($\approx$ 45 nm in edge size) with in vitro experiments on cancer cells. Nanocages resemble nanoshells made by the Halas' group, except they have a cubic shape, a hollow body (no silica core) and truncated corners. The authors compared the efficacy of their nanocages with nanorods and nanoshells previously reported in the literature. Using nanocages, they measure a photothermal cell damage power density threshold of 1.5 $W{\cdot}cm^{-2}$, compared to 35 $W{\cdot}cm^{-2}$ for nanoshells [79] and 10 $W{\cdot}cm^{-2}$ for nanorods [83]. However, it is difficult to make comparisons from one experiment to another because the laser power threshold may depend on the nanoparticle concentration, nanoparticle agglomeration, beam diameter (due to collective thermal effects, which is certainly occurring in this article due to the large beam diameter of 2 mm and the cell confluence) and cell line (as demonstrated in Reference [63]).

The in vivo implementation of *active* targeting was first reported by Melancon *et al.* in 2008 [132, 121]. Biodistribution studies following 24 h of intravenous circulation found significantly enhanced tumor accumulation by the active targeting strategy (13% versus 5%) [121].

The years 2008 and 2009 saw a burst of review articles on PPTT, most of them being the most cited reviews on the thematics today [94, 27, 71, 105, 92, 85, 86, 93]. It was the time when an "impending clinical impact" was announced [105] by Halas and coworkers. Indeed, in 2008, the company founded in 2002 by Halas and West, Nanospectra Biosciences,[1] undertook the first clinical trials, but the results were not disseminated.

---

[1] www.nanospectra.com

After 2009, most of the proofs of principle and preclinical challenges were demonstrated: in vitro and in vivo experiments, passive and active targeting, etc. So the activities rather focused on refining the technique by developing, for instance, more exotic nanoparticle shapes such as nanocubes [182], nano-popcorn structures [122], hybrid structures [146, 163, 97] and non-plasmonic structures such as carbon-based nanoparticles [183, 67, 187, 152].

In 2010, a collaboration between the groups of El-Sayed and Nie re-examined the interest of *active* targeting in PPTT. They concluded that for PPTT, direct administration of particles to the tumor can be more effective than intravenous injection. This conclusion corroborates previous experiments with liposomes labeled with anti-Her2 [102] and gold nanoparticles labeled with transferrin [46], where it was shown that functionalization improves nanoparticle penetration into cells but produces no appreciable increase in particle accumulation in tumors [59].

As a conclusion, Figure 6.2 displays statistics on the number of articles published each year related to the field of PPTT. Still today, this field is an active and increasing area of research.

Clinical trials. Clinical trials have been conducted by only one private company, Nanospectra Biosciences, founded in 2002 by Jennifer West and Naomi Halas, from Rice University. The company launched an FDA[2]-approved pilot study named Aurolase® aimed at using silica-gold core–shell nanoparticles for photothermal therapy. The elected approach involved intravenous injection in the patient's bloodstream and passive targeting. From 2008 clinical trials have been conducted on patients suffering from lung cancer. Although trials are complete the results have not yet been disseminated [145].

—— **Webofknowledge statistics – Plasmonic photothermal cancer therapy** ——

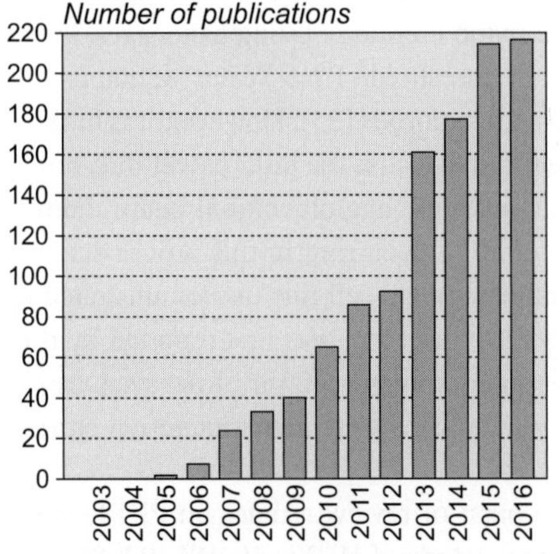

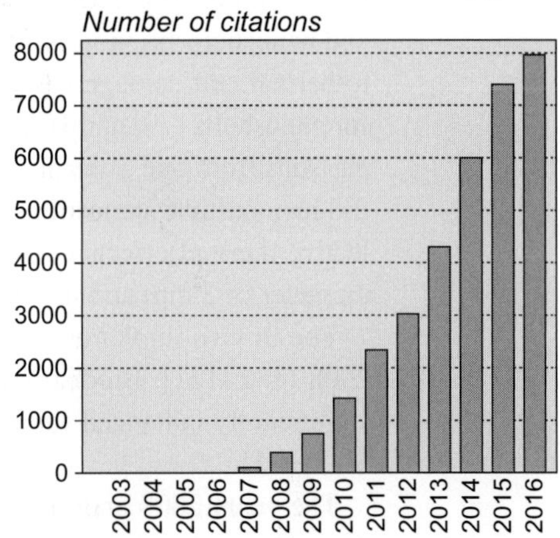

Fig. 6.2 · Web of Knowledge statistics [3] (January 2017), corresponding to the search criterion "TOPIC: (plasmon* and (thermal or photothermal) and (therapy or cancer))." Results found: 1134 / Average citations per item: 30.26 / h-index: 84.

---

[2] Food and Drug Administration

Note that clinical trials on hyperthermia for cancer therapy using *magnetic* nanoparticles for a magnetothermal approach are, conversely, much more advanced. For instance, clear results of Phase-II clinical trials have been reported in 2007 [124].

6.2.4 Remaining Current Challenges in PPTT of Cancer

Although PPTT certainly is the most promoted application of thermoplasmonics, the fact that the results of the clinical trials on cancer therapy have not been communicated may be a sign that the targeted challenge is bigger than expected, and may be out of reach. Here are the current problems that may limit the applicability of this therapy.

Restriction to subcutaneous tumors. Preclinical trials have been successfully conducted only on *subcutaneous* tumors, *i.e.*, tumors that are easily accessible, removable using simple surgery, and that do not need therapy. Surprisingly, no preclinical trials have been demonstrated on tumors of interest that were more deeply located in the body of living animals. One can hardly believe that nobody has tried, so this lack of demonstration may be a sign that PPTT can treat superficial tumors but cannot be used to treat deeper tumors.

Temperature spreading. Many approaches are based on a global photothermal effect under cw illumination. In such a case, the spatial distribution of the temperature will not be localized around each nanoparticle but rather delocalized throughout the whole tumor (due to thermal collective effects, see Subsection 2.2.5 on page 50), and even outside due to heat diffusion. For this reason, it will be difficult to confine the temperature increase to the sole tumor. As all the preclinical trials were involving subcutaneous tumors, far from any important organ, a temperature spreading out of the area of interest has never been detrimental. But for real applications, where tumors are more deeply located in the human body, temperature spreading out of the area of interest could be highly hazardous, like heating a lung cancer at 50°C with the heart in proximity. And it will be all the more hazardous that it will be difficult to ensure a precise temperature increase. For this reason, approaches based on pulsed illumination seem wiser as the temperature rises are confined in the vicinity of the nanoparticles (see Figure 2.21 on page 75).

Temperature nonuniformity. Another problem occurs under cw illumination. There is no reason for the nanoparticle distribution to be uniform in the tumor. Consequently, the temperature distribution throughout the tumor is not supposed to be uniform. It is thus difficult to understand how the temperature of 45–46°C could be applied through the whole tumor. Some parts of the tumor may be destroyed while some other parts (that remained at a lower temperature) may survive, and even worse may induce the cancer to metastasize (to different organs).

Temperature monitoring. Ideally, to make sure any part of the tumor reaches the desired temperature and no injury is caused to nearby organs, a three-dimensional map of the temperature increase would be required. But no imaging technique enables this performance, except MRI.

Opacity of human body. Finally, a very common problem is the fact that the human body is not transparent. This may explain why magnetothermal treatments have reached phase-II clinical trials [124], unlike PPTT approaches (human body is fully "transparent" to magnetic fields). Even in the infrared, it is difficult to reach a light penetration larger than one centimeter, not only because of water and blood absorption [44], but also due to tissue scattering. Using an optical fiber as a means to carry light in the tumor location has been proposed to overcome this issue. However, the resulting temperature distribution may be highly nonuniform as the light intensity pattern would rapidly decay from the fiber tip. A strong temperature increase will be confined at the fiber extremity, making unclear the interest of an optical fiber compared with the use of a resistive heating wire located at the tip of a micrometric probe, which would roughly create the same temperature gradient in the tumor.

6.2.5 Plasmonic Photothermal Therapy of Atheroma

Noteworthily, in 2015, effective phase-II clinical trials were disseminated regarding the PPTT of atheroma [99]. Atheroma consists of an abnormal local accumulation of cells, lipids, and calcium in artery walls, leading to a restriction of blood flow. In most cases, atheroma most commonly results in heart attack and ensuing debility. It was shown that PPTT using silica-gold nanoparticles led to significant regression of coronary atherosclerosis. This kind of PPTT looks more promising compared to PPTT of cancer because it does not suffer from most of the drawbacks mentioned above. In particular, unlike with a tumor, destructing the whole atheroma is not a stringent requirement.

6.2.6 Plasmonic Photothermal Therapy of Acne Vulgaris

Apart from tumors and atheroma, another possible target of PPTT was identified in 2015: acne vulgaris [143]. The idea was proposed by D. Y. Paithankar and (many) coworkers. The authors used gold-coated silica nanoparticles and delivered them into sebaceous glands. By illuminating the glands using millisecond pulses of light, a local injury to sebaceous follicles and glands was performed resulting in a reduction in inflammatory lesion burden on the cheeks of patients (Figure 6.3). This is also a promising application that does not suffer from all the limitation encountered with PPTT of cancer.

6.3 Drug and Gene Delivery (DGD)

6.3.1 Principle of DGD

Transport and release of drugs or oligonucleotides (D&O) in a specific location in vivo is a crucial challenge for the improvement of therapies for human diseases [57, 172]. Although it barely existed some 30 years ago, DGD is now being utilized by tens of millions of patients every year. In this context, a delivery of D&O remotely triggered by an external stimulus offers strong advantages over a passive release or an internally triggered release (*e.g.*, by a chemical stimulus). In particular, the timing of drug delivery could be finely adjusted. For example, insulin is most effective when delivered to a diabetic in short bursts,

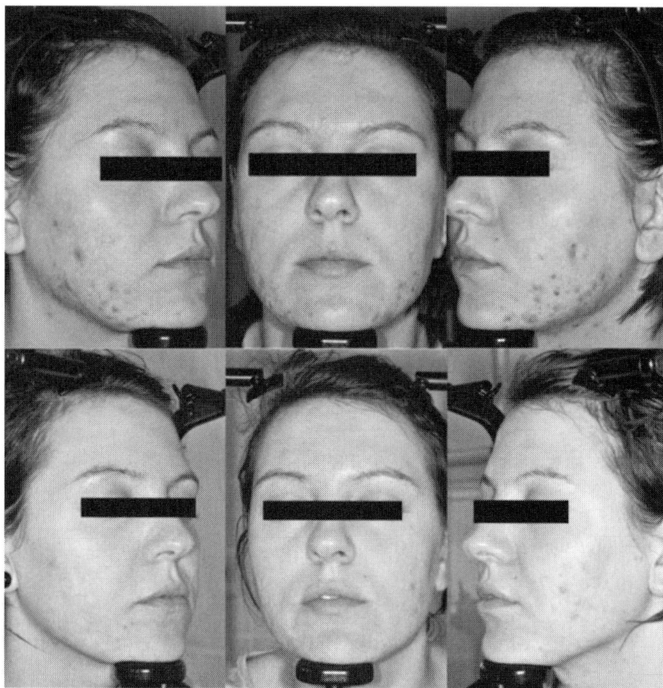

Fig. 6.3   Baseline (top row) and 24-week post-baseline (bottom row) photographs of a subject showing a reduction in inflammatory lesion burden on the cheeks. Reproduced from Reference [143]. Copyright 2015, The Society for Investigative Dermatology, Inc.

whereas an anesthetic should be delivered in a steady, continuous fashion [172]. The possible remote stimuli are light (ideally in the near-infrared), ultrasounds and magnetic fields. In any case, an efficient delivery system must fulfill several requirements. First, the active compounds must be protected against the surrounding biochemical conditions during transport. Second, it must remain inactive (mute) outside the target. Third, the delivery system must be nontoxic and biodegradable if it is given parenterally.

The initial principle of *plasmonic-assisted* delivery of D&O was proposed in 2000 [162]. Therapeutic compounds, functionalized to the surface of metal nanoparticles, are supposed to be released only under illumination due to a temperature increase inducing a bond breakage. The following 15 years of investigation mainly consisted in improving this basic scheme by proposing different variants, such as the drug release from capsule-like vehicles (nanocages, liposomes, micelles, …). The nature of the active compounds was also multiple, namely pharmaceutical drugs, nucleic acids, proteins or plasmid DNA.

The next subsection chronologically lists the articles that introduced all the variants of the basic principle of plasmonic photothermal D&O delivery.

6.3.2 History of Plasmonic Photothermal DGD

The idea of plasmon-assisted delivery of active molecules was initially proposed and patented by the group of J. W. West in 2000 [162]. In this seminal work, the molecular

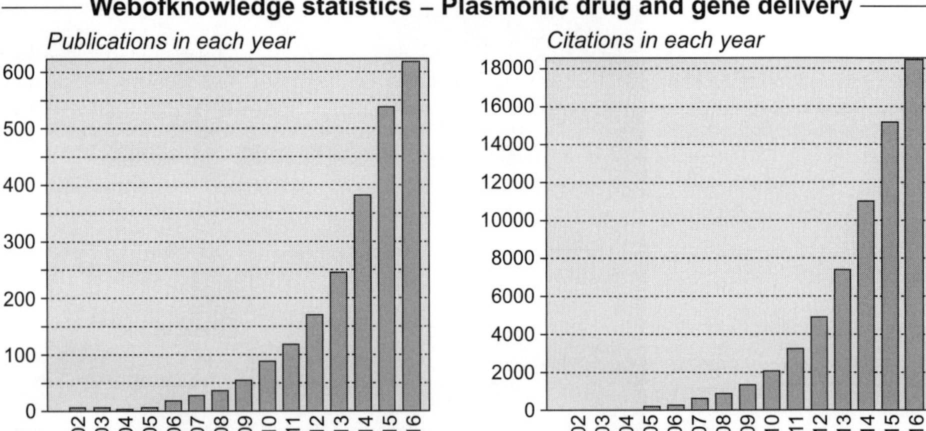

Fig. 6.4 TOPIC: ((drug or gene) and (delivery or release) and nanoparticle* and (plasmon* or photothermal*)) (January 2017). Results found: 2334 / Average citations per item: 28.46 / h-index: 115.

release resulted from the thermal-induced collapse of an hydrogel functionalized to gold nanoshells and containing the active molecules. As a proof of concept, the compounds of interest were methylene blue molecules, which are strongly absorbing dyes, easing the release monitoring via absorbance measurements. Experiments were also conducted on proteins.

In 2005, after some time of weak activity, this thematic demonstrated a renewed interest with the work of Takahashi *et al.* [170]. The authors released plasmid DNA immobilized on gold nanorods and introduced the use of a ns-pulsed illumination (repetition rate: 10 Hz). The use of a pulsed illumination confined the temperature increase around each nanoparticle and avoid a global, extended heating of the whole medium (see Subsection 2.21 on page 75). The authors discussed the mechanism of molecular release. According to them, it was due to a thermal-induced reshaping of the nanorods into nanospheres, favored by the pulsed nature of the illumination. Although the authors demonstrated the effective release of DNA, it was not carried out in a biological environment.

In 2006, Chen *et al.* performed the first cell transfection using a thermoplasmonic-assisted release of DNA [36]. The gene of enhanced green fluorescence protein (EGFP) was attached to gold nanorods. This enabled the authors to easily visualize the effective cell transfection by fluorescence measurements. Basically, HeLa cells that were illuminated using fs-pulsed NIR illumination became fluorescent. The proposed mechanism of DNA release was a nanoparticle reshaping into a nanosphere.

In 2006, the group of M. A. El-Sayed proposed a new mechanism of molecular release from plasmonic nanoparticles [91], different from the other admitted mechanisms that mainly consisted in thermal-induced bond breakage or nanoparticle reshaping. Here, the authors explained that a fs-pulsed illumination could lead to a sulfur–gold bond breakage involving the energy of the photoexcited electrons of the nanoparticles (today referred to as hot electrons).

Still in 2006, which was definitely the year were the burgeoning activity started in this thematic, Skirtach *et al.* introduced the concept of release of encapsulated compounds [165]. The authors used the layer-by-layer method[3] to produce nano-capsules. These capsules were decorated with silver nanoparticles in their walls, which could be remotely activated to release encapsulated material inside living cells. Fluorescently labeled polymers were chosen as a model system to evidence efficient molecular release. Successful experiments were conducted in living cells.

In 2007, Paasonen and coworkers pioneered the use of liposomes in plasmonic photothermal DGD [142]. Liposomes are artificial nanometric vesicles consisting of a hollow-body, spherical lipid bilayer. They are commonly used to carry pharmaceutical drugs or DNA (for cell transfection) until a target location. The content of a liposome can be subsequently released *e.g.*, by cell endocytosis, pH variations, enzymatic activity, ultrasounds, light or temperature variations. But some of these approaches may be invasive. The authors introduced a new concept of liposomal release based on a photothermal liposome permeabilization using gold nanoparticles. Indeed, a moderate temperature increase can induce a gel-to-liquid crystalline phase transition of lipid layers. To establish the concept, the authors investigated the release of a fluorescent compound (calcein), using cw UV light, encapsulated at self-quenching concentrations within the temperature sensitive liposomes. This way, the release was monitored by the apparition of fluorescence.

In 2008, the group of K. Hamad-Schifferli introduced the interesting concept of selective release of different molecules [180]. By attaching two different DNA oligonucleotides respectively to nanorods of distinct morphologies featuring distinct resonance wavelength, the authors were able the heat and thus release one type of DNA oligonucleotide at a time by adjusting the illumination wavelength (800 nm and 1100 nm). The same concept applied to liposomes was demonstrated later in 2011 by Leung *et al.* [111].

In 2008, the group of C. Burda demonstrated the use of plasmonic photothermal drug delivery in vivo [43]. The authors delivered drug for photodynamic therapy by passive targeting.

In 2010, the group of Lapotko introduced a new mechanism for plasmonic photothermal release [7]. Using liposomes decorated with spherical gold nanoparticles and containing fluorescent proteins, the authors demonstrated the effective release stemming from a bubble formation leading to a disruption of the liposome membrane. The bubble was created by heating the nanoparticles using a pulsed laser illumination. The delivery is thus mechanically induced by the formation of a bubble, not by a direct photothermal effect.

In 2012, the group of W. J. Parak addressed the problem of delivering macromolecules within the cell cytosol [32]. Crossing the cellular membrane to achieve efficient nanoparticle uptake is not straightforward and sometimes not convincingly demonstrated in the literature, as pointed out by Lévy and coworkers [112]: "several papers claim intracellular delivery of nanoparticles, but show pictures of cells that are, to the expert biologist, evidently dead (and therefore permeable)."

---

[3] The layer-by-layer method produces capsules by alternately adsorbing oppositely charged polymers on colloidal core templates, followed by template dissolution to end up with a hollow core.

In 2012, the group of Skirtach introduced an original approach [54]. The capsule-like vehicles for transport and release of molecules were red blood cells, decorated with gold nanoparticles. The red blood cells were filled with fluorescent molecules and the effective release was monitored by wide-field fluorescence microscopy imaging.

6.4 Heat-Assisted Magnetic Recording (HAMR)

6.4.1 History of HAMR

Magnetic recording, or magnetic storage, consists in storing binary information on a ferromagnetic film. Each bit value is spatially coded by the orientation of the magnetic dipole of ferromagnetic grains in one direction or the other (up/down or side to side). This concept, at the basis of current hard disk drives present in most computers, was invented more than 100 years ago and has enjoyed an exponential growth since the 1950s. The storage density of hard disk drives has doubled every three years since their introduction in 1956 by IBM. At the origin, one bit was recorded on unit areas of $(500\ \mu m)^2$, which corresponded to an areal density of $2\ \text{Kbit·in}^{-2}$ [137]. In 2014, one bit was recorded on a unit area smaller than $(100\ \text{nm})^2$, *i.e.*, at an areal density of $\sim 800\ \text{Gbit·in}^{-2}$.[4] During the last decade, by replacing longitudinal magnetic recording (LMR), *perpendicular* magnetic recording (PMR) was able to keep on increasing areal density, where the magnetic elements are aligned perpendicular to the disk surface.

However, nothing remains exponential for ever. The estimated limit of PMR is 1 Tbit·in^{-2}. Here are the two main reasons. (i) First, within the magnetic film, a sufficient number of magnetic grains (currently 10–20) have to be recorded per bit in order to increase the signal-to-noise ratio. A single grain would not be sufficient. Moreover, the grain distribution is random, not periodic, within the film. This prevents us from shrinking the size of a unit area at will. (ii) Second, since the 2000s, the technology is facing the problem of the stability of small magnetic grains undergoing thermal agitation. Below a certain size, the magnetic dipole of a ferromagnetic grain may become unstable, even below the Curie temperature, and the grain is said to be in a superparamagnetic state. The magnetization curve of an assembly of superparamagnetic grains shows no hysteresis and behaves like paramagnetic material with giant magnetic moments, which lends the effect its name. This prevents from shrinking the unit areas on the magnetic films by shrinking the grain size. More precisely, the stability of the magnetic dipole of a single grain of ferromagnetic material is given by a criterion that reads[5] [130]

$$\eta_0 = \frac{KV}{k_B T} \geq 60 \tag{6.1}$$

---

[4] There is a confusion in the literature where the unit Gb·in^{-2} sometimes means Gbit·in^{-2} and sometimes Gbytes·in^{-2}, where 1 byte equals 8 bits.

[5] The standard Arrhenius Néel theory [150] shows that an energy barrier of KV about $40\ k_B T$ is required to keep the magnetization stable for roughly 10 years at room temperature. For practical applications, the minimum required $KV/k_B T$ has to be increased to at least 60, because some margin has to be allowed for external demagnetization fields and other factors.

where V is the volume of the magnetic grain, KV its magnetic anisotropy energy and T the temperature. This is the so-called *superparamagnetic limit*. Thus, the smaller the grain volume, the higher the probability for spontaneous demagnetization and loss of information in magnetic storage. Logic suggests that shrinking the volume can be compensated by increasing K. For instance, materials with a large K value, such as FePt [179], could support stable grains as small as 2 to 3 nm in diameter and storage densities [103] up to 155 $Pb \cdot m^{-2}$. However, K cannot be increased at will. Indeed, the magnitude of the magnetic field required for data writing scales in proportion to K. Too high a value of K would require non-realistic magnetic field amplitudes to act on the magnetic dipole and write data. For this reason, the maximum achievable areal density is 1 $Tbit \cdot in^{-2}$. This fundamental barrier, also coined the superparamagnetic limit, is almost reached today (~ 800 $Gbit \cdot in^{-2}$ in 2014). To go beyond this limit, new technologies have to be developed.

A promising approach to circumvent this limitation consists in benefiting from the temperature dependence of $K(T)$, which is a decreasing function. K can be high at room temperature to optimize η_0 and the thermal stability of the storing magnetic medium, and can be made transiently smaller by locally increasing the temperature of individual unit areas only when recording information. This strategy has been investigated since the 1980s and is called heat-assisted magnetic recording (HAMR) by the American company Seagate Technology (Bloomington, MN, USA) [34], or thermally assisted magnetic recording (TAR) by the Japanese company Hitachi [169], or thermomagnetic recording. It is, however, not commercialized yet. To be effective, heating and cooling of unit areas of the magnetic medium have to be fast, and the heat deposit has to be spatially confined. Reference [130] derives some rough estimations of the required length, timescales and heat power. According to this article, for a targeted spatial scale on the order of 50 nm, which corresponds to an areal density of 1.5 $Pb \cdot m^{-2}$ (1 $Tb \cdot in^{-2}$), the transit time should be on the order of 0.1 ns and the delivered heat power on the order of 5 mW.

Statistics on the articles published in HAMR are displayed in Figure 6.5.

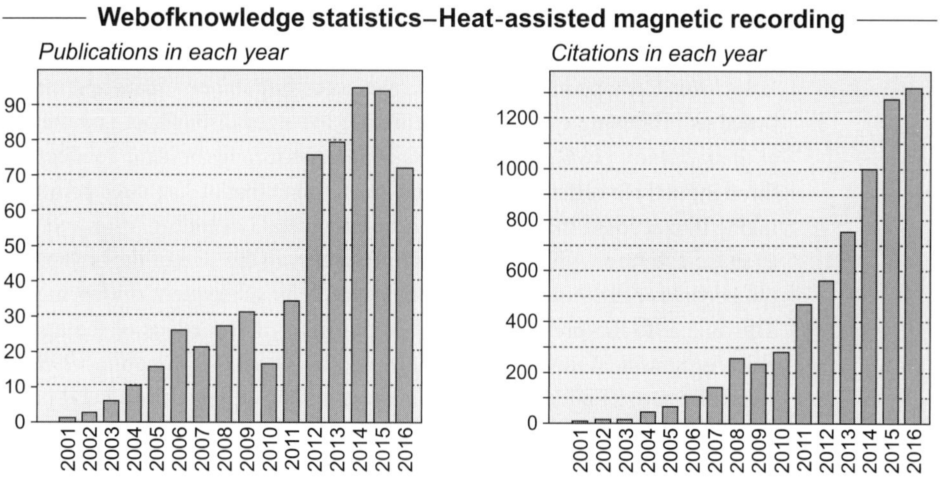

Webofknowledge statistics–Heat-assisted magnetic recording

Publications in each year

Citations in each year

Fig. 6.5   Web of Knowledge statistics [3] (Mars 2017), corresponding to the search criterion "`TOPIC:` `(heat-assisted magnetic recording)`." Results found: 619 / Average citations per item: 10.82 / h-index: 34.

Various strategies have been proposed to achieve HAMR, for instance heating using a contact with a hot metal tip [76] and by proximate resistive heating [98]. But a strategy that has been receiving increased attention recently consists in heating using optical means [33]. Such a contact-less approach is promising as it could achieve the requirements of delivered power (a few mW) and temporal dynamics (sub-nanosecond range). However, the diffraction limit does not allow heating over a scale as small as 50 nm using conventional laser wavelengths and simple laser focusing. For this reason, most of the proposed photothermal approaches involve the optical near-field of light nanosources (apertures, antennae, waveguides, solid immersion lenses) [33].

Noble metals are generally involved in such approaches (not necessarily for their plasmonic properties). They are often considered as perfect conductors and used to confine the light field. Indeed, in order to squeeze the spatial extension of an optical field, common sense would suggest drilling a nanometric aperture in a metal film and send light through it. This was basically done in a pioneer article in 1992 [25]. Erik Betzig[6] and coworkers used an NSOM (near-field scanning optical microscope) to image and record domains on a magnetic film. The probe of a NSOM consists of a tapered and metallized optical fiber in which light escapes via a nanoaperture drilled in the metal coating at the tip of the fiber. Although this approach is supposed to yield a light confinement as small as the size of the nano-aperture, a reduction of the size of the aperture comes along with a strong damping of the light throughput: the smaller the aperture size, the weaker the power delivery. Although the authors demonstrated a data recording density of ~ 45 Gb·in^{-2}, a throughput of only 10^{-6} for a 50 nm aperture is clearly too small for a practical storage device.

In theory, for a perfect conductor (no loss, acting as a perfect "mirror"), the throughput of a small, circular hole of diameter d ($d \ll \lambda$) scales as the diameter to the power of 4:

$$\frac{I_{\text{out}}}{I_{\text{in}}} = \frac{64\pi^2}{27} \left(\frac{d}{\lambda}\right)^4 \tag{6.2}$$

where λ is the wavelength of the incoming light, I_{in} the power per unit area of a plane wave that is normally incident upon the aperture and I_{out} the power per unit area transmitted through the aperture. Hence, shrinking an aperture size at will is not the magic solution.

In 2002, C-shape metal apertures, instead of circular or square apertures, have been proposed as promising candidates to enhance the light throughput and the light confinement at their vicinity [164]. Indeed, the C-shape aperture in an ideal conductor film can have a throughput three orders of magnitude greater than that of a square aperture. However, when taking into account the true properties of the metal (including loss), and when comparing a C-shape aperture with a rectangular aperture, both the size of the hot spot and the throughput becomes nearly identical, as demonstrated by subsequent studies with a gold film [33]. This is due to the presence of surface plasmons at the edges of the aperture, which only exist for non-ideal metals, as demonstrated by Thomas Ebbesen in 1998 [60]. For this reason, Equation (6.2) underestimates the energy deposit, which is a good point. However, the drawback of considering a real metal is the resulting light absorption by the metal itself

---

[6] Erik Betzig is better known for his Nobel Prize in Chemistry, awarded in 2015, for a totally different activity related to high-resolution microscopy using single molecules.

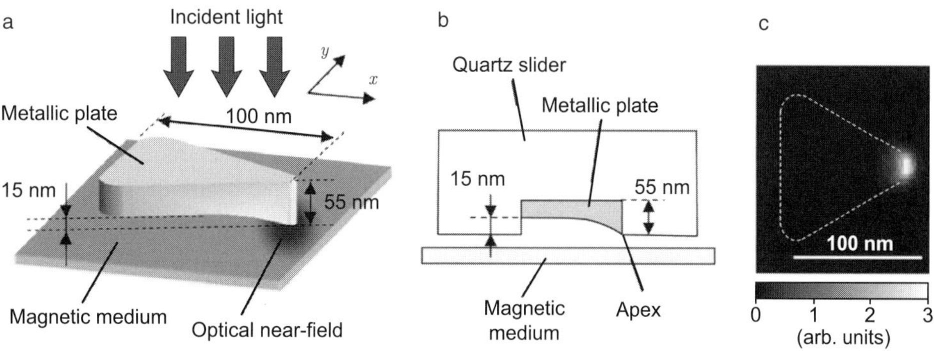

Fig. 6.6 Schematic of the probe used by Matsumoto and coworkers [127, 128, 129]: (a) 3D view and (b) cross-sectional view. (c) Intensity distribution of the optical near-field calculated on the surface of the recording medium. Reproduced with permission from Reference [128]. Copyright 2006, The Optical Society.

and the subsequent temperature increase of the writing head, which turns out to be the real limitation.

In 2006, Takuya Matsumoto and coworkers from the Hitachi company introduced for the first time the use of metallic finite-size nanostructures in HAMR as optical near-field enhancers [129]. Their idea was to benefit from the ability of metal nanotips to create a strong and confined optical field at its vicinity, which can be used to very locally heat the substrate over an area below the diffraction limit. Thus, in this application, it is not the temperature increase within the metal nanostructure itself that is involved in the mechanism, but rather the optical near-field. In this pioneer work, the metal structure, acting as a near-field transducer (NFT), consisted of a triangular plate endowed with a sharp beak, as represented in Figure 6.6. As a general rule in HAMR, the enhanced near-field from the metallic structure does not really originate from a plasmonic resonance, but rather from a lightning rod effect [52]. Marks as small as 40 nm, far below the diffraction limit (using a 780 nm laser and an objective lens of 0.8 NA), were successively obtained on a $Ge_2Sb_2Te_5$ magnetic layer. The distance between the metal tip and magnetic films was estimated to be 6 nm. This was a first proof of concept, but still far from practical applications as the magnetic film was not rotating at high speed and the dynamic aspect of data recording was not investigated. This pioneer article gave birth to the field of research named plasmon-induced HAMR (see statistics in Figure 6.7).

In 2008, the same group performed NFT-HAMR using their beaked triangle with a bit-patterned magnetic medium (BPM) rather than a conventional granular medium [129]. With such a medium, the unit cell area for bit recording is no longer constituted of multiple magnetic grains. It consists of single magnetic islands typically as small as 20 nm in size. Such a scheme is called bit-patterned recording (BPR) and enables a much higher areal density. By combining HAMR, NFT and BPR, the Hitachi company achieved an areal density of 830 Gbits·in^{-2} [129]. Magnetic-force-microscope images of the medium showed that the reverse magnetizations of single grains had been achieved. Merging NFT-HAMR and BPR is ultimately supposed to extend data storage density by up to a factor of 100 beyond the limits of conventional recording [130].

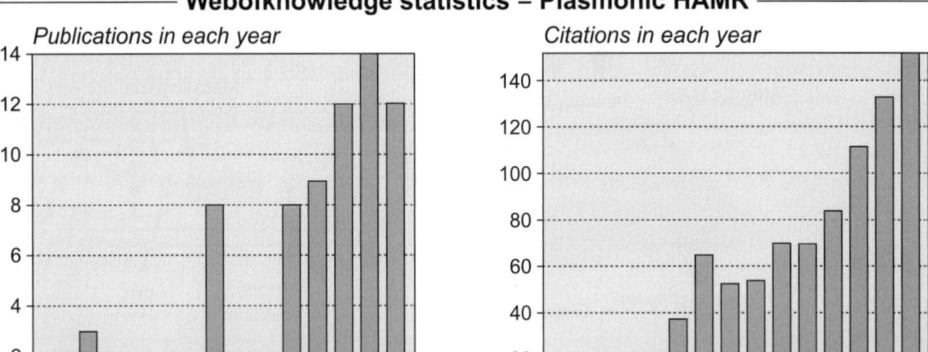

Fig. 6.7 Web of Knowledge statistics [3] (January 2017), corresponding to the search criterion "TOPIC: (`heat-assisted magnetic recording' and plasmon*)." Results found: 73 / Average citations per item: 12.05 / h-index: 13.

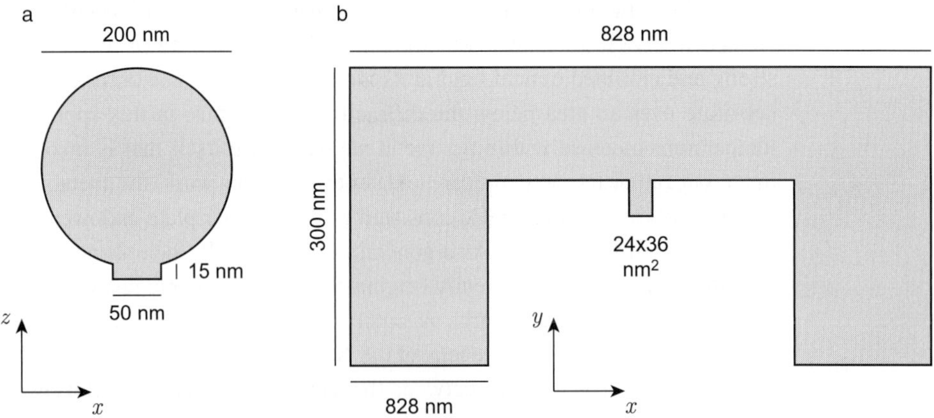

Fig. 6.8 Schematics of the e-beam metal nanostructures used as near-field transducers (NFTs) for HAMR in Reference [34] (a) and in Reference [169] (b).

In 2009, William A. Challenger [34], from the Seagate company (the other main actor in the field of HAMR), implemented the use of a HAMR metallic head flying over a magnetic recording medium on a rotating disk. The metal structure, called a NFT (near-field transducer) by the authors, consisted of a planar structure made by e-beam lithography, composed of a gold disc, 200 nm in diameter, endowed with a nanobeak (called a peg in the article) on its edge, 50 nm wide and 15 nm long (see Figure 6.8a). The whole structure was 22 nm thick. Such a lollipop-like structure, very similar to the beaked triangle of Matsumoto, was meant to produce an enhanced optical near-field around the nanobeak and heat the FePt magnetic film that was located 15 nm below. The authors achieved a data recording at an areal density of ~ 375 Tb·m^{-2}. However, no clear information was given regarding the recording speed, and on how much data can be recorded before the metal NFT gets damaged.

In 2010, a year after the seminal work of the Seagate company, the Hitachi company went one step further [169]. NFT-HAMR was performed with a much higher areal density, stemming from the use a BPM sample, which consisted in patterning the recording medium with lithographically defined magnetic islands, 20 nm in size, arranged in a hexagonal lattice, instead of randomly distributed magnetic grains. The use of a BPM sample in NFT-HAMR had already been pioneered by the Hitachi company a couple of years before [129], as mentioned above, but not on a rotating disc, and not with a nanosecond pulsed laser. Here, binary information was coded with a 24 nm track pitch, which led to them to reach the milestone of $\sim$ 1 Tbit·in^{-2} (1.5 Pbit·m^{-2}). 4 ns laser pulses were required with a 30 mW power incident on the antenna. The data rate was only about 1 Kbit·s^{-1}. Note that in this work, the authors are no longer using a beaked triangle, but rather an E-shape gold structure, as represented in Figure 6.8b. The notch at the centre was meant to concentrate in a small volume the surface charges generated through a plasmonic resonance in the body. With the assistance of the lightning rod effect produced by the notch, the FWHM spot size was reduced from more than 200 nm to $\sim$ 40 nm, along with increased peak intensity within the spot by 7 times.

In 2013, Seagate reached 1 Tbit·in^{-2} using HAMR, [181] and introduced a prototype of a fully integrated and functioning HAMR drive at CEATEC 2013 [1]. Western Digital (Irvine, CA, USA) also demonstrated its HAMR technology at the 2013 China International Forum on Advanced Materials and Commercialization, where a PC powered by a 2.5-inch HAMR hard drive was presented [2].

Today, each company in the hard disk drive industry has plans to introduce HAMR technology into the market within the next few years. Figure 6.9 proposes some scenarios depending on the implementation of the HAMR and BPM technologies.

Although HAMR is the subject of a full section of this last chapter, it is not, strictly speaking, an application of thermoplasmonics. Indeed, the heat generation of the metal

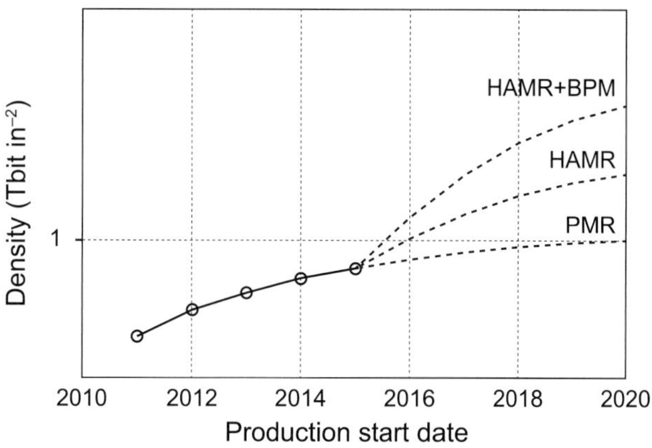

Fig. 6.9   Different scenarios in the magnetic recording industry: parallel magnetic recording (PMR), heat-assisted magnetic recording (HAMR) and HAMR on a bit-patterned medium (BPM). Inspired from Reference [189].

nanostructure itself is not the process of interest in such an application. It is rather the optical near field of the metal structures and the subsequent heat generation in the neighboring magnetic medium. Moreover, one can even wonder if what was discussed in this section stands for an application of plasmonics. Indeed, in most developments, the enhanced near-field of the metallic near-field transducer results from a simple lightning rod effect rather than from a plasmonic resonance. The lightning rod effect is a non-resonant phenomenon, also called "non-resonant amplification" to better describe the mechanism [52].

6.4.2 Remaining Issues in Near-Field Assisted HAMR

One of the major problems is the heat generation within the metal near-field transducer itself. One can hardly imagine that NFT could heat up a neighboring solid via its near-field, while remaining cold. Gold is the material of choice for NFTs as it features a melting point (1064°C) much higher than the Curie temperature of the magnetic medium in HAMR. However, gold nanoparticles are known to reshape at temperatures much weaker than the melting point of gold (typically between 100°C and 400 °C, see Section 5.11 on page 210). The lifetime of the NFT may suffer from this problem. For this reason, efforts are made to find new materials in plasmonics that can sustain higher temperatures, such as metal nitrides (TiN, ZrN) or refractory metals (W, Mo). TiN and ZrN have been shown promising as their plasmonic resonance are similar to gold's resonance (in wavelength and magnitude), and because they feature melting points close to 3000°C. However, they are not supposed to be good near-field enhancers [106].

Apart from this problem of heating of the NFT, the future challenges will be to achieve large chip-scale areas of bit-patterned medium inexpensively; optimizing the efficiency of the transducer to reduce wasted light energy; preventing the plasmonic particle and the magnetic medium from degrading owing to photothermal and thermo-mechanical effects arising from the enormous amount of heat being channeled through the write head; and ensuring fast write speeds (which will be limited by how fast the medium can be heated and cooled) [137].

6.5 Photoacoustic Imaging (PAI)

6.5.1 Principle of Photoacoustic Imaging

Photoacoustic imaging refers to a family of biomedical imaging modalities based on the photoacoustic effect, which consists of the generation of acoustic waves produced by the absorption of pulses of light (or of radio-frequency waves in some cases). Photoacoustic imaging (PAI) uses optical illumination and ultrasonic detection to produce deep tissue images based on their light absorption, and uses endogenous or exogenous contrast agents. This technique enables imaging in real time, with a high-spatial resolution (~ 5 µm), deep inside tissues (5–6 cm), on the anatomical functional and molecular content of biological tissues in the absence of ionizing radiation. Two main imaging modalities exist:

photoacoustic microscopy and photoacoustic tomography. As explained in Reference [75], "photoacoustic microscopy employs a coupled, focused ultrasonic detector–confocal optical illumination system to generate multidimensional tomographic images without the need for reconstruction algorithms, whereas the detectors in photoacoustic tomography scan the laser-illuminated object in a circular path and use inverse algorithms to construct three-dimensional images."

First proofs of concept of PAI dates from 1979–1980. But it was necessary to wait until the early 2000s to observe a rapid expansion of this area of research (see Figure 6.10), presumably due to progresses in microscopy and laser technologies. This technique proved to be a powerful tool to image both endogenous (blood vessels, tumors, etc.) and exogenous (nanoparticles, dyes, etc.) agents, with good spatial resolution and good contrast. Statistics on research carried out in PAI are displayed in Figure 6.10.

There has been tremendous effort devoted to the development of exogenous photoacoustic contrast agents [114], and gold nanoparticles are naturally very good candidates because of their strong light absorption properties in the infrared and their biocompatibility. The idea to benefit from gold nanoparticles as contrast agents in PAI dates from 2001 [140]. The use of nanoparticle-based contrast agents greatly extended PAI applications [185]. The benefit is three-fold: (i) It allows deeper imaging within tissue with enhanced contrast. Metal nanoparticles are highly absorbing and their absorption properties can be tuned in biological transparency windows. (ii) It allows active targeting of specific locations in living organisms using metal nanoparticles conjugated with antibodies. This way, systems endowed with weak endogenous photoacoustic contrast can be made highly visible using PAI. (iii) PAI can be coupled with photothermal therapy using gold nanoparticles acting both as photoacoustic and photothermal agents in tumors. Statistics on research carried out in PAI using plasmonic nanoparticles are displayed in Figure 6.11. Note that a new gain of

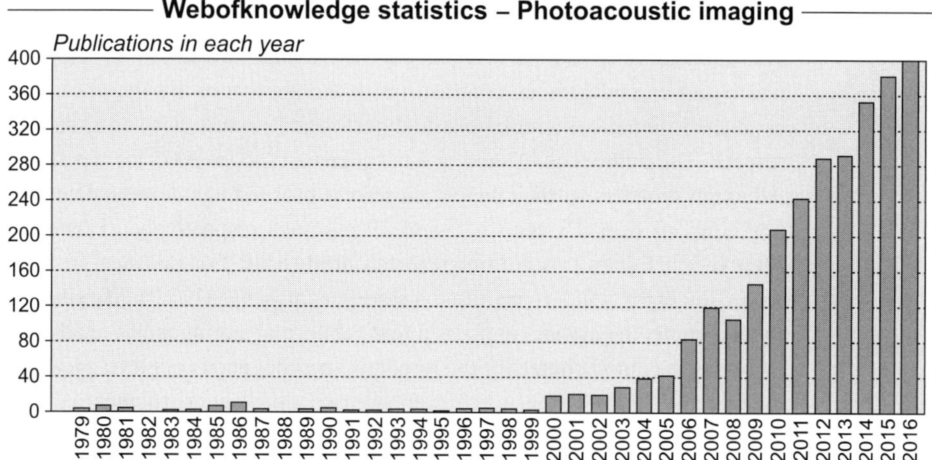

Fig. 6.10 Web of Knowledge statistics [3] (January 2017), corresponding to the search criterion "`TITLE: ((photoacoustic or optoacoustic) and (imaging or tomography or microscopy))`." Results found: 2879 / Average citations per item: 14.14 / h-index: 93.

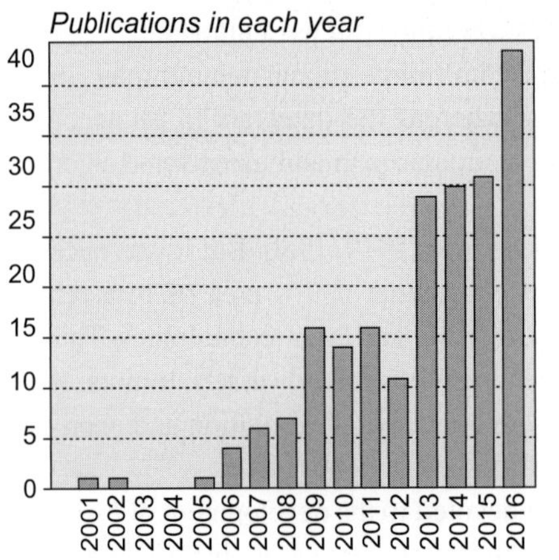
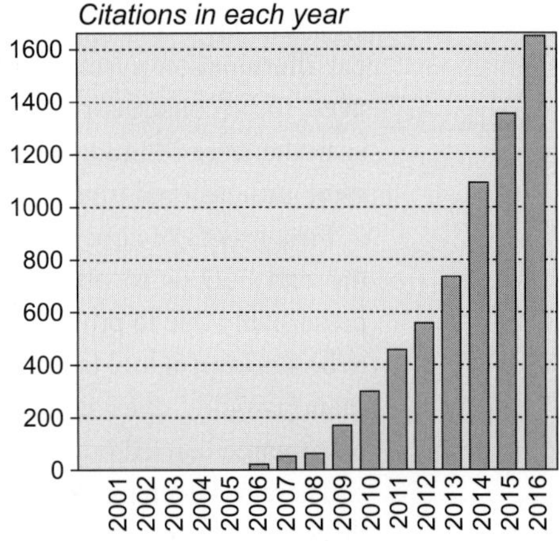

Fig. 6.11 Web of Knowledge statistics [3] (January 2017), corresponding to the search criterion "`TOPIC: ((photoacoustic or optoacoustic) and (imaging or tomography) and nano* and plasmon*)`." Results found: 210 / Average citations per item: 31.05 / h-index: 41.

interest has been observed since 2013, which can be correlated with the gain of interest observed in stress wave emission from plasmonic nanoparticles under pulsed illumination (see statistics presented in Figure 5.35 on page 203.)

6.5.2 History of Photoacoustic Imaging

In this subsection, the most important articles in PAI using plasmonic nanoparticles are reviewed.

In 2001, Alexander A. Oraevsky and coworkers proposed for the first time the use of gold (and carbon) nanoparticles as contrast agents in PAI [140]. Prior to this work, PAI was based on endogenous contrast, or on the absorption by dye molecules. The authors conducted numerical simulations and performed experiments on tissue phantoms in this respect. The authors published a subsequent article in 2003 on the same message [62]. Although pioneers in their fields, these two proceedings communications have gathered few citations over 15 years (16 and 19 citations respectively). However, Oraevsky has been very active in this field of research until today. For instance, in 2004, he published a peer-reviewed article with a much stronger impact [51] (210 citations in January 2017). In this article, the authors explained that "abnormal angiogenesis in advanced tumors, that increases the blood content of the tumor, is an endogenous contrast agent for photoacoustic tomography. In early stages, however, angiogenesis is not sufficient to differentiate a tumor from normal tissue; justifying the application of an exogenous contrast agent." For this purpose, they used spherical gold nanoparticles conjugated with antibodies (Herceptin) that specifically target the membrane of breast cancer cells. The nanoparticle targeted cancer cells were imaged using PAI in a gel phantom that optically resembled breast tissue (see Figure 6.12). A resolution of 0.5 mm was demonstrated.

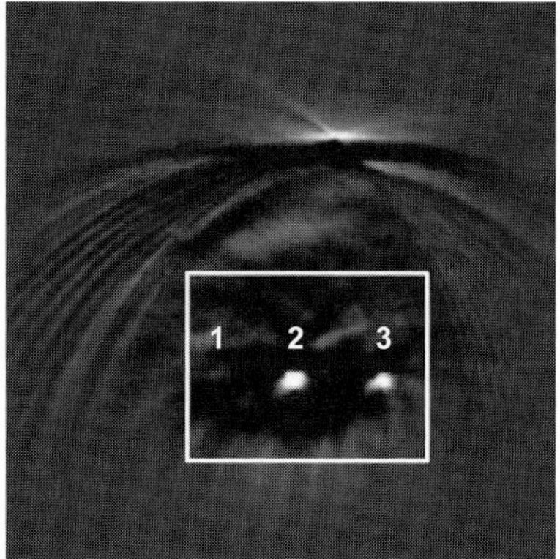

Fig. 6.12 (a) Photoacoustic image of a gel phantom with three embedded tubes: Tube 1 (1) was filled with the same media as the phantom gel (negative control), tube 2 (2) was filled with NPs (1.3×10^9 nanoparticle/ml) and tube 3 (3) was filled with the SK-BR-3 cells treated with nanoparticles. Tubes 2 and 3 can be clearly visualized. Reproduced with permission from Reference [51]. Copyright 2004, Elsevier.

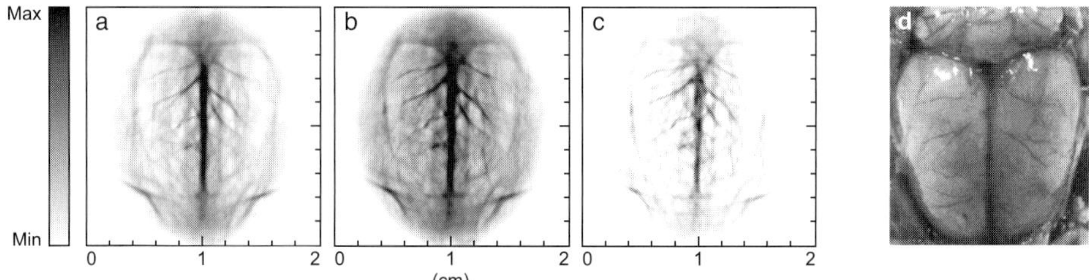

Fig. 6.13 Noninvasive PAT of a rat brain in vivo employing the nanoshell contrast agent and NIR light at a wavelength of 800 nm. (a) Open-skull photograph of the rat brain cortex obtained after the data acquisition for photoacoustic tomography. (a) Photoacoustic image acquired before the administrations of nanoshells. (b) Photoacoustic image obtained 20 min after the third administration of nanoshells. (c) Differential image that was obtained by subtracting the pre-injection image (a) from the post-injection image (b). Reproduced with permission from Reference [178]. Copyright 2004, American Chemical Society.

Still in 2004, in parallel with the seminal work of Oraevsky, the group of Lihong V. Wang reported on the use of gold nanoshells as photoacoustic agents to perform photoacoustic tomography in rat brain, in vivo [178]. This work in vivo was more advanced compared to the work of Oraevsky who focused on measurements in gel phantoms. However, the gain of using nanoparticles is not obvious from the images presented in Figure 6.13: The authors started from a situation where the endogenous contrast was already excellent (Figure 6.13a).

In 2007, after the use of nanoshells of the group of L. V. Wang and of nanosphere by Oraevsky *et al.*, Agarwal and coworkers implemented the use of gold nanorods in PAI [6]. The authors conducted two types of experiments, using antibody conjugated gold nanorods: experiments in vitro on LNCaP prostate cancer lines, and experiments in mice by implanting two small objects made of gelled nanoparticle solution in the upper part of

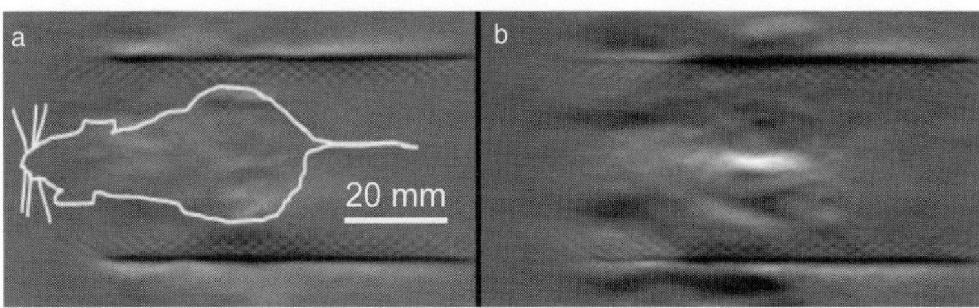

Fig. 6.14 A typical photoacoustic image of a nude mouse before (a) and after (b) subcutaneous injection of gold nanoparticles in the abdominal area. Drawing in (a) depicts the approximate position of the nude mouse during experiment. Reproduced with permission from Reference [61]. Copyright 2007, American Chemical Society.

a mouse hind limb. The authors evidenced that the presence of gold nanorods can enhance the photoacoustic contrast above the endogenous background by a factor of five with a gold nanorod concentration of about 1.5×10^{11} particles/ml. With this concentration, the authors managed to get a good signal-to-noise ratio with a fluence four times below the ANSI Z136 safety limit of 20 mJ·cm^{-2}.

In 2007 also, Eghtedari, Oraevsky and coworkers also reported the use of gold nanorods for in vivo applications of PAI [61]. The authors performed a subcutaneous injection of 100 µL of gold nanorods at a concentration of 7.5×10^{10} NRs per mL in the abdominal area, a concentration very similar to the work of L V Wang *et al.* [6]. The photoacoustic images obtained by Eghtedari are displayed in Figure 6.14.

The same year, Kim *et al.* also used nanorods to image early inflammatory response using PAI [101]. And Mallidi *et al.* used 50 nm gold nanospheres conjugated with Anti-EGFR monoclonal antibodies to study the benefit of active targeting for PAI of tumors [125]. Experiments were carried out on three-dimensional tissue phantoms prepared using a human keratinocyte cell line.

Still in 2007, using the same outline as in their previous work on gold nanoshells, the group of L. V. Wang published an important article introducing the use of gold nanocages [184]. They worked on rat brains and published similar images as the ones reproduced in Figure 6.13. Compared to their gold nanoshells, the authors concluded that gold nanocages seem to have small advantages due to their absorption-dominant extinction and their smaller size, advantageous for in vivo delivery.

In 2008, Chamberland *et al.* [35] used PAI with gold nanorods to image articular tissues in rat tail joints, in order to monitor anti-rheumatic drug delivery

In 2010, the L. V. Wang's group published one of the most cited article in plasmon-assisted PAI [100], certainly because their three-dimensional reconstruction of the morphology of a melanoma is particularly impressive, as represented in Figure 6.15. The main objective of this study was to quantitatively compare the photoacoustic contrast enhancement in vivo provided by active and passive targeting of gold nanoparticles. The authors used gold nanocages as photoacoustic contrast agents. They managed to separately image a melanoma and the surrounding blood vessels by using two different illumination wavelengths. The active targeting approach led to a three-fold photoacoustic signal enhancement (as observed when comparing Figures 6.15b and 6.15d). Note that the tumor visible in Figure

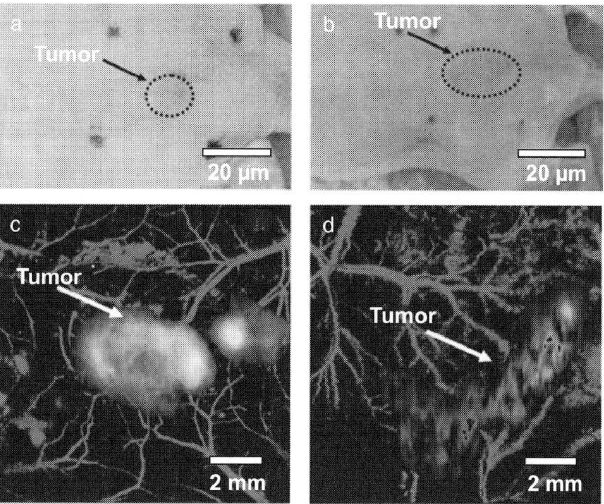

Fig. 6.15 In vivo noninvasive PA time-course coronal MAP images of B16 melanomas using [Nle4,D-Phe7]- -MSH- and PEG-nanocages. (a,e) Photographs of nude mice transplanted with B16 melanomas before injection of (a) [Nle4,D-Phe7]- -MSH- and (e) PEG-nanocages. Photoacoustic images of the B16 melanomas after intravenous injection with (b) [Nle4,D-Phe7]- -MSH- and (d) PEG-nanocages through the tail vein. The background vasculature images were obtained using the photoacoustic microscope at 570 nm (ultrasonic frequency 50 MHz), and the melanoma images were obtained using a photoacoustic macroscope at 778 nm (ultrasonic frequency 10 MHz). Color schemes: red (dark grey) for blood vessels and yellow (light grey) for the increase in photoacoustic amplitude. Reproduced with permission from Reference [100]. Copyright 2010, American Chemical Society.

6.15 is not the true signal. In the photoacoustic images, the color scale codes the increase of the photoacoustic signal as a percentage compared to the endogenous signal (which was already quite satisfying).

In 2009 and 2011, the same group published two timely review articles dedicated to the use of gold nanoparticles as contrast agents in PAI [185, 115]. The first one was a general review while the second one was more oriented toward the use of gold nanocages (mainly highlighting the two articles published by these authors on this subject [184, 100]).

In 2010, McLaughlan and coworkers introduced a refined strategy to enhance the photoacoustic signal [131]. By coupling laser illumination and ultrasound fields, nanobubbles formed around the nanoparticles produced strong acoustic emissions. As explained by the authors, the threshold laser fluence for bubble formation around metal nanoparticles can be reduced if the nanoparticles are simultaneously subjected to an ultrasound field and a laser pulse, and if the laser pulse is timed to illuminate the nanoparticles during the rarefaction phase of the ultrasound. Experiments were conducted with 80 nm gold nanospheres embedded in a phantom tissue.

In 2010, the Emilianov's group reported the use of silica-coated gold nanorods [40]. The idea was to show that a silica coating could stabilize the nanorods and avoid their thermal reshaping under pulse illumination. In 2011, the same group published a highly cited article giving a powerful advice to enhance the photoacoustic signal from gold nanoparticles [41]. Using their silica-coated nanoparticles again, the authors evidenced a much stronger signal compared to bare gold nanoparticles. After having shown that the enhanced signal could not be ascribed to an enhanced absorption cross section, the authors concluded that this effect came from a decrease of the Kapitza resistance on the nanoparticle surface, leading to an increase of the temperature in the surrounding medium. Experiments were conducted on phantoms with inclusions containing PEG-ylated gold nanorods, and gold-silica core–shell nanorods with silica shells of various thicknesses.

Subsequently, during the 2010s, most efforts have been devoted to

- trying to find new nanoparticle morphologies. Besides nanoshells, nanospheres, nanorods, nanocages and silica-coated nanoparticles, other morphologies have been introduced such as nanoprisms [17], stars [136] or tripods [42].
- expanding the range of applications, from cancer diagnosis to imaging of atherosclerotic plaques, brain function and image-guided therapy [114].

6.6 Nanochemistry

When gold nanoparticles are dispersed in a chemical reaction medium and illuminated at their plasmonic resonance, an increase of the chemical yield of the reaction can be observed (see Section 5.1 on page 143). This observation gave birth to the area of research that I shall name PINC for *plasmonics induced nanochemistry*. There are at least four mechanisms leading to the enhancement of chemical reaction yields in plasmonics [12]:

1. The *optical near-field* enhancement in the case of photochemical reactions.
2. The local *temperature increase* due to light absorption and subsequent heat generation (named T-PINC, the subject that will be developed in this section).
3. *Hot electron transfer* to surrounding oxidizing chemical species
4. A *catalytic activity* of the nanoparticle due to its nanometric size and which is not observed with its bulk counterpart [5, 78, 45]. Unlike the three other mechanisms, this one is not related to plasmonic properties.

The fourth mechanism can be coupled to the previous ones. All the mechanisms are schematized in Figure 6.16.

In the context of this chapter devoted to thermoplasmonics, we will focus only on the developments that involved the second mechanism (related to a local temperature increase), although other mechanisms (such as the catalytic activity) are also concomitantly involved in some applications.

Most chemical reactions are temperature-dependent according to the Arrhenius law (see Chapter 5, Section 5.1). In brief, the chemical yield is a strictly increasing function of temperature. Thus, nanoparticle heating can contribute to activate chemical reactions or boost reaction kinetics.

A first important question has to be clarified before going further. Heating as a means to boost a chemical reaction is a trivial approach in chemistry. Thus, one can wonder what could be the benefits of heating using nanoparticles compared to heating using a hot plate.

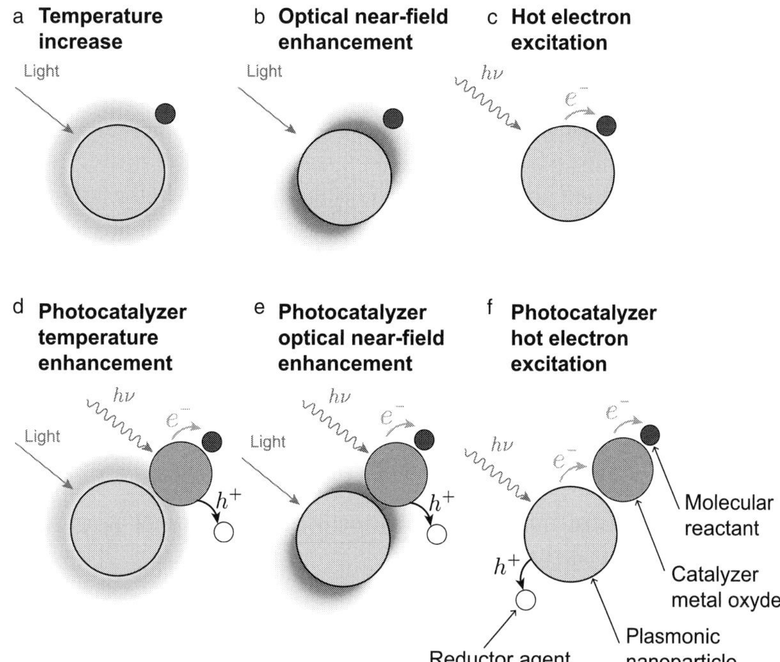

Fig. 6.16 Different mechanisms proposed for enhancing a chemical reaction around plasmonic nanoparticles. Reproduced with permission from Reference [12]. Copyright 2014, Royal Society of Chemistry.

This issue is not always clearly addressed upon reading some reported works. The benefits of using plasmonic nanoparticles compared with the use of a regular hot plate are a priori as follows:

- Heating a small region makes it possible to make the thermal dynamics faster due to a reduced thermal inertia (typically below the microsecond timescale; see Section 2.3 on page 55). In other words, it is much faster to heat (or let cool) a small volume than a large volume.
- Heating a micrometric area makes it possible to easily superheat the fluid above its boiling point (up to around 240°C for water; see Section 5.4 on page 162), with possible applications in solvothermal chemistry without using an autoclave [14].
- Heating on the nanoscale enables the formation of products with a nanometric spatial resolution.

If heating is performed upon illuminating a macroscopic assembly of nanoparticles, the three above-mentioned benefits no longer hold and heating with a hot plate or a Peltier cell would be rather equivalent (except if the laser illumination is pulsed). Indeed, collective effects (temperature homogeneization) will make the temperature completely uniform even at the nanoscale, despite the nanometric scale of the sources of heat (see Section 2.2.5, page 50). This situation typically occurs with plasmonic nanoparticles dispersed in solution, where it is impossible to heat a small number of particles even by focalizing a laser beam.

Consequently, the interest of T-PINC is not to produce industrial amounts of chemicals [31, 5] and T-PINC only makes sense when the temperature distribution remains localized, confined. The exception is the use of a pulsed illumination. Under pulsed illumination, the temperature increase can be very high and confined at the vicinity of the nanoparticles even when shining a large population of nanoparticles (see Figure 2.21 on page 75).

The second important question is the following. In the case of T-PINC applications for *surface chemistry*, one can also wonder what the benefit of using plasmonic nanoparticle is, *e.g.*, compared to a light-absorbing layer, like a thin metal film ... or anything black! This should also enable a strong temperature increase. The interest is not really that plasmonic nanoparticles are much more absorbent. A substrate can be made nearly 100% absorbent without requiring nanoparticles and nanoparticles will never do better than 100% absorption. And even if the absorption is not total, one just has to increase the laser power or focus a bit more tightly the sun light to reach the desired temperature. The benefits of using a layer of nanoparticles instead of a plain absorbing layer are rather as follows:

- The heated area can be much smaller than the size of a focused laser beam, down to the size of a single nanoparticle.
- Metal films are highly reflective. With metal nanoparticles, it is possible to make the light-sample interaction only absorbent (no scattering, no reflection, only absorption).
- Depositing a metal layer requires bulky and expensive apparatus, while metal nanoparticle substrates can be obtained using chemical means (like block copolymer lithography [96, 148]).
- Plasmonic nanoparticles exhibit an absorption resonance in a finite region of the spectrum, which makes it possible to heat the sample at a given wavelength range, while the sample remains transparent at another wavelength range. This eases the parallel observation of a sample using optical microscopy means, for example.

These seven points demarcate a finite area of interest to conduct investigation in PINC. Any investigation outside this area of interest would be useless as other simpler approaches would exist (*e.g.*, heating with a hot plate or heating a black surface with light).

All the applications of T-PINC have been based on a two-dimensional geometry where plasmonic nanoparticles are dispersed on a solid, planar substrate, and where the chemicals are freely dispersed on top of the substrate, either in a gas or a liquid phase. More precisely, applications in T-PINC can be divided into three main categories:

1. plasmonic-assisted chemical vapor deposition (PACVD)
2. thermoplasmonic-assisted catalysis in gas phase
3. thermoplasmonic-assisted catalysis in liquids

Let us successively review these three categories.

6.6.1 Plasmon-Assisted Chemical Vapor Deposition (PACVD)

In 2006, Goodwin's group conducted the seminal experiment introducing the concept of PINC [29], which was based on chemical vapor deposition (CVD). CVD is a method of thin film deposition on a substrate under vacuum from volatile precursors, which react

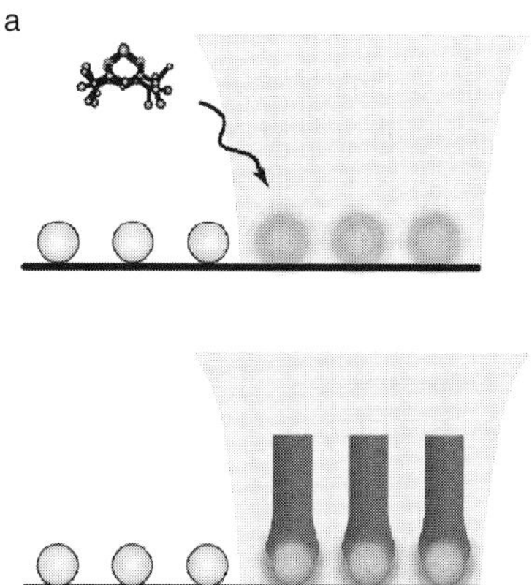 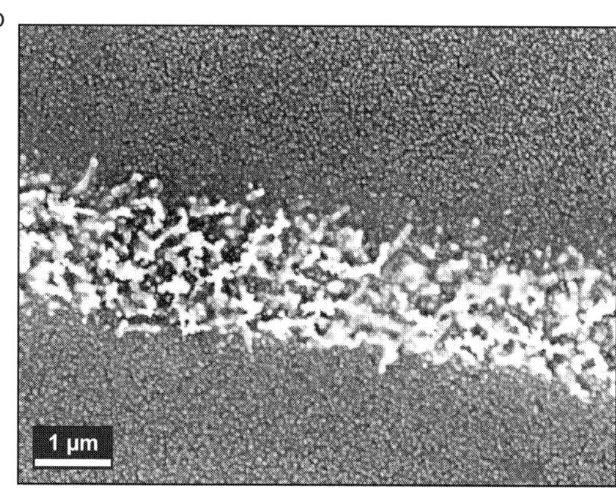

Fig. 6.17 (a) Schematic of PACVD process. A substrate of gold nanodots is exposed to a gaseous environment containing the CVD precursor in a carrier gas, and a laser (green) is focused on the surface heating the nanoparticles (top). Growth of nanowires is initiated only on the heated nanoparticles (bottom). (b) SEM image of a line of orthorombic PbO nanowires deposited by PACVD at a scan rate of 1 μm s^{-1}. Reproduced with permission from Reference [29]. Copyright 2006, American Chemical Society.

or decompose on the substrate. It produces high-quality solid materials across extended regions. The authors introduced a new CVD process in which the local heating necessary to induce the reaction was performed by local laser heating on the micrometric scale of gold nanoparticles deposited on a substrate. They named this technique PACVD (plasmon-assisted CVD). The author could achieve the microscale patterning of metal oxides such as PbO and TiO_2 on a glass substrate by heating up to 150°C (Figure 6.17). Prior to this advance, the approach of local laser heating for CVD, named laser-assisted CVD (LACVD), had already been actively explored for a few decades [19, 110]. The benefit of PACVD, compared to LACVD, is that plasmonic nanoparticles make the required power of the heating laser much smaller and potentially enables pattering on the subdiffraction limit in the case of single nanoparticle heating. Quoting the authors, "this technique is general and can be used to spatially control the deposition of virtually any material for which a CVD process exists."

A year later, in 2007, the group of Brongersma [31] (who was involved in the seminal work of Goodwin as well) achieved on its side subdiffraction PACVD around single plasmonic nanoparticles, leading to the growth of single nanowires. Several reactions were conducted to illustrate the message: the formation of silicon nanowires (from SiH_4 in Ar and with gold nanoparticles), germanium (from GeH_4 in Ar) or carbon nanowires (from methane and ethylene as a source gas and Ni nanoparticles as the catalyst). Although the catalytic role of Ni nanoparticles for the growth of carbon nanowires is established, the catalytic role of gold nanoparticle is not evident, as Ni, Ti and Fe nanoparticles seem to be working as well. Gold nanoparticles presumably just act as sources of heat in this study.

In 2008 [81], the group of Cronin renamed this technique PRCVD (plasmon-resonant CVD), with little credit to the seminal work of Goodwin.

In 2011, Hwang *et al.* also conducted a very detailed study of PACVD by investigating the effects of parameters such as the illumination orientation, exposure time and temperature increase [88] on the growth of silicon nanowires from SiH_4 on gold nanoparticles.

6.6.2 Thermoplasmonic-Assisted Catalysis in Gas Phase

Producing new catalytic compounds that are highly active under solar illumination is an important current challenge. With the pioneering works of Haruta starting in 1987 [78, 45], gold is known to feature strong catalytic activity provided it is used in the form of nanoparticles supported by titania (or other oxides), due to changes in the electronic structure, catalyst-substrate interactions, and morphological features, compared to its bulk counterpart. Using this concept, oxidation of various volatile organic compounds, such as CO, CH_3OH, HCHO, have been demonstrated using gold nanoparticles on metal oxides, at moderately elevated temperatures. But at that time, heating was performed using conventional means (no photothermal plasmonic effects involved). The idea of Chen and coworkers in 2008 [39] was to further enhance the heterogeneous catalytic activity of gold nanoparticles by heating the nanoparticles themselves under laser illumination. A new application of plasmonics was born. The authors used gold nanoparticles on ZrO_2 and SiO_2, *i.e.*, wide band gap semiconductors to preclude any photocatalytic effect, to generate the oxidation of HCHO in CO_2. According to the authors, the benefit was that heating was limited to the catalytically active area of the system, hence a gain in energy consumption compared to a global heating approach. In 2010, the group of Cronin borrowed the same idea to generate the exothermic oxidation of carbon monoxide in carbon dioxide, using gold nanoparticles on Fe_2O_3 [82]. The authors quantified the benefit of heating using a laser approach compared to a global heating. A gain of two to three orders of magnitude was evidenced.

6.6.3 Thermoplasmonic-Assisted Catalysis in Liquids

In 2009, Psaltis, in collaboration with the group of Goodwin, proposed to apply the same idea in a liquid environment (not in gas phase). The technique was named PAC for *plasmon-assisted catalysis* [5]. The authors chose to investigate the thermal-induced reforming of a liquid mixture of ethanol and water, leading to the formation of CO_2, CO and H_2. The experiments were conducted in a microfluidic channel in order to more easily collect the gas products (Figure 6.18). The authors stressed the fact that their plasmon-assisted catalysis approach was general and could be used with a variety of endothermic catalytic processes. Noteworthily, their theoretical estimations of the temperature increase are not properly conducted. Eq. (2) of Reference [5] giving the temperature threshold for bubble formation is not appropriate as it discards superheating (see Section 5.4, page 162). Then, Eq. (3) underestimates the temperature increase because it does not take into account collective effects (see Section 2.2.5). This is why they end up with a theoretical temperature increase of only 9 K.

Along the same lines, in 2011, Fasciani *et al.* investigated the thermal decomposition of peroxides [64], a reaction supposed to occur at more than 140°C. The key parameter of this

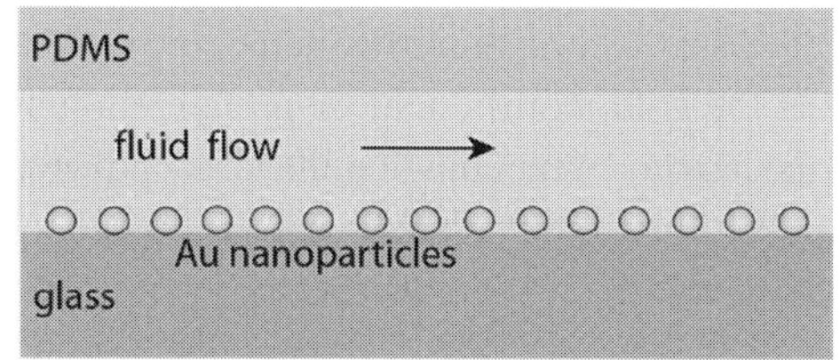

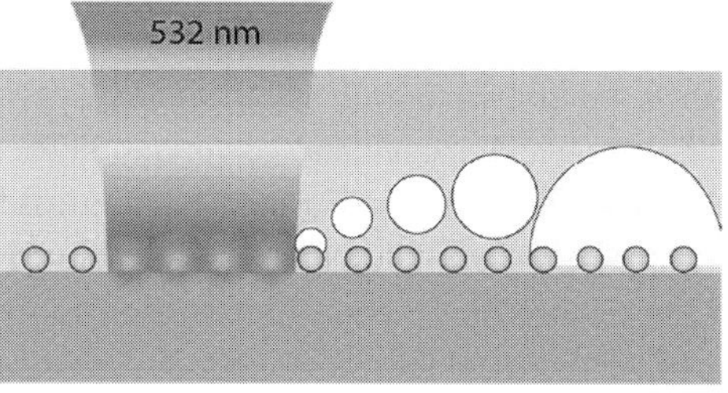

Fig. 6.18 Schematic of the PAC process (side view). A microfluidic channel with gold nanoparticles attached to a glass support; fluid flows from left to right. (Top) A laser is focused on the top of the support, and the subsequent heat generated in the nanoparticles is transferred to the surrounding fluid and forms vapor. The vapor phase components react on the catalyst and the resulting gas bubbles are carried downstream. The channel height is 40 µm and the radius of the nanoparticles is around 10 nm. Reproduced with permission from Reference [5]. Copyright 2009, American Chemical Society.

Fig. 6.19 Laser drop pictures before, during, and a few milliseconds after 532 nm laser excitation. Note a slight blueish tint during excitation. A 532 nm notch filter was used in front of the camera lens, to make it "blind" to the green laser beam and to prevent camera damage. Reproduced with permission from Reference [64]. Copyright 2011, American Chemical Society.

work was the use of a nanosecond-pulsed illumination (1 Hz repetition rate, 8 ns, 50 mJ per pulse) to generate rapid and intense temperature increases, without boiling the fluid. This way, the authors evidenced that organic chemistry could be performed both at relatively high energy and at near-ambient temperature. The geometry of the system consisted of a droplet of solution suspended at the very end of a needle (Figure 6.19). The laser beam was focused on the droplet. After 75 laser pulses, the chemical transformation of the reactant to products was observed to be total in the droplet.

In 2012, the group of J. Feldmann, an active researcher in thermoplasmonics, observed the oxidation of gold nanorods that were optically trapped [135]. Oxidation tends to decrease the size of a nanorod by removing gold atoms. Interestingly, the authors have shown that the gold atoms could be preferentially removed from the tip or from the side walls of the nanorod depending on the laser-induced temperature increase achieved during laser trapping. Small temperatures led to a reduction of the aspect ratio (oxidation of the tips of the nanorod), while superheating conditions (up to 140°C; see Section 5.4) led to a strong destabilization of the CTAB molecules on the whole surface of the nanorod and a preferential oxidation of the side walls, hence an increase of the aspect ratio. As water superheating is not possible under bulk heating, the authors stressed the fact that such an unusual increase of the nanorod aspect ratio is specific to single nanoparticle trapping.

In 2013, Vázquez Vázquez and coworkers [176] presented an original approach where thermal induced PAC was performed on the inside of hollow nanometric silica shells containing the reactants. The shell was decorated with gold nanoparticles intended to generate the required heat upon illumination to generate the reaction. The interest was to demonstrate that local chemical reactions can be generated in a confined volume with a surrounding solvent at ambient temperature, and that the extent of reaction can be followed by SERS.

In 2013, the group of J. Feldmann introduced a nanofabrication method based on the polymerization of PDMS by a single trapped nanoparticle [65]. By driving a single nanoparticle along a predefined path, the authors could fabricate PDMS nanoparticles and nanowires, as illustrated in Figure 6.20.

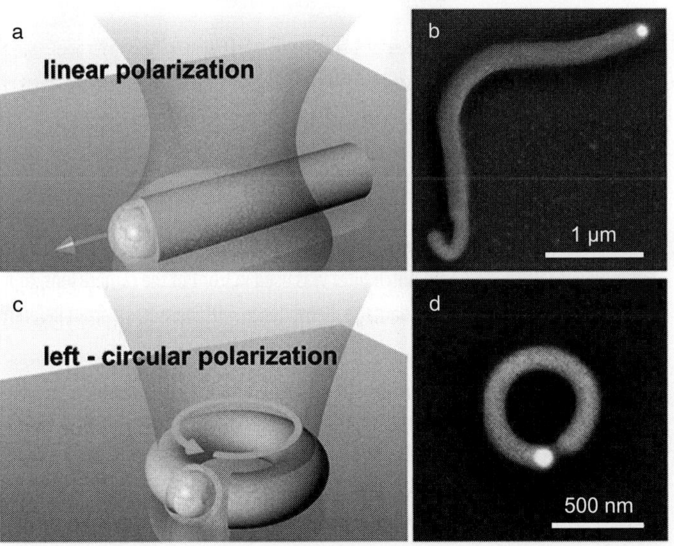

Fig. 6.20 Fabrication of PDMS nanowires. (a) Schematic illustration of the nanowire formation with a linear polarized laser beam. The focused beam is pushing the gold nanoparticle over the surface of the substrate along a straight line. (b) SEM image of the PDMS nanowire structure. (c) Schematic illustration of the nanowire formation with circularly polarized light. (e) SEM image of the PDMS nanowire made with left circularly polarized light. Reproduced with permission from Reference [65]. Copyright 2013, American Chemical Society.

In 2016, Robert *et al.* introduced the concept of hydrothermal chemistry at ambient pressure using gold nanoparticles [151]. The authors conducted chemical reactions under hydrothermal conditions, *i.e.*, using liquid water at more than 100°C as a solvent. The authors benefited from the ability of gold nanoparticles to superheat water without boiling even at ambient pressure (see Section 5.4 on page 162). They managed to synthesize $In(OH)_3$ microcrystals from a solution of $InCl_3$, a reaction that normally occurs in autoclaves at 200°C. The reaction kinetics have been observed to be several orders of magnitude higher compared with the use of autoclaves.

6.6.4 Unidentified Thermal Effects in PINC

As explained at the beginning of the section, photothermal effects are not always the targeted mechanism in PINC. Pure optical effects or hot electron injection processes are also considered and intensively studied as a means to enhance the yield of chemical reactions. However, since any approach in PINC consists in illuminating assemblies of nanoparticles, photothermal effects are likely to occur anyway. The occurrence of photothermal effects in PINC studies, and more generally in plasmonics, are not always easy to address.

In 2009, this problem was honestly pointed out by the group of El Sayed [186]. The group studied the reduction of ferricyanide by thiosulfate and concluded that the origin of the observed enhanced chemical yield was purely due to a laser-induced temperature increase of the solution, not due to an enhanced photocatalytic effect, as anticipated. This conclusion questioned the interest of using gold nanoparticles compared to a regular hot plate.

In 2010 during their experiments of PAC in gas phase, the group of Cronin [82] monitored the temperature rise of their sample simply using an infrared camera, which revealed a significant temperature increase. Using an infrared camera is an excellent way to evidence thermal effects. One can regret that so few researchers are using infrared measurements in this context.

In 2016, Bora *et al.* dedicated a full article to address the effect of plasmonic heating during plasmon-induced nanochemistry experiments [28]. Using a hybrid substrate composed of gold nanoparticles and ZnO nanowires, the authors demonstrated a temperature rise of the sample up to around 300°C, which partly explained the high chemical yield observed in photochemical processes.

One can regret that many reported works did not perform rigorous tests to quantify the significance of thermal effects in PINC applications. For instance, the group of Linic conducted several studies in PINC [47, 48] where a purely photocatalytic effect was claimed, but they never monitored the temperature of their samples. Yet, several of their results are consistent with a purely thermal effect. The data sets in their figure 3b in Reference [48] – reproduced in Figure 6.21a – and the associated linear fit were intended to rule out any thermal contribution to the yield enhancement. But these data sets could be nicely fitted with the Arrhenius law as well[7], as evidenced in Figure 6.21b. All the results presented in

---

[7] $K = A \exp(-E_a/(R(T_0 + cI)))$ where K is the reaction rate, E_a the activation energy, R the ideal gas constant, $T_0 = 450°C$, I is the laser intensity and $A = 2.9 \times 10^{10}$ the fitting parameter. $c = 8 \times 10^{-2}$ $K \cdot cm^2 \cdot mW^{-1}$

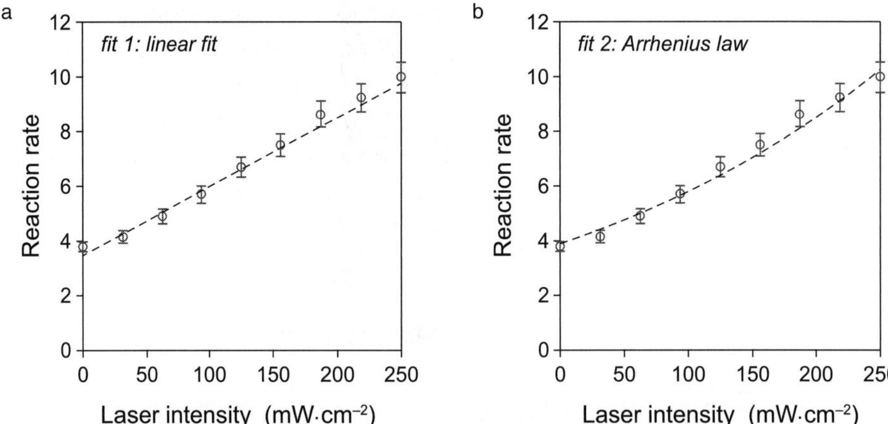

Fig. 6.21 (a) The rate of ethylene epoxidation as a function of laser intensity. The dashed line shows a linear fit to the experimental data, consistent with a photocatalytic mechanism (data taken from Reference [48]). Same data points that have been fitted with the Arrhenius law, showing that a purely photothermal mechanism could also explain the observations.

this article are consistent with a temperature rise of around 20 K, which casts some doubt about the validity of this article.

6.7 Photothermal Imaging (PTI)

Introduced in 2002 in plasmonics [30], photothermal microscopy enables detection of nano-objects solely based on their absorption, notably gold nanoparticles [50, 177]. The good sensitivity of the technique and the stability of the signal enabled advances in nano-object spectroscopy (absorption spectroscopy and correlation spectroscopy) and optical detection in living cells (localization and tracking of biomolecules and organelles). Figure 6.22 gives an example of the type of image that can be acquired using this technique.

The main interest of detecting nano-objects via absorption (and not via fluorescence, for instance) is that they behave as ideal labels: they are small enough to remain noninvasive and, more importantly, they do not suffer from photobleaching, or blinking like common fluorescent probes.

Although this application is historically important, photothermal imaging has not become a widely used technique, even today. Most of the developments have been mainly reported by only three groups all over the world, led by Michel Orrit (Leiden, Neitherlands), Brahim Lounis (Bordeaux, France) and Frank Cichos (Leipzig, Germany).

is a constant that was determined to have a maximum temperature variation of 20 K, consistently with the temperature shift observed in their figure 1b.

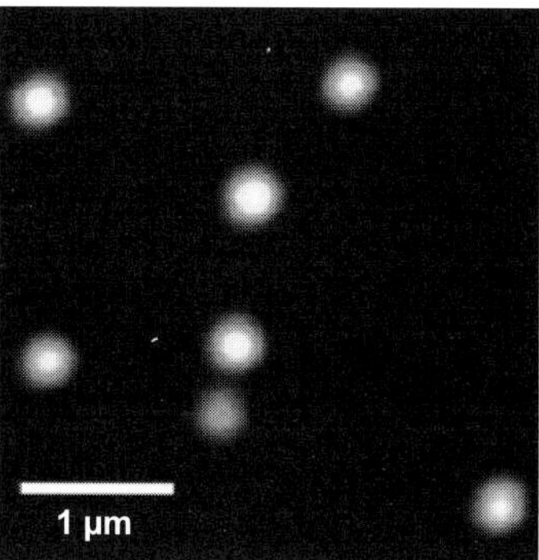

Fig. 6.22 Photothermal microscopy image of 10 nm gold nanoparticles deposited on a glass slide. Reproduced with permission from Reference [177]. Copyright 2014, Royal Microscopical Society

This section introduces the principle of PTI, depicts the different reported experimental configurations, mentions the underlying theories and chronologically reviews the most important works.

6.7.1 Principle of Photothermal Imaging

Photothermal imaging (PTI) [177] is one of the first advances (with photothermal cancer therapy) that initiated the interest of the community for photothermal effects in plasmonics. PTI was introduced and demonstrated by Michel Orrit and Brahim Lounis in 2002 in an article published in *Science* [30]. This technique was originally named *photothermal interference contrast* (PIC) microscopy by the authors. It is now more commonly named *photothermal imaging* or *photothermal microscopy*.

The principle is simple, albeit subtle. Detecting and tracking nano-objects (typically smaller than 40 nm) using optical means is fundamentally difficult. Such small objects do not scatter the incident light as Rayleigh scattering decreases as the sixth power of their diameter, and the scattering signal must be discriminated from the background. In brief, they look invisible. Most of the interaction between the incoming light and a nano-object rather results in absorption and subsequent heat dissipation. Scattering becomes dominant only for larger particles, typically above 100 nm (depending on the nature of the particle). The idea of Orrit and Lounis was to detect very small nanoparticles via their light-absorption, as it is the main transduction pathway. For this reason, the authors considered highly absorbent nano-objects and naturally chose plasmonic nanoparticles. In their seminal work, gold nanoparticles a few nanometer big were randomly deposited on a glass substrate and immersed in water [30]. The gold nanoparticles were heated by a few kelvins using a focused laser beam, which resulted in a decay of the refractive index of the surrounding water (see Section 5.9 on page 205). Interestingly, such a refractive index variation spreads over a distance from the particle much larger than the particle size itself, according to the thermal diffusion law. For this reason, the authors ended up with effective objects that became much larger than the nanoparticle size as soon as it is heated. This larger volume of liquid undergoing a refractive index variation was sufficiently big to scatter an incident probe beam and make the presence of the nano-object detectable using any phase imaging technique. Using an angular modulation Ω of the heating laser (typically $\Omega \sim 0.1 - 1$ MHz) along with a synchronous detection in order to get rid of the background signal, they could achieve the detection of spherical nanoparticles as small as 5 nm in diameter with a signal to noise ratio better than 10, and 2.5 nm spherical nanoparticles with a signal to noise ratio on the order of 2, for an integration time of 10 ms per pixel.

6.7.2 Experimental Configurations

PTI is based on the detection of phase objects. All the experimental setups were thus naturally based on the use of phase imaging techniques. In any case, two laser illuminations were implemented:

- A laser beam (usually at $\lambda = 532$ nm, a few mW or less) intended to heat the nanoparticle. This laser was mechanically or acousto-optically modulated to enable a synchronous detection of the signal.

- A low-intensity laser beam (in the near-infrared) to build a phase contrast image.

Three main experimental configurations have been reported so far.

DIC-like microscopy. The initial configuration reported by Orrit and Lounis was a slightly modified DIC (differential interference contrast) microscope, according to the configuration proposed by Gleyzes, Boccara and Saint-Jalmes in 1997 [72]. DIC microscopy involves the use of a Wollaston prism creating two separate beams illuminating the sample at different angles (with orthogonal polarizations). The two beams are supposed to experience different optical paths through the sample of interest, before interfering. Although the configuration of Orrit and Lounis involved a Wollaston prism, the interference did not result from two incoming beams widely illuminating the sample at different illumination angles like in DIC. On the contrary, the two beams originating from the Wollaston prism were *focused* on the sample, but separated by a few microns. Only one beam (called the reference beam) were overlapped with the heating beam, while the other was considered as a reference beam. The heating beam beating of the angular frequency Ω was modulating the optical path experienced of the probe beam. This way, three laser beams were focused on the sample and the signal resulted from the interference between the probe and reference beams, back reflected/scattered by the sample, and measured at the angular frequency Ω using a lock-in amplifier. Figure 6.23(a) details the experimental configuration. This configuration implies a raster scan of the sample to construct an image point by point.

DIC-like microscopy in transmission. A slight modification of the previous configuration can be implemented. The previous configuration was relying on the portion of the reference beam back-reflected by the water/glass interface, and the portion of the probe beam back-scattered by the heated volume. As these amounts of light are weak, the idea of this second configuration was to back-reflect both beams using the configuration represented in Figure 6.23(a) with afocal system in order to collect more light.

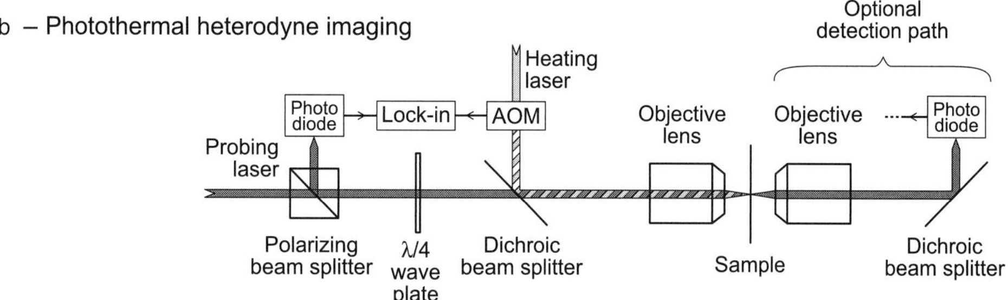

Fig. 6.23 Main experimental configurations used in photothermal imaging. (a) Original configuration that resembles a DIC microscope. An afocal system can be implemented on top of the sample to back-reflect the probe and reference beams and increase the amount of collected photons. (b) Simpler, albeit more sensitive, configuration involving a single focus point on the sample. The signal can be measured either in reflexion or in transmission.

Photothermal heterodyne imaging. A more sensitive configuration, named photothermal heterodyne imaging (PHI), is based on the use of a single beam and does not require a Wollaston prism and the presence of a reference beam (Figure 6.23(b)).

In this configuration, a polarizing cube and a quarter wave plate were used to extract the interference between the reflected field (so-called reference field) and the back-scattered field. Reflected red beams are collected on fast photodiodes and fed into a lock-in amplifier to detect the beat signal at the angular frequency of the heating beam modulation (Ω).

Thanks to this simpler configuration that involves a single beam path and gets rid of a Wollaston prism, a raster scan of the beams can be possibly implemented instead of scanning the sample.

More importantly, this configuration, albeit much simpler than the DIC-like configuration, features a much higher sensitivity by two orders of magnitude. It is able to detect temperature variations smaller than 1 K, compatible with bio-related studies. This gain in sensitivity comes from the fact that this technique neither suffers from the quality of the overlap of reference and probe beams on the detector, nor from their relative phase fluctuations.

This configuration was implemented by Berciaud *et al.* in 2004 [21], although it is very similar to the photothermal microscopy configuration introduced by Harada *et al.* in 1993 [77].

Single-beam photothermal imaging. Their also exists a more recent technique, introduced in 2014 by Selmke and coworkers [161] (not represented in Figure 6.23). This variant is very similar to the PHI technique, except it uses a single beam acting both as a heating beam and a probe beam. So far, this technique has been only reported in a single article.

6.7.3 Theories of Photothermal Imaging

The measured signal stems from the interaction of a light beam with a time-dependent refractive index profile around the nanoparticle, due to the heating beam modulation at the angular frequency Ω, that writes (see Equations (2.109) and (5.85)):

$$\delta n(r,t) = \frac{\mathrm{d}n}{\mathrm{d}T}\delta T(r,t) = \frac{\mathrm{d}n}{\mathrm{d}T}\frac{Q_0}{4\pi\kappa_{\mathrm{s}}r}\left[1 + \exp\left(-\frac{r}{\delta_{\mathrm{th}}}\right)\cos\left(\Omega t - \frac{r}{\delta_{\mathrm{th}}}\right)\right] \tag{6.3}$$

where

$$\delta_{\mathrm{th}} = \sqrt{2a_{\mathrm{s}}/\Omega} \tag{6.4}$$

is the attenuation length of the thermal "wave." Several theoretical descriptions of this system, involving different models and approximations, have been proposed [157], namely:

- Heterodyne Imaging theory [23, 133]
- Equivalent dipole model [69, 144]
- Generalized Lorenz–Mie theory [160, 159]
- Photonic Rutherford scattering [156, 155]
- Fresnel Diffraction [49]
- Gaussian ABCD method [158]

Among the most important models, the heterodyne imaging theory developed by Berciaud *et al.* in 2006 considered the theory of light scattering from a fluctuating medium [23]. The generalized Lorenz–Mie theory described by Selmke *et al.* in 2012 rather assumed a nano-lens effect considering the long-range variation of the refractive index [160, 159]. A detailed review of all these mechanisms is provided in Reference [157].

An interesting outcome of these theories is the derivation of a fundamental limitation to the signal to noise ratio that can be achieved in photothermal imaging. Berciaud [23] and subsequently Gaiduk [69] derived similar expressions that read

$$\mathrm{SNR} \approx \frac{1}{\pi r_0 \lambda^2 \Omega} n \frac{\mathrm{d}n}{\mathrm{d}T}\frac{1}{C_{\mathrm{p}}}\frac{\sigma_{\mathrm{abs}}}{A}P_{\mathrm{heat}}\sqrt{\frac{P_{\mathrm{probe}}\Delta t}{h\nu}} \tag{6.5}$$

where r_0 is the probe beam focal radius (beam waist), n is the refractive index of the surrounding medium, A is the area of the heating laser beam waist, C_{p} is the heat capacity per unit volume of the photothermal medium, λ and P_{probe} are the wavelength and power of the probe beam, and Δt is the integration time of the lock-in amplifier.

According to this expression, the efficiency of photothermal imaging can be optimized in different manners [69]:

- Increase the dissipated power Q_0, in the limit of small temperature increase to avoid damage of the sample, especially for biological samples. In practice, photothermal imaging is efficient even for very small temperature rises (typically a few K).
- Increase the power of the probe laser, but one should avoid heating the nanoparticle with the probe laser. This is made easier by choosing a probe wavelength in the infrared far from the plasmon resonance wavelength of the nanoparticle.
- Adjust the heat diffusion length δ_{th} with the spot size of the probe beam by choosing the right modulation frequency Ω [69]. In other words, $1/\Omega$ has to be on the order of the characteristic time for thermal relaxation of the nanoparticle (see Equation (2.134) on page 73).
- Optimize the thermal properties of the liquid, such as the refractive index, its derivative with respect to temperature, specific heat capacity and thermal conductivity. In particular, glycerol is leading to better results than water [70] and the use of liquids near the critical point enables a 100-fold enhancement of the signal to noise ratio [56].
- Increase the integration time Δt of the lock-in amplifier. This parameter can be arbitrarily increased, provided the mechanical stability of the experimental setup and the photostability of the sample allow it. Typically, Δt ranges from 10 ms [30] to 300 ms [70].

6.7.4 History of Photothermal Imaging

Statistics of the articles published in photothermal imaging and microscopy are displayed in Figure 6.24.

Thermal lens detection techniques were introduced long ago, in the 1970s, to perform absorbance spectroscopy of weakly absorbing *solutions*. For instance, Long *et al.* managed to record the absorption spectrum of a solution featuring an absorbance as weak as 10^{-6} $M^{-1}{\cdot}cm^{-1}$ [119]. But the measurements were not performed on nanoscale objects and were not using microscopy means.

The study of light-absorbing *objects* based on a photothermal-induced variation of refractive index was introduced by A. Claude Boccara back in the 1980s via a dozen articles [26, 90, 68, 153]. He named the technique "photothermal deflection spectroscopy/detection" or "photothermal imaging using the mirage effect." Boccara and coworkers, however, focused mainly on the study of solids and surfaces, not on single nano-objects and not on microscale measurements. Moreover, Boccara was rather using a deflection beam technique, not an interferometric technique.

In 1993, Harada and coworkers [77] reported on a photothermal microscopy technique suited to study single absorbing microparticles (dielectric microbeads decorated with absorbing Fe-oxine particles). The experimental configuration was very similar to the setups developed from 2002 in plasmonics. At that time, however, experiments were restricted to beads larger than ~ 50 μm and the typical heating laser power was 100 mW. Moreover, the thermal induced variations of the refractive index of the 100 μm dielectric beads was involved, not the variations of the refractive index of the surrounding medium.

In 2000, Uchiyama and coworkers published an article entitled *Thermal Lens Microscope*. The authors reported on a technique capable of mapping the distribution of

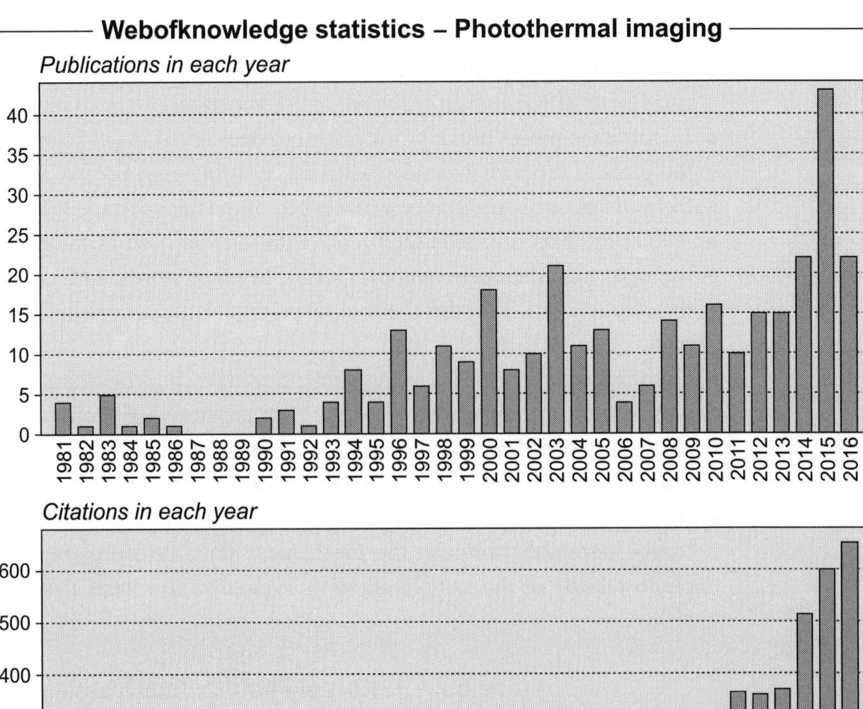

Fig. 6.24   Web of Knowledge statistics [3] (January 2017), corresponding to the search criterion "TOPIC: ('photothermal imaging' or "photothermal microscopy")". Results found: 334 / Average citations per item: 15.00 / h-index: 35.

absorbing molecules dispersed on a substrate via *photothermal effects*. But once again the measurements were not scaled down to nanoscale single objects.

The progress of Orrit and Lounis in 2002 compared with the state of the art was to implement a photothermal imaging technique down to the study of single nano-objects, opening the path for detection and tracking experiments *e.g.*, in cell biology.

Following the pioneer experiment of group of Orrit and Lounis in 2002 [30], refinements of the technique and practical applications in cell biology have been continuously demonstrated.

In 2003, the group of Lounis introduced a second PIC experimental configuration [49]. Measurements were no longer performed in reflexion through the objective lens, but in *transmission* using another objective lens on top of the sample. Using such a transmission-like detection, the group of Lounis could detect and localize single individual

10 nm gold particles in the plasma membrane of COS7 cells. However, the estimated temperature increase was still 15 K, fairly prohibitive for practical application in cell biology.

In 2004, still in the same group, Berciaud *et al.* refined the PIC technique by implementing a single beam, heterodyne detection [21]. Albeit much more simple, this new approach led to an improvement of the signal-to-noise ratio by two orders of magnitude. The signal no longer resulted from the interference between a probe and a reference beam separated on the sample, but from the transmitted (or reflected) and scattered fields [23]. In the reported experiments, 5 nm gold nanoparticles were detected with a signal to noise ratio larger than 100 at a heating power of 1 mW, which is supposed to correspond to a nanoparticle temperature increase of 4 K in aqueous solution. Such characteristics offer the possibility to detect nanoparticles with a temperature increase smaller than 1 K, opening the path for safe biological applications. After this work, the PTI technique was no longer coined *photothermal interference contrast* (PIC) microscopy, but rather *photothermal heterodyne imaging* (PHI) or *photothermal heterodyne microscopy*.

In 2005, Berciaud *et al.* demonstrated how PHI could be used as an absorption spectroscopy technique of weakly absorbing nano-objects such as semiconductor nanocrystals [22] and carbon nanotubes [24].

In 2006, the group of Lounis demonstrated the first practical application of PHI for cell biology [108]. The authors managed to track individual 5 nm nanoparticles attached to membrane proteins of live neurons. In order to avoid scanning the whole area of interest and achieve efficient tracking, a method of triple-point measurement was implemented. The authors called this technique LISNA for *laser induced scattering around a nanoabsorber*. An example of a LISNA implementation is given in Figure 6.25. This technique was further extended in a label-free version where the photothermal contrast in living cells was not due to metal nanoparticles but to intrinsic light absorption by mitochondria [109]. This way, the authors managed to follow the morphological states of mitochondria under different physiological treatments.

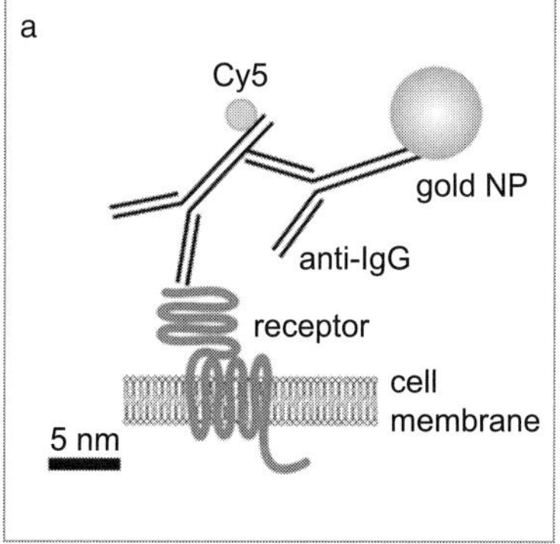

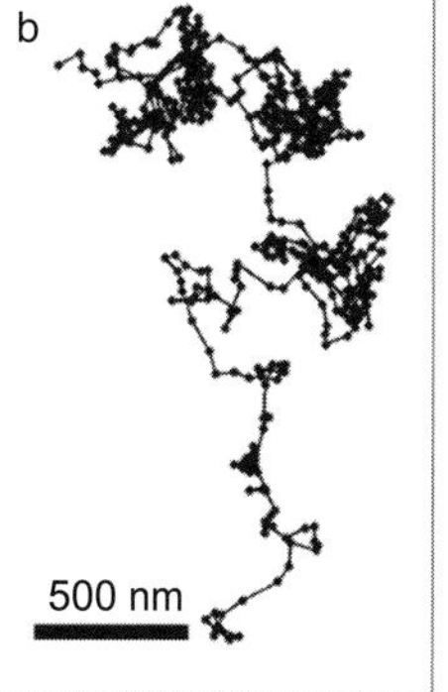

Fig. 6.25 (a) Biological construct. (b) LISNA trajectory over a portion of a transfected COS7 cell labeled with 5 nm gold NPs detected with a SNR $\sim$ 30. Reproduced with permission from Reference [108]. Copyright 2006, The Biophysical Society.

In 2008, Michael Atlan and coworkers introduced a photothermal microscopy technique that combined a photothermal approach with a digital heterodyne holography [9]. This wide-field technique was demonstrated with 50 nm gold particles. The method involved a trade-off between large observation areas and video rate acquisition speed [4]. Compared to single-point detection, the signal-to-noise of wide-field detection is about an order of magnitude lower.

In 2009, two groups, the group of Lounis [138] (still in Bordeaux, France) and the group or Orrit [144] (who moved to Leiden, Neitherlands), concomitantly introduced the concept of photothermal correlation spectroscopy. The idea was to perform fluorescence correlation spectroscopy (FCS) using PTI on gold nanoparticles, ruling out the possible artifacts occurring when performing FCS using fluorescent molecules and stemming from photobleaching, blinking, and saturation. The aim was to perform more reliable FCS measurements of the diffusion of single nanoparticles in fluid media.

In 2010, Gaiduk, Orrit and coworkers published a detailed theoretical investigation of the fundamental detection limits in photothermal microscopy. The idea was to derive an expression of the expected signal-to-noise ratio as a function of all the parameters of the experiments. The authors ended up with the expression (6.5). The authors further carried out experimental measurements with the concern of optimizing each parameter of this expression. This way, they managed to detect a dissipated power of 3 nW with a signal-to-noise ratio of 8, and an integration time of 10 ms. This corresponds to a less than 0.1 K surface temperature rise for a 20 nm-diameter gold nanosphere. This optimization work even led them to detect single dye molecules [70] (featuring an absorption cross section of 4 Å^2) with a signal to noise ratio of $\sim$ 10 and 300 ms of integration time.

In 2012, a new group joined the playground of photothermal imaging, which was so far almost exclusively occupied by the two groups of Bordeaux and Leiden. Markus Selmke and Franck Cichos investigated in detail the underlying physics governing the signal measured in PTI via several articles published in 2012 and 2013 [156, 158, 159, 160]. Their main idea was that the physics could be described by a *nanolens effect*. This vision contrasts with the previous models rather considering a scattering description due to the small size of the objects that prevents from using geometrical optics a priori.

In 2014, Selmke and coworkers introduced a variant of the PHI technique using a single beam instead of two (heating and probe) beams [161]. In this configuration, the detection was performed via the measurement of the self-induced phase delay of the transmitted heating beam modulation.

In 2015, Selmke and Cichos deposited on Arxiv a review article on the different theories and experimental setups related to photothermal microscopies [157].

6.7.5 Conclusion

After more than a decade of investigation, the interest of the photothermal imaging technique is not obvious, compared to other tracking techniques based for instance on the use of fluorescent quantum dots. In the original *Science* article, it was noted that, although fluorescent quantum dots "resist bleaching longer than dyes, their luminescence brightness is liable to blinking, and they are difficult to functionalize in a controlled way." The fact is that, today, more tracking experiments are based on the use of fluorescent quantum dots while the photothermal imaging technique remains sparingly employed. Even researchers who were using the photothermal techniques some time ago are now involved in works based on the use of fluorescent quantum dots labels [175].

6.8 Thermoplasmonics for Cell Biology

Gold nanoparticles as nanosources of heat have been proved efficient to perform local thermodynamic investigation on nanoscale and microscale biosystems, such as proteins, DNA, lipid membranes, vesicles or single living cells. This last section deals with this kind of application.

In general, the state of most biosystems is highly dependent on temperature. Studies of the effect of temperature on living cells in culture are often carried out by heating the whole petri dish with a temperature-controlled sample holder, or even by heating the whole microscope. The major drawback of such an approach is the inherent large thermal inertia. One has to wait around several 10s of minutes to reach the desired temperature, which prevents from studying the dynamics of fast thermal-induced processes.

To circumvent this limitation, shrinking the size of the heated region appears as the natural approach. The smaller the system, the smaller the thermal inertia. For this reason, heating using a laser and an absorbing medium seems ideal. However, the main limitation of this approach is the difficulty to reliably measure a temperature distribution at the microscale. Many temperature microscopy techniques have been developed since the early 2010s, as detailed in Chapter 4. While most of them proved efficient in reliably mapping the temperature of metal nanoparticles in simple environments, things are more complicated in a medium as complex as the cytosol of a living cell. Many parameters such as pH, ionicity, microviscosity, gradients of refractive index, etc., can affect the measured signal, in particular when fluorescence is used. This problem led to several artifactual temperature measurements in living cells, as explained in polemic Commentaries published in Nature Methods in 2014, 2015 [13] [15].

Certainly for this reason, not so many thermal-induced processes have been investigated at the single living cell level. Yet, heating single cells with a laser has the following benefits:

- The temperature dynamics can lie on the microsecond to millisecond time-scale, which would enable the study of the dynamics of fast thermal-induced processes.
- One cell at a time can be heated at a given temperature, which should ease the studies aimed at investigating a wide range of temperature.
- Heating can be performed at the subcellular level. Heating specific organelles is possible.

For this purpose, gold nanoparticles seem ideal sources of heat: they can be designed to efficiently absorb in the infrared (a requirement to avoid phototoxicity of the high power laser used for heating), they are biocompatible and they can lead to nano- and microscale heating, depending on the number of nanoparticles under illumination.

6.8.1 History of Thermoplasmonics for Cell Biology and Biosystems

Here is an overview of the articles reporting on the heating of small biosystems using gold nanoparticles.

Oddly enough, while most of the articles published in this field of research date from the 2010s, the very first work devoted to the study of biosystems using gold nanoparticle heating dates from 1999. As mentioned in the first section of this chapter, two researchers, Hüttmann and Birngruber [87] studied the thermal-induced denaturation of proteins using a pulsed laser to heat gold nanoparticles. The authors evidenced the denaturation of chymotrypsin proteins within 300 ps at temperatures below 380 K. This visionary work was carried out in the context of photothermal treatment of vessels or pigmented cells.

A decade after, this domain of research benefited from a renewed interest thanks to the active research group led by J. Feldmann. In 2008, this group studied the melting of DNA on the microsecond timescale using DNA strands attached to gold nanoparticles and 300 ns laser pulses [167, 80]. This is much faster than conventional DNA melting achieved upon overall heating where the temperature is slowly ramped up. The authors illustrated this way the gain in dynamics when using gold nanoparticles and pulsed illumination.

In 2009, the same group carried out other in vitro experiments, based this time on the local heating of artificial phospholipid membranes using single gold nanoparticles [173, 10]. The system consisted of a giant unilamellar vesicle (GUV), which is basically a spherical phospholipid bilayer, 15 to 50 µm big. The authors evidenced a reversible and controlled gel–fluid phase transition of GUVs.

In 2010, the group of L. B. Oddershede performed similar experiments on the phase transition of lipid membranes [20, 104]. But in this work, the idea was to estimate the temperature increase by looking at the size of the melted area on the membrane using fluorescence means (see Figure 4.5 on page 109).

In 2011, the Feldmann's group pursued his effort on artificial phospholipid membranes [174]. By trapping a single gold nanoparticles at the beam waist of a focused infrared laser beam, and moving it in three dimensions, the authors managed to make a gold nanoparticle cross the interface of a GUV. The nanoparticle couldn't cross the membrane unless it is heated in order to induce a local phase transition on the membrane and make it locally permeable (see Figure 6.26).

In 2012, the group of J. Polleux performed experiments on single living cells [190]. The authors managed to control the migration of living cells by creating temperature gradients at their vicinity, as illustrated by Figure 6.27.

In 2015, following his work on GUVs, the group of Feldmann managed to make single trapped nanoparticles cross the membrane of living cells [113]. The process combined plasmonic heating and optical force and reached a cell survival rate of more than 70%.

In 2015, Iwaki and coworkers managed to accelerate the motion of myosin V proteins along actin filaments by attaching a gold nanoparticle to it [89]. Experiments were conducted on artificial actin filaments grown in vitro and adsorbed onto a cover glass. Heating the nanoparticle using a laser light resulted in a local heating of the myosin molecule and a faster motion, up to 4.5 times faster, followed by dark field imaging.

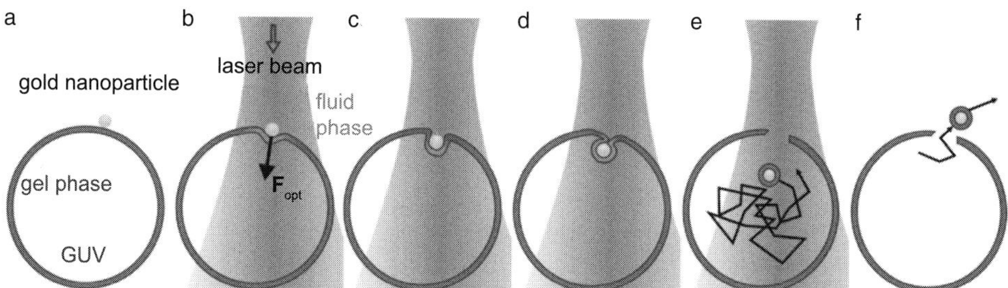

Fig. 6.26   Mechanism of optical injection. (a) The gold nanoparticles are attached to a GUV in the gel phase. (b d) The gold nanoparticle is irradiated by the laser, inducing a phase change in a membrane region around the nanoparticle into the fluid phase. The optical force pushes the gold nanoparticle into the vesicle and the elastic fluid phase membrane bends inward until the membrane pinches off and (e) the nanoparticle diffuses through the vesicle. The vesicle membrane cools down quickly after the injection and the phospholipids retract, forming a pore in the membrane (f) enabling the nanoparticle to leave the vesicle again. Reproduced with permission from Reference [174]. Copyright 2011, American Chemical Society.

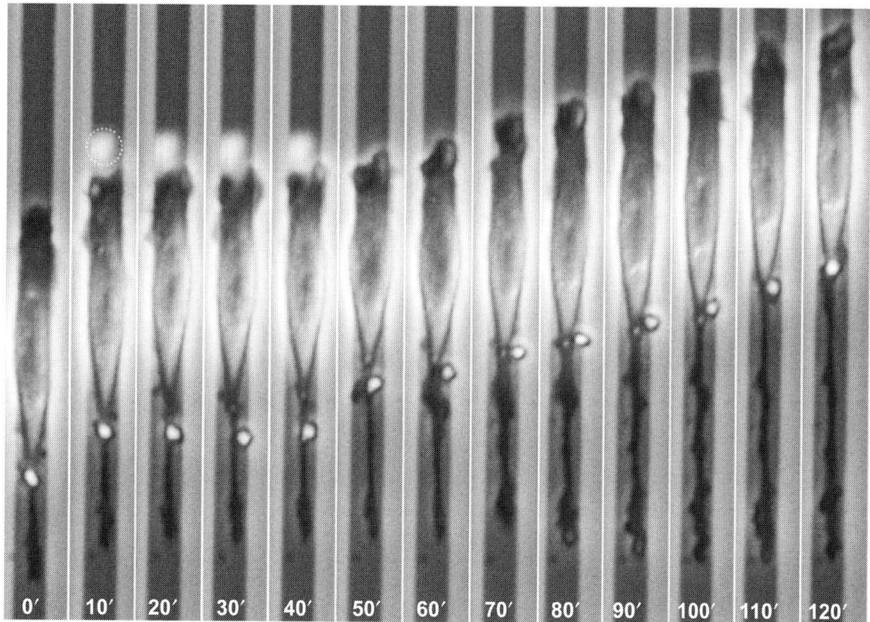

Fig. 6.27   Successive phase contrast images of a cell (fibroblast), every 10 min, migrating along a stripe of RGD-coated gold nanoparticles. A time $t = 10$ min, a laser beam is focused in front of the cell to increase the temperature by a few K. The direct consequence is a stop of cell migration. The cell migration starts again only when the laser is stopped ($t = 50$ min). Reproduced with permission from Reference [190]. Copyright 2012, American Chemical Society.

In 2017, Bahadori, Oddershede Thend Bendix reported on the thermoplasmonic-assisted fusion of living cells [16]. The authors used a gold nanoparticle placed at the interface between two optically trapped living cells in contact (see Figure 6.28). The local temperature increase created by the nanoparticle induced the fusion of the cell membranes. Fusion of

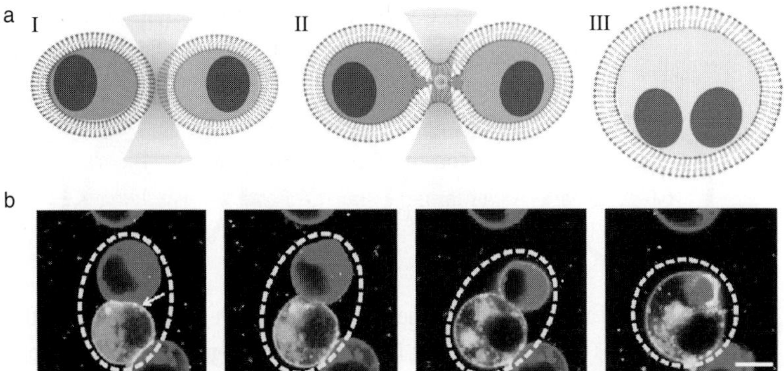

Fig. 6.28   Cell–cell fusion mediated by optically heated gold nanoparticles. (a) Schematic illustration of the fusion process, nuclei are drawn as black ellipsoids: (I) Two cells of interest are selected and placed next to each other using optical traps, (II) a 150 nm gold nanoparticle diffuses into the optical trap located at the contact zone between the two cells, the released heat induces the fusion process, (III) the cells fully fuse whereby their cell membranes and cytoplasmic contents mix. (b) Time series of confocal images of fusion between two selected cells (inside the yellow dashed ellipse), laser power at the sample was 350 mW. The 150 nm gold nanoparticle is visible as a bright spot in the contact zone (white arrow). Scale bar is 10 µm. Reproduced with permission from Reference [16]. Copyright 2016, Tsinghua University Press and Springer-Verlag Berlin Heidelberg

cells had already been achieved using other means, and is even observable in nature. Here, the novelty consisted in benefiting from the use of gold nanoparticles as nanosources of heat. Fusion between a cell and a GUV was also demonstrated in this work.

6.8.2 Conclusion

Heating at the single cell level is certainly one of the future active fields of research in thermoplasmonics, favored by the development of more efficient and reliable temperature imaging techniques. It will enable the study of fast thermal-induced processes and enable the heating at the cellular or sub-cellular level.

References

[1] 2013. *Seagate Press Release. Seagate To Demo Its Revolutionary Heat Assisted Magnetic Recording Storage Technology At CEATEC 2013. www.seagate.com/ about/newsroom/press-releases/HMR-demo-ceatec-2013-pr-master/.*

[2] 2013. *WD Press Release. WD Demonstrates Heat Assisted Magnetic Recording Hard Drive Technology at 2013 China (Ningbo) International Forum on Advanced Materials and Commercialization.*

[3] apps.webofknowledge.com, 2016.

[4] Absil, E., Tessier, G., Gross, M., Atlan, M., Warnasooriya, N., Suck, S., Coppey-Moisan, M., and Fournier, D. 2010. Photothermal Heterodyne Holography of Gold Nanoparticles. *Opt. Express*, **18**, 780–786.

[5] Adleman, J. R., Boyd, D. A., Goodwin, D. G., and Psaltis, D. 2009. Heterogenous Catalysis Mediated by Plasmon Heating. *Nano Lett.*, **9**(12), 4417–4423.

[6] Agarwal, A., Huang, S. W., O'Donnell, M., Day, K. C., Day, M., Kotov, N., and Ashkenazi, S. 2007. Targeted Gold Nanorod Contrast Agent for Prostate Cancer Detection by Photoacoustic Imaging. *J. Appl. Phys.*, **102**, 064701.

[7] Anderson, L. J. E., Hansen, E., Lukianova-Hleb, E. Y., Hafner, J. H., and Lapotko, D. O. 2010. Optically Guided Controlled Release from Liposomes with Tunable Plasmonic Nanobubbles. *J. Control. Release*, **144**, 151–158.

[8] Arnfield, M. R., Mathew, R. P., Tulip, J., and McPhee, M. S. 1992. Analysis of Tissue Optical Coefficients Using an Approximate Equation Valid for comparable absorption and scattering. *Phys. Med. Biol.*, **37**, 1219.

[9] Atlan, M., Gross, M., Desbiolles, P., Absil, E., Tessier, G., and Coppey-Moisan, M. 2008. Heterodyne Holographic Microscopy of Gold Particles. *Opt. Lett.*, **33**, 500–502.

[10] Ba, H., Rodríguez-Fernández, J., Stefani, F. D., and Feldmann, J. 2010. Immobilization of Gold Nanoparticles on Living Cell Membranes upon Controlled Lipid Binding. *Nano Lett.*, **10**, 3006–3012.

[11] Baffou, G., and Quidant, R. 2013. Thermo-Plasmonics: Using Metallic Nanostructures as Nano-Sources of Heat. *Laser & Photon. Rev.*, **7**(2), 171–187.

[12] Baffou, G., and Quidant, R. 2014. Nanoplasmonics for Chemistry. *Chem. Soc. Rev.*, **43**, 3898–3907.

[13] Baffou, G., Rigneault, H., Marguet, D., and Jullien, L. 2014a. A Critque of Methods for Temperature Imaging in Single Cells. *Nature Methods*, **11**, 899–901.

[14] Baffou, G., Polleux, J., Rigneault, H., and Monneret, S. 2014b. Super-Heating and Micro-Bubble Generation around Plasmonic Nanoparticles under cw Illumination. *J. Phys. Chem. C*, **118**, 4890.

[15] Baffou, G., Rigneault, H., Marguet, D., and Jullien, L. 2015. Reply to: "Validating Subcellular Thermal Changes Revealed by Fluorescent Thermosensors" and "The 10^5 Gap Issue Between Calculation and Measurement in Single-Cell Thermometry." *Nature Methods*, **12**, 803.

[16] Bahadori, A., Oddershede, L. B., and Bendix, P. M. 2017. Hot-Nanoparticle-Mediated Fusion of Selected Cells. *Nano Research*, doi: 10.1007/s12274–016–1392–3.

[17] Bao, C, Beziere, N, del Pino, P, Pelaz, B, Estrada, G, Tian, F. R., Ntziachristos, V., de la Fuente, J. M., and Cui, D. X. 2013. Gold Nanoprisms as Optoacoustic Signal Nanoamplifiers for In Vivo Bioimaging of Gastrointestinal Cancers. *Small*, **9**(1), 68–74.

[18] Barreto, J. A., O'Malley, W., Kubeil, M., Graham, B., and Stephan, H. 2011. Nanomaterials: Applications in Cancer Imaging and Therapy. *Adv. Mater.*, **23**, H18–H40.

[19] Bäuerle, D., Irsigler, P., Leyendecker, G., Noll, H., and Wagner, D. 1982. Ar$^+$ Laser Induced Chemical Vapor Deposition of Si from SiH$_4$. *Appl. Phys. Lett.*, **40**, 819–821.

[20] Bendix, P. M., Reihani, S. N. S., and Oddershede, L. B. 2010. Direct Measurements of Heating by Electromagnetically Trapped Gold Nanoparticles on Supported Lipid Bilayers. *ACS Nano*, **4**(4), 2256.

[21] Berciaud, S., Cognet, L., Blab, G. A., and Lounis, B. 2004. Photothermal Heterodyne Imaging of Individual Nonfluorescent Nanoclusters and Nanocrystals. *Phys. Rev. Lett.*, **93**, 257402.

[22] Berciaud, S., Cognet, L., and Lounis, B. 2005. Photothermal Absorption Spectroscopy of Individual Semiconductor Nanocrystals. *Nano Lett.*, **5**(11), 2160–2163.

[23] Berciaud, S., Lasne, D., Blab, G. A., Cognet, L., and Lounis, B. 2006. Photothermal Heterodyne Imaging of Individual Metallic Nanoparticles: Theory Versus Experiment. *Phys. Rev. B*, **73**, 045424.

[24] Berciaud, S., Cognet, L., Poulin, P., Weisman, R. B., and Lounis, B. 2007. Absorption Spectroscopy of Individual Single-Walled Carbon Nanotubes. *Nano Lett.*, **7**(5), 1203–1207.

[25] Betzig, E., Trautman, J. K., Wolfe, R., Gyorgy, E. M., Finn, P. L., Kryder, M. H., and Chang, C. H. 1992. Near-Field Magneto-Optics and High Density Data Storage. *Appl. Phys. Lett.*, **61**, 142–144.

[26] Boccara, A. C., Fournier, D., and Badoz, J. 1980. Thermooptical Spectroscopy: Detection by the "Mirage Effect." *Appl. Phys. Lett.*, **36**, 130–132.

[27] Boisselier, E., and Astruc, D. 2009. Gold Nanoparticles in Nanomedicine: Preparations, Imaging, Diagnostics, Therapies and Toxicity. *Chem. Soc. Rev.*, **38**, 1759–1782.

[28] Bora, T., Zoepfl, D., and Dutta, J. 2016. Importance of Plasmonic Heating on Visible Light Driven Photocatalysis of Gold Nanoparticle Decorated Zinc Oxide Nanorods. *Sci. Rep.*, **6**, 26913.

[29] Boyd, D. A., Greengard, L., Brongersma, M., El-Naggar, M. Y., and Goodwin, D. G. 2006. Plasmon-Assisted Chemical Vapor Deposition. *Nanolett.*, **6**(11), 2592–2597.

[30] Boyer, D., Tamarat, P., Maali, A., Lounis, B., and Orrit, M. 2002. Photothermal Imaging of Nanometer-Sized Metal Particles Among Scatterers. *Science*, **297**, 1160.

[31] Cao, L., Barsic, D. N., Guichard, A. R., and Brongersma, M. L. 2007. Plasmon-Assisted Local Temperature Control to Pattern Individual Semiconductor Nanowires and Carbon Nanotubes. *Nano Lett.*, **7**(11), 3523–3527.

[32] Carregal-Romero, S., Ochs, M., Rivera-Gil, P., Ganas, C., Pavlov, A. M., Sukhorukov, G. B., and Parak, W. J. 2012. NIR-Light Triggered Delivery of Macromolecules into the Cytosol. *J. Control. Release*, **159**, 120–127.

[33] Challener, W. A., McDaniel, T. W., Mihalcea, C. D., Mountfield, K. R., Pehlos, K., and Sendur, I. K. 2003. Light Delivery Techniques for Heat-Assisted Magnetic Recording. *Jpn. J. Appl. Phys.*, **42**, 981–988.

[34] Challener, W. A., Peng, C., Itagi, A. V., Karns, D., Peng, W., Peng, Y., Yang, X. M., Zhu, X., Gokemeijer, N. J., Hsia, Y.-T., Ju, G., Rottmayer, R. E., Seigler, M. A., and

Gage, E. C. 2009. Heat-Assisted Magnetic Recording by a Near-Field Transducer with Efficient Optical Energy Transfer. *Nat. Photon.*, **3**, 220–224.

[35] Chamberland, D. L., Agarwal, A., Kotov, N., Fowlkes, J. B., Carson, P. L., and Wang, X. 2008. Photoacoustic Tomograph of Joints Aided by an Etanercept-Conjugated Gold Nanoparticle Contrast Agent – an *ex vivo* Preliminary Rat Study. *Nanotechnology*, **19**, 095101.

[36] Chen, C. C., Lin, Y. P., and Wang, C. W. 2006. DNA-Gold Nanorod Conjuates for Remote Control of Localized Gene Expression by Near Infrared Irradiation. *J. Am. Chem. Soc.*, **11**(3709–3715).

[37] Chen, J., Wang, D., Xi, J., Au, L., Siekkinen, A., Warsen, A., Li, Z. Y., Zhang, H., Xia, Y., and Li, X. 2007. Tailored Optical Properties for Targeted Photothermal Destruction of Cancer Cells. *Nano Lett.*, **7**(5), 1318–1322.

[38] Chen, J., Glaus, C., Laforest, R., Zhang, Q., Yang, M., Gidding, M., Welch, M. J., and Xia, Y. 2010a. Gold Nanocages as Photothermal Transducers for Cancer Treatment. *Small*, **6**(7), 811.

[39] Chen, X., Zhu, H. Y., Zhao, J. C., Zheng, Z. F., and Gao, X. P. 2008. Visible-Light-Driven Oxidation of Organic Contaminants in Air with Gold Nanoparticle Catalysts on Oxide Supports. *Angew. Chem. Int. Ed.*, **47**, 5353–5356.

[40] Chen, Y. S., Frey, W., Kim, S., Homan, K., Kruizinga, P., Sokolov, K., and Emilianov, S. 2010b. Enhanced Thermal Stability of Silica-Coated Gold Nanorods for Photoacoustic Imaging and Image-Guided Therapy. *Opt. Express*, **18**(9), 8867–8877.

[41] Chen, Y. S., Frey, W., Kim, S., Kruizinga, P., Homan, K., and Emilianov, S. Y. 2011. Silica-Coated Gold Nanorods as Photoacoustic Signal Nanoamplifiers. *Nano Lett.*, **11**(2), 348–354.

[42] Cheng, K., Kothapalli, S. R., Liu, H., Leen Koh, A., Jokerst, J. V., Jiang, H., Yang, M., Li, J., Levi, J., Wu, J. C., Gambhir, S. S., and Cheng, Z. 2014. Construction and Validation of Nano Gold Tripods for Molecular Imaging of Living Subjects. *J. Am. Chem. Soc.*, **136**, 3560–3571.

[43] Cheng, Y., Samia, A. C., Meyers, J. D., Panagopoulos, I., Fei, B., and Burda, C. 2008. Highly Efficient Drug Delivery with Gold Nanoparticle Vectors for in Vivo Photodynamic Therapy of Cancer. *J. Am. Chem. Soc.*, **130**, 10643–10647.

[44] Cherukuri, P., Glazer, E. S., and Curley, S. A. 2010. Targeted Hyperthermia Using Metal Nanoparticles. *Adv. Drug Deliv. Rev.*, **62**, 339–345.

[45] Cho, A. 2003. Connecting the Dots to Custom Catalysts. *Science*, **299**(5613), 1684–1685.

[46] Choi, C. H. J., Alabi, C. A., Webster, P., and Davis, M. E. 2010. Mechanism of Active Targeting in Solid Tumors with Transferrin-Containing Gold Nanoparticles. *Proc. Natl. Acad. Sci. U.S.A.*, **107**(3), 1235–1240.

[47] Christopher, P., Ingram, D. B., and Linic, S. 2010. Enhancing Photochemical Activity of Semiconductor Nanoparticles with Optically Active Ag Nanostructures: Photochemistry Mediated by Ag Surface Plasmons. *J. Phys. Chem. C*, **114**, 9173–9177.

[48] Christopher, P., Xin, H., and Linic, S. 2011. Visible-Light-Enhanced Catalytic Oxidation Reactions on Plasmonic Silver Nanostructures. *Nature Chem.*, **3**, 467–472.

[49] Cognet, L., Tardin, C., Boyer, D., Choquet, D., Tamarat, P., and Lounis, B. 2003. Single Metallic Nanoparticle Imaging for Protein Detection in Cells. *Proc. Natl. Acad. Sci. U.S.A.*, **100**, 11350–11355.

[50] Cognet, L., Berciaud, S., Lasne, D., and Lounis, B. 2008. Photothermal Methods for Single Nonluminescent Nano-Objects. *Anal. Chem.*, **80**(7), 2288–2294.

[51] Copland, J.A., Eghtedari, M., Popov, V. L., Kotov, N., Mamedova, N., Motamedi, M., and Oraevsky, A. 2004. Bioconjugated Gold Nanoparticles as a Molecular Based Contrast Agent: Implications for Imaging of Deep Tumors Using Optoacoustic Tomography. *Mol. Imaging Biol.*, **6**(5), 341–349.

[52] Crozier, K. B., Sundaramuerthy, A., Kino, G. S., and Quate, C. F. 1993. Surface Enhancement of Local Optical Fields and the Lightning-Rod Effect. *Quantum Electron*, **23**, 435–440.

[53] Day, E. C., Thompson, P. A, Zhang, L., Lewinski, N. A., Ahmed, N., Drezek, R. A., Blaney, S. M., and West, J. L. 2011. Nanoshell-Mediated Photothermal Therapy Improves Survival in a Murine Glioma Model. *J. Neuro-Oncol.*, **104**(1), 55–63.

[54] Delcea, M, Sternberg, N., Yashchenok, A. M., Georgieva, R., Bäumler, H., Möhwald, H., and Skirtach, A. G. 2012. Nanoplasmonics for Dual-Molecule Release through Nanopores in the Membrane of Red Blood Cells. *ACS Nano*, **6**(5), 4169–4180.

[55] Dickerson, E. B., Dreaden, E. C., Huang, X., El-Sayed, I. H., Chu, H., Pushpanketh, S., McDonald, J. F., and El-Sayed, M. A. 2008. Gold Nanorod Assisted Near-Infrared Plasmonic Photothermal Therapy (PPTT) of Squamous Cell Carcinoma in Mice. *Cancer Letters*, **269**, 57–66.

[56] Ding, T. X., Hou, L., van der Meer, H., Alivisatos, A. P., and Orrit, M. 2016. Hundreds-fold Sensitivity Enhancement of Photothermal Microscopy in Near-Critical Xenon. *J. Phys. Chem. Lett.*, **7**, 2524–2529.

[57] Doane, T. L., and Burda, C. 2012. The Unique Role of Nanoparticles in Nanomedicine: Imaging, Drug Delivery and Therapy. *Chem. Soc. Rev.*, **41**, 2885–2911.

[58] Dou, Y., Zhigilei, L. V., Winograd, N., and Garrison, B. J. 2001. Explosive Boiling of Water Films Adjacent to Heated Surfaces: A Microscopic Description. *J. Phys. Chem. A*, **105**, 2748–2755.

[59] Dykman, L., and Khlebtsov, N. 2012. Gold Nanoparticles in Biomedical Applications: Recent Advances and Perspectives. *Chem. Soc. Rev.*, **41**, 2256–2282.

[60] Ebbesen, T. W., Lezec, H. J., Ghaemi, H. F., Thio, T., and Wolff, P. A. 1997. Extraordinary Optical Transmission through Sub-Wavelength Hole Arrays. *Nature*, **391**, 667–669.

[61] Eghtedari, M., Oraevsky, A., Copland, J. A., Kotov, N. A., Conjusteau, A., and Motamedi, M. 2007. High Intensity of *in vivo* Detection of Gold Nanorods using a Laser Optoacoustic Imaging System. *Nano Lett.*, **7**(7), 1914–1918.

[62] Eghtedari, M. A., Copland, J.A., Popov, V. L., Kotov, N. A., Motamedi, M., and Oraevsky, A. A. 2003. Bioconjugated Gold Nanoparticles as a Contrast Agent for Optoacoustic Detection of Small Tumors. *Proc. SPIE*, **4960**, 76–85.

[63] El-Sayed, I. H., Huang, X., and El-Sayed, M. A. 2006. Selective Laser Photo-Thermal Therapy of Epithelial Carcinoma Using anti-EGFR Antibody Conjugate Gold Nanoparticles. *Cancer Letters*, **239**, 129–135.

[64] Fasciani, C., Bueno Alejo, C. J., Grenier, M., Netto-Ferreira, J. C., and Scaiano, J. C. 2011. High-Temperature Organic Reactions at Room Temperature Using Plasmon Excitation: Decomposition of Dicumyl Peroxide. *Org. Lett.*, **2**, 204–207.

[65] Fedoruk, M., Meixner, M., Carretero-Palacios, S., Lohmüller, T., and Feldmann, J. 2013. Nanolithography by Plasmonic Heating and Optical Manipulation of Gold Nanoparticles. *ACS Nano*, **7**, 7648–7653.

[66] Field, S. B., and Bleehen, N. M. 1979. Hyperthermia in the Treatment of Cancer. *Cancer Treat. Rev.*, **6**, 63–94.

[67] Fisher, J. W., Sarkar, S., Buchanan, C. F., Szot, C. S., Whitney, J., Hatcher, H. C., Torti, S. V., Rylander, C. G., and Rylander, M. N. 2010. Photothermal Response of Human and Murine Cancer Cells to Multiwalled Carbon Nanotubes After Laser Irradiation. *Cancer Res.*, **70**(23), 9855–9864.

[68] Fournier, D., Lepoutre, F., and Boccara, A. 1983. Tomographic Approach for Photothermal Imaging Using the Mirage Effect. *J. Phys. Coll.*, **44**, C6–479–C6–482.

[69] Gaiduk, A., Ruijgrok, P. V., Yorulmaz, M., and Orrit, M. 2010a. Detection Limits in Photothermal Microscopy. *Chem. Sci.*, **1**, 343.

[70] Gaiduk, A., Yorulmaz, M., Ruijgrok, P. V., and Orrit, M. 2010b. Room-Temperature Detection of a Single Molecule's Absorption by Photothermal Contrast. *Science*, **330**, 353–356.

[71] Ghosh, P., Han, G., De, M., Kim, C. K., and Rotello, V. M. 2008. Gold Nanoparticles in Delivery Applications. *Adv. Drug Deliv. Rev.*, **60**, 1307–1315.

[72] Gleyzes, P., Boccara, A. C., and Saint-Jalmes, H. 1997. Multichannel Nomarski Microscope with Polarization Modulation: Performance and Applications. *Opt. Lett.*, **22**(20), 1529–1531.

[73] Gobin, A. M., Lee, M. H., Halas, N. J., James, W. D., Drezek, R. A., and West, J. L. 2007. Near-Infrared Resonant Nanoshells for Combined Optical Imaging and Photothermal Cancer Therapy. *Nano Lett.*, **7**(7), 1929–1934.

[74] Gu, Frank X., Karnik, Rohit, Wang, Andrew Z., Alexis, Frank, Levy-Nissenbaum, Etgar, Hong, Seungpyo, Langer, Robert S., and Farokhzad, Omid C. 2007. Targeted Nanoparticles for Cancer Therapy. *Nano Today*, **2**(3), 14.

[75] Hahn, M. A., Singh, A. K., Sharma, P., Brown, S. C., and Moudgil, B. M. 2011. Nanoparticles as Contrast Agents for in-vivo Bioimaging: Current Status and Future Perspectives. *Anal. Bioanal. Chem.*, **399**, 3–27.

[76] Hamann, H. F., Martin, Y. C., and Wickramasinghe, H. K. 2004. Thermally Assisted Recording Beyond Traditional Limits. *Appl. Phys. Lett.*, **84**, 810.

[77] Harada, M., Iwamoto, K., Kitamori, T., and Sawada, T. 1993. Photothermal Microscopy with Excitation and Probe Beams Coaxial under the Microscope and its Application to Microparticle Analysis. *Anal. Chem.*, **65**, 2938–2940.

[78] Haruta, M., Kobayashi, T., Sano, H., and Yamada, N. 1987. Novel Gold Catalysts for the Oxidation of Carbon Monoxide at a Temperature far Below 0°C. *Chem. Lett.*, **16**(2), 405–408.

[79] Hirsch, L. R., Stafford, R. J., Bankson, J. A., Sershen, S. R., Rivera, B., Price, R. E., Hazle, J. D., Halas, N. J., and West, J. L. 2003. Nanoshell-Mediated Near-Infrared Thermal Therapy of Tumors under Magnetic Resonance Guidance. *Proc. Natl. Acad. Sci. U.S.A.*, **100**(23), 13549–13554.

[80] Hrelescu, C., Stehr, L., Ringler, M., Sperling, R. A., Parak, W. J., Klar, T. A., and Feldmann, J. 2010. DNA Melting in Gold Nanostove Clusters. *J. Phys. Chem. C*, **114**, 7401–7411.

[81] Hsuan Hung, W., Hsu, I. K., Bushmaker, A., Kumar, R., Theiss, J., and Cronin, S. B. 2008. Laser Directed Growth of Carbon-Based Nanostructures by Plasmon Resonant Chemical Vapor Deposition. *Nanolett.*, **8**(10), 3278–3282.

[82] Hsun Hung, W., Aykol, M., Valley, D., Hou, W., and Cronin, S. B. 2010. Plasmon Resonant Enhancement of Carbon Monoxide Catalysis. *Nanolett.*, **10**, 1314–1318.

[83] Huang, X., El-Sayed, I. H., Qian, W., and El-Sayed, M. A. 2006a. Cancer Cell Imaging and Photothermal Therapy in the Near-Infrared Region by Using Gold Nanorods. *J. Am. Chem. Soc.*, **128**(6), 2115.

[84] Huang, X., Jain, P. K., El-Sayed, I. H., and El-Sayed, M. A. 2006b. Determination of the Minimum Temperature Required for Selective Phothermal Destruction of Cancer Cells with the Use of Immunotargeted Gold Nanoparticles. *Photochemistry and photobiology*, **82**, 412–417.

[85] Huang, X., Jain, P. K., El-Sayed, I. H., and El-Sayed, M. A. 2008. Plasmonic Photothermal Therapy (PPTT) Using Gold Nanoparticles. *Laser Med. Sci.*, **23**, 217–228.

[86] Huang, X. H., Jain, P. K., El-Sayed, I. H., and El-Sayed, M. A. 2007. Gold Nanoparticles: Interesting Optical Properties and Recent Applications in Cancer Diagnostic and Therapy. *Nanomedecine*, **2**(5), 681–693.

[87] Hüttmann, G., and Birngruber, R. 1999. On the Possibility of High-Precision Photothermal Microeffects and the Measurement of Fast Thermal Denaturation of Proteins. *IEEE J. Sel. Top. Quant.*, **5**(4), 954–962.

[88] Hwang, D. J., Ryu, S. G., and Grigoropoulos, C. P. 2011. Multi-Parametric Growth of Silicon Nanowires in a Single Platform by Laser-Induced Localized Heat Sources. *Nanotechnology*, **22**, 385303.

[89] Iwaki, M., Iwane, A. H., Ikezaki, K., and Yanagida, T. 2015. Local Heat Activation of Single Myosins Based on Optical Trapping of Gold Nanoparticles. *Nano Lett.*, **15**, 2456–2461.

[90] Jackson, W. B., Amer, N. M., Boccara, A. C., and Fournier, D. 1981. Photothermal Deflection Spectroscopy and Detection. *Appl. Opt.*, **20**(8), 1333–1344.

[91] Jain, P. K., Qian, W., and El-Sayed, M. A. 2006. Ultrafast Cooling of Photoexcited Electrons in Gold Nanoparticle-Thiolated DNA Conjugates Involves the Dissociation of the Gold-Thiol Bond. *J. Am. Chem. Soc.*, **128**(7), 2426–2433.

[92] Jain, P. K., El-Sayed, I. H., and El-Sayed, M. A. 2007a. Au Nanoparticles Target Cancer. *Nano Today*, **2**(1), 18.

[93] Jain, P. K., Huang, X., El-Sayed, I. H., and El-Sayed, M. A. 2007b. Review of Some Interesting Surface Plasmon Resonance-Enhanced Properties of Noble Metal Nanoparticles and their Applications to Biosystems. *Plasmonics*, **2**(3), 107–118.

[94] Jain, P. K., Huang, X., El-Sayed, I. H., and A., M. 2008. Noble Metals on the Nanoscale: Optical and Photothermal Properties and Some Applications in Imaging, Sensing, Biology, and Medicine. *Acc. Chem. Res.*, **41**(12), 1578–1586.

[95] Jaque, D., Martinez Maestro, L., del Rosal, B., Haro-González, P., Benayas, A., Plaza, J. L., Martín Rodriguez, E., and García Solé, J. 2014. Nanoparticles for Photothermal Therapies. *Nanoscale*, **6**, 9494–9530.

[96] Jaramillo, T. F., Baeck, S. H., Roldan Cuenya, B., and McFarland, E. W. 2003. Catalytic Activity of Supported Au Nanoparticles Deposited from Block Copolymer Micelles. *J. Am. Chem. Soc.*, **125**, 7148–7149.

[97] Ji, X., Shao, R., Elliott, A. M., Stafford, J., Esparza-Coss, E., Bankson, J. A., Liang, G., Luo, Z. P., Park, K., Markert, J. T., and Li, C. 2007. Bifunctional Gold Nanoshells with a Superparamagnetic Iron Oxide–Silica Core Suitable for Both MR Imaging and Photothermal Therapy. *J. Phys. Chem. C*, **111**, 6245–6251.

[98] Kasiraj, P., Robertson, N.L., and Wickramasinghe, H.K. 2002. *Thermally-Assisted Magnetic Recording System with Head Having Resistive Heater in Write Gap*. US Patent 6,493,183 B1.

[99] Kharlamov, A. N., Tyurnina, A. E., Veselova, V. S., Kovtun, O. P., Shur, V. Y., and Gabinsky, J. L. 2015. Silica-Gold Nanoparticles for Atheroprotective Management of Plaques: Results of the NANOM-FIM Trial. *Nanoscale*, **7**, 8003–8015.

[100] Kim, C., Cho, E. C., Chen, J., Song, K. H., Au, L., Favazza, C., Zhang, Q., Cobley, C. M., Gao, F., Xia, Y., and Wang, L. V. 2010. *In Vivo* Molecular Photoacoustic Tomography of Melanomas Targeted by Bioconjugated Gold Nanocages. *ACS Nano*, **4**, 4559–4564.

[101] Kim, K., Huang, S. W., Ashkenazi, S., O'Donnell, M., Agarwal, A., and Kotov, N. A. and. 2007. Photoacoustic Imaging of Early Inflammatory Response Using Gold Nanorods. *Appl. Phys. Lett.*, **90**, 223901.

[102] Kirpotin, D. B., Drummond, D. C., Shao, Y., Shalaby, M. R., Hong, K., Nielsen, U. B., Marks, J. D., Benz, C. C., and Park, J. W. 2006. Antibody Targeting of Long-Circulating Lipidic Nanoparticles does not Increase Tumor Localization but Does Increase Internalization in Animal Models. *Cancer Res.*, **66**(13), 6732–6740.

[103] Kryder, M. H., Gage, E. C., McDaniel, T. W., Challener, W. A., Rottmayer, R. E., Ju, G., Hsia, Y.-T., and Erden, M. F. 2008. Heat Assisted Magnetic Recording. *Proc. IEEE*, **96**, 1810–1835.

[104] Kyrsting, A., Bendix, P. M., Stamou, D. G., and Oddershede, L. B. 2011. Heat Profiling of Three-Dimensionally Optically Trapped Gold Nanoparticles using Vesicle Cargo Release. *Nano Lett.*, **11**, 888–892.

[105] Lal, S., Clare, S. E., and Halas, N. J. 2008. Nanoshell-Enabled Photothermal Cancer Therapy: Impending Clinical Impact. *Acc. Chem. Res.*, **41**, 1842.

[106] Lalisse, A., Tessier, G., Plain, J., and Baffou, G. 2016. Plasmonic Efficiencies of Nanoparticles Made of Metal Nitrides (TiN, ZrN) Compared with Gold. *Sci. Rep.*, **6**, 38647.

[107] Lapotko, D. 2011. Plasmonic Nanobubbles as Tunable Cellular Probes for Cancer Theranostics. *Cancers*, **3**(1), 802–840.

[108] Lasne, D., Blab, G. A., Berciaud, S., Heine, M., Groc, L., Choquet, D., Cognet, L., and Lounis, B. 2006. Single Nanoparticle Photothermal Tracking (SNaPT) of 5-nm Gold Beads in Live Cells. *Biophys. J.*, **91**, 4598–4604.

[109] Lasne, D., Blab, G. A., De Giorgi, F., Ichas, F., Lounis, B., and Cognet, L. 2007. Label-Free Optical Imaging of Mitochondria in Live Cells. *Opt. Express*, **15**(21), 14184–14193.

[110] Lehmann, O., and Stuke, M. 1995. Laser-Driven Movement of Three-Dimensional Microstructures Generated by Laser Rapid Prototyping. *Science*, **270**, 1644–1646.

[111] Leung, S. J., Kachur, X. M., Bobnick, M. C., and Romanowski, M. 2011. Wavelength-Selective Light-Induced Release from Plasmon Resonant Liposomes. *Adv. Funct. Mater.*, **21**, 1113–1121.

[112] Lévy, R., Shaheen, U., Cesbron, Y., and Sée, V. 2010. Gold Nanoparticles Delivery in Mammalian Live Cells: a Critical Review. *Nano Reviews*, **1**, 4889.

[113] Li, M., Lohmüller, T., and Feldmann, J. 2015. Optical Injection of Gold Nanoparticles into Living Cells. *Nano Lett.*, **15**, 770–775.

[114] Li, W., and Chen, X. 2015. Gold Nanoparticles for Photoacoustic Imaging. *Nanomedicine*, **10**(2), 299–320.

[115] Li, W., Brown, P. K., Wang, L. V., and Xia, Y. 2011. Gold Nanocages as Contrast Agents for Photoacoustic Imaging. *Contrast Media Mol. Imaging*, **6**, 370–377.

[116] Lim, Z. Z. J., Li, J. E. J., Ng, C. T., Yung, L. Y. L., and Bay, B. H. 2011. Gold Nanoparticles in Cancer Therapy. *Acta Pharmacol. Sin.*, **32**, 983–990.

[117] Lin, C. P., and Kelly, M. W. 1998. Cavitation and Acoustic Emission around Laser-Heated Microparticles. *Appl. Phys. Lett.*, **72**(22), 2800–2802.

[118] Lin, C. P., Kelly, M. W., Sibayan, S. A. B., Latina, M. A., and Anderson, R. R. 1999. Selective Cell Killing by Microparticle Absorption of Pulsed Laser Radiation. *IEEE J. Sel. Top. Quant.*, **5**, 963–968.

[119] Long, M. E., Swofford, R. L., and Albrecht, A. C. 1976. Thermal Lens Technique: A New Method of Absorption. *Science*, **191**, 183–185.

[120] Loo, C., Lowery, A., Halas, N., West, J., and Drezek, R. 2005. Immunotargeted Nanoshells for Integrated Cancer Imaging and Therapy. *Nano Lett.*, **5**(4), 709–711.

[121] Lu, W., Xiong, C., Zhang, G., Huang, Q., Zhang, R., Zhang, J. Z., and Li, C. 2009. Targeted Photothermal Ablation of Murine Melanomas with Melanocyte-Stimulating Hormone Analog – Conjugated Hollow Gold Nanospheres. *Clin. Cancer Res.*, **15**(3), 876–886.

[122] Lu, W., Kumar Singh, A., Afrin Khan, S., Senapati, D., Yu, H., and Chandra Ray, P. 2010. Gold Nano-Popcorn-Based Targeted Diagnosis, Nanotherapy Treatment, and In Situ Monitoring of Photothermal Therapy Response of Prostate Cancer Cells Using Surface-Enhanced Raman Spectroscopy. *J. Am. Chem. Soc.*, **132**(51), 18103–18114.

[123] Lukianova-Hleb, E., Ren, X., Townley, D., Wu, X., Kupferman, M. E., and Lapotko, D. O. 2012. Plasmonic Nanobubbles Rapidly Detect and Destroy Drug-Resistant Tumors. *Theranostics*, **2**(10), 976–987.

[124] Maier-Hauff, K., Rothe, R., Scholz, R., Gneveckow, U., Wust, P., Thiesen, B., Freussner, Annelie, von Deimling, Andreas, Waldoefner, N., Felix, R., and Jordan, A. 2007. Intracranial Thermotherapy Using Magnetic Nanoparticles Combined with External Beam Radiotherapy: Results of a Feasibility Study on Patients with Glioblastoma Multiforme. *J. Neuro-Oncol.*, **81**, 53.

[125] Mallidi, S., Larson, T., Aaron, J., Sokolov, K., and Emelianov, S. 2007. Molecular Specific Optoacoustic Imaging with Plasmonic Nanoparticles. *Opt. Express*, **15**, 6583.

[126] Mallidi, S., T., Larson, Tam, J., Joshi, P. P., Karpiouk, A., Sokolov, K., and Emelianov, S. 2009. Multiwavelength Photoacoustic Imaging and Plasmon Resonance Coupling of Gold Nanoparticles for Selective Detection of Cancer. *Nano Lett.*, **9**(8), 2825.

[127] Matsumoto, T., Shimano, T., Saga, H., Sukeda, H., and Kiguchi, M. 2004. Highly Efficient Probe with a Wedge-Shaped Metallic Plate for High Density Near-Field Optical Recording. *Appl. Phys. Lett.*, **95**(8), 3901–3906.

[128] Matsumoto, T., Anzai, Y., Shintani, T., Nakamura, K., and Nishida, T. 2006. Writing 40 nm Marks by Using a Beaked Metallic Plate Near-Field Optical Probe. *Opt. Lett.*, **31**(2), 259–261.

[129] Matsumoto, T., Nakamura, K., Nishida, T., Hieda, H., Kikitsu, A., Naito, K., and Koda, T. 2008. Thermally Assisted Magnetic Recording on a Bit-Patterned Medium by Using a Near-Field Optical Head with a Beaked Metallic Plate. *Appl. Phys. Lett.*, **93**, 031108.

[130] McDaniel, T. W. 2005. Ultimate Limits of Thermally Assisted Magnetic Recording. *J. Phys.: Condens. Matter*, **17**, R315–R332.

[131] McLaughlan, J. R., Roy, R. A., Ju, H., and Murray, T. W. 2010. Ultrasonic Enhancement of Photoacoustic Emissions by Nanoparticle-Targeted Cavitation. *Opt. Lett.*, **35**(13), 2127–2129.

[132] Melancon, M. P., Lu, W., Yang, Z., Zhang, R., Cheng, Z., Elliot, A. M., Stafford, J., Olson, T., Zhang, J. Z., and Li, C. 2008. *In Vitro* and *In Vivo* Targeting of Hollow Gold Nanoshells Directed at Epidermal Growth Factor Receptor for Photothermal Ablation Therapy. *Mol. Cancer Ther.*, **7**(6), 1730–1739.

[133] Miyazaki, J., Tsurui, H., Kawasumi, K., and Kobayashi, T. 2014. Optimal Detection Angle in sub-Diffraction Resolution Photothermal Microscopy: Application for High Sensitivity Imaging of Biological Tissues. *Opt. Express*, **22**(16), 18833–18842.

[134] Nam, J., Won, N., Jin, H., Chung, H., and Kim, S. 2009. pH-Induced Aggregation of Gold Nanoparticles for Photothermal Cancer Therapy. *J. Am. Chem. Soc.*, **131**(38), 13639.

[135] Ni, W., Ba, H., Lutich, A. A., Jäckel, F., and Feldmann, J. 2012. Enhancing Single-Nanoparticle Surface-Chemistry by Plasmonic Overheating in an Optical Trap. *Nano Lett.*, **12**, 4647–4650.

[136] Nie, L., Wang, S., Wang, X., Rong, P., Bhirde, A., Ma, Y., Liu, G., Huang, P., Lu, G., and Chen, X. 2014. In Vivo Volumetric Photoacoustic Molecular Angiography and Therapeutic Monitoring with Targeted Plasmonic Nanostars. *Small*, **10**(8), 1585–1593.

[137] O'Connor, D., and Zayats, A. V. 2010. The Third Plasmonic Revolution. *Nature Nanotech.*, **5**, 482.

[138] Octeau, V., Cognet, L., Duschene, L., Lasne, D., Schaeffer, N., Fernig, D. G., and Lounis, B. 2009. Photothermal Absorption Correlation Spectroscopy. *ACS Nano*, **3**(2), 345–350.

[139] Oleson, J. R., and Dewhirst, M. W. 1983. Hyperthermia: An Overview of Current Progress and Problems. *Current Problems in Cancer*, **8**(6), 1–62.

[140] Oraevsky, A. A., Karabutov, A. A., and Savateeva, E. V. 2001. Enhancement of Optoacoustic Tissue Contrast with Absorbing Nanoparticles. *Proc. SPIE*, **4434**, 60–69.

[141] Overgaard, J., and Suit, H. D. 1979. Time-Temperature Relationship in Hyperthermic Treatment of Malignant and Normal Tissue *in Vivo*. *Cancer Res.*, **39**, 3248–3253.

[142] Paasonen, L., Laaksonen, T., Johans, C., Yliperttula, M., Kontturi, K., and Urtti, A. 2007. Gold Nanoparticles Enable Selective Light-Induced Contents Release from Liposomes. *J. Control. Release*, **122**, 86–93.

[143] Paithankar, D. Y., Sakamoto, F. H., Farinelli, W. A., Kositratna, G., Blomgren, R. D., Meyer, T. J., Faupel, L. J., Kauvar, A. N. B., Lloyd, J. R., Cheung, W. L., Owczarek, W. D., Suwalska, A. M., Kochanska, K. B., Nawrocka, A. K., Paluchowska, E. B., Podolec, K. M., Pirowska, M. M., Wojas-Pelc, A. B., and Anderson, R. R. 2015. Acne Treatment Based on Selective Photothermolysis of Sebaceous Follicles with Topically Delivered Light-Absorbing Gold Microparticles. *J. Invest. Dermatol.*, **135**, 1727–1734.

[144] Paulo, P. M. R., Gaiduk, A., Kulzer, F., Krens, S. F. G., Spaink, H. P., Schmidt, T., and Orrit, M. 2009. Photothermal Correlation Spectroscopy of Gold Nanoparticles in Solution. *J. Phys. Chem. C*, **113**, 11451–11457.

[145] Pedrosa, P., Vinhas, R., Fernandes, A., and Baptista, P. V. 2015. Gold Nanotheranostics: Proof-of-Concept or Clinical Tool? *Nanomaterials*, **5**, 1853–1879.

[146] Peng Qian, L., Han Zhou, L., Too, H. T., and Chow, G. M. 2011. Gold Decorated $NaYF_4$:Yb,Er/$NaYF_4$/Silica (Core/Shell/Shell) Upconversion Nanoparticles for Photothermal Destruction of BE(2)-C Neuroblastoma Cells. *J. Nanopart. Res.*, **13**, 499–510.

[147] Pitsillides, C. M., Joe, E. K., Anderson, R. R., and Lin, C. P. 2003. Selective Cell Targeting with Light Absorbing Microparticles and Nanoparticles. *Biophys. J.*, **84**, 4023.

[148] Polleux, J., Rasp, M., Louban, I., Plath, N., Feldhoff, A., and Spatz, J. P. 2011. Benzyl Alcohol and Block Copolymer Micellar Lithography: A Versatile Route to Assembling Gold and in Situ Generated Titania Nanoparticles into Uniform Binary Nanoarrays. *ACS Nano*, **5**(8), 6355–6364.

[149] Quek, C. H., and Leong, K. W. 2012. Near-Infrared Fluorescent Nanoprobes for *in Vivo* Optical Imaging. *Nanomaterials*, **2**, 92–112.

[150] Richter, H. R., Lyberatos, A., Nowak, U., Evans, R. F. L., and Chantrell, R. W. 2012. The Thermodynamic Limits of Magnetic Recording. *J. Appl. Phys.*, **111**, 033909.

[151] Robert, H., Kundrat, F., Bermúdez Ureña, E., Rigneault, H., Monneret, S., Quidant, R., Polleux, J., and Baffou, G. 2016. Light-Assisted Solvothermal Chemistry Using Plasmonic Nanoparticles. *ACS Omega*, **1**, 2–8.

[152] Robinson, J. T., Tabakman, S. M., Liang, Y., Wang, H., Sanchez Casalongue, H., Vinh, D., and Dai, H. 2011. Ultrasmall Reduced Graphene Oxide with High Near-Infrared Absorbance for Photothermal Therapy. *J. Am. Chem. Soc.*, **133**, 6825–6831.

[153] Roger, J., Fournier, D., Boccara, A., and Lepoutre, F. 1989. Coatings Characterization by the Mirage Effect and the Photothermal Microscope. *J. Phys. Coll.*, **50**, C5–295–C5–310.

[154] Sapareto, S. A., and Dewey, W. C. 1984. Thermal Dose Determination in Cancer Therapy. *Int. J. Radiation Oncology Biol. Phys.*, **10**, 787–800.

[155] Selmke, M., and Cichos, F. 2013a. Photonic Rutherford Scattering: A Classical and Quantum Mechanical Analogy in Ray and Wave Optics. *Am. J. Phys.*, **81**, 405–413.

[156] Selmke, M., and Cichos, F. 2013b. Photothermal Single Particle Rutherford Scattering Microscopy. *Phys. Rev. Lett.*, **110**, 103901.

[157] Selmke, M., and Cichos, F. 2015. The Physics of the Photothermal Detection of Single Absorbing Nano-Objects: A Review. *Arxiv*, 1510.08669v1.

[158] Selmke, M., Braun, M., and Cichos, F. 2012a. Gaussian Beam Photothermal Single Particle Microscopy. *J. Opt. Soc. Am. A*, **29**(10), 2237–2241.

[159] Selmke, M., Braun, M., and Cichos, F. 2012b. Nano-Lens Diffraction around a Single Distribution Analysis. *Opt. Express*, **20**(7), 8055–8070.

[160] Selmke, M., Braun, M., and Cichos, F. 2012c. Photothermal Single Particle Microscopy: Detection of a Nanolens. *ACS Nano*, **6**, 2741–2749.

[161] Selmke, M., Heber, A., Braun, M., and Cichos, F. 2014. Photothermal Single Particle Microscopy Using a Single Laser Beam. *Appl. Phys. Lett.*, **105**, 013511.

[162] Sershen, S. R., Westcott, S. L., Halas, N. J., and West, J. L. 2000. Temperature-Sensitive Polymer–Nanoshell Composites for Photothermally Modulated Drug Delivery. *J. Biomed. Mater. Res.*, **51**, 293–298.

[163] Sharma, P., Brown, S. C., Singh, A., N., Iwakuma, Pyrgiotakis, G., Krishna, V., Knapik, J. A., Barr, K., Moudgil, B. M., and Grobmyer, S. R. 2010. Near-Infrared Absorbing and Luminescent Gold Speckled Silica Nanoparticles for Photothermal Therapy. *J. Mater. Chem.*, **20**, 5182–5185.

[164] Shi, X., Thornton, R. L., and Hesselink, L. 2002. A Nano-Aperture with $1000\times$ Power Throughput Enhancement for Very Small Aperture Laser System (VSAL). *Proc SPIE*, **4342**, 320–327.

[165] Skirtach, A. G., Munoz Javier, A., Kreft, O., Köhler, K., Piera Alberola, A., Möhwald, H., Parak, W. J., and Sukhorukov, G. B.—. 2006. Laser-Induced Release of Encapsulated Materials inside Living Cells. *Angew. Chem. Int. Ed.*, **45**, 4612–4617.

[166] Smith, A. M., Mancini, M. C., and Nie, S. 2009. Bioimaging: Second Window for *in vivo* Imaging. *Nature Nanotech.*, **4**, 710–711.

[167] Stehr, J., Hrelescy, C., Sperling, R. A., Raschke, G., Wunderlich, M., Nichtl, A., Heindl, D., Kurzinger, K., Parak, W. J., Klar, T. A., and Feldmann, J. 2008. Gold Nanostoves for Microsecond DNA Melting Analysis. *Nano Lett.*, **8**(2), 619.

[168] Stern, J. M., Stanfield, J., Kabbani, W., Hsieh, J. T., and Cadeddu, J. A. 2008. Selective Prostate Cancer Thermal Ablation with Laser Activated Gold Nanoshells. *J. Urol.*, **179**, 748–753.

[169] Stipe, B. C., Strand, T. C., Poon, C. C., Balamane, H., Boone, T. D., Katine, J. A., Li, J. L., Rawat, V., Nemoto, H., Hirotsune, A., Hellwig, O., Ruiz, R., Dobisz, E., Kercher, D. S., Robertson, N., Albrecht, T. R., and Terris, B. D. 2010. Magnetic Recording at 1.5 Pb m^{-2} Using an Integrated Plasmonic Antenna. *Nat. Photon.*, **4**, 484–488.

[170] Takahashi, H., Niidome, Y., and Yamada, S. 2005. Controlled Release of Plasmid DNA from Gold Nanorods Induced by Pulsed Near-Infrared Light. *Chem. Comm.*, 2247–2249.

[171] Takahashi, H., Niidome, T., Nariai, A., Niidome, Y., and Yamada, S. 2006. Gold Nanorod-sensitized Cell Death: Microscopic Observation of Single Living Cells Irradiated by Pulsed Near-infrared Laser Light in the Presence of Gold Nanorods. *Chemistry Letters*, **35**(5), 500–501.

[172] Timko, B. P., Whitehead, K., Gao, W., Kohane, D. S., Farokhzad, O., Anderson, D., and Langer, R. 2011. Advances in Drug Delivery. *Annu. Rev. Mater. Res.*, **41**, 1–20.

[173] Urban, A. S., Fedoruk, M., Horton, M. R., Rädler, J. O., Stefani, F. D., and Feldmann, J. 2009. Controlled Nanometric Phase Transitions of Phospholipid Membranes by Plasmonic Heating of Single Gold Nanoparticles. *Nano Lett.*, **9**(8), 2903–2908.

[174] Urban, A. S., Pfeiffer, T., Fedoruk, M., Lutich, A.A., and Feldmann, J. 2011. Single Step Injection of Gold Nanoparticles Through Phospholipid Membranes. *ACS Nano*, **5**(5), 3585–3590.

[175] Varela, J. A., Dupuis, J. P., Etchepare, L., Espana, A., Cognet, L., and Groc, L. 2016. Targeting neurotransmitter receptors with nanoparticles in vivo allows single-molecule tracking in acute brain slices. *Nature Commun.*, **7**, 10947.

[176] Vásquez Vásquez, C., Vaz, B., Giannini, V., Pérez-Lorenzo, M., Alvarez-Puebla, R. A., and Correa-Duarte, M. A. 2013. Nanoreactors for Simultaneous Remote Thermal Activation and Optical Monitoring of Chemical Reactions. *J. Am. Chem. Soc.*, **135**, 13616–13619.

[177] Vermeulen, P., Cognet, L., and Lounis, B. 2014. Photothermal Microscopy: Optical Detection of Small Absorbers in Scattering Environments. *J. Microsc.*, **254**(3), 115–121.

[178] Wang, Y., Xie, X., Wang, X., Ku, G., Gill, K. L., O'Neal, D. P., Stoica, G., and Wang, L. V. 2004. Photoacoustic Tomography of a Nanoshell Contrast Agent in the in Vivo Rat Brain. *Nano Lett.*, **4**(9), 1689–1692.

[179] Weller, D., and Moser, A. 1999. Thermal Effect Limits in Ultrahigh-Density Magnetic Recording. *IEEE Trans. Magn.*, **35**, 4423–4439.

[180] Wijaya, A., Schaffer, S. B., Pallares, I. G., and Hamad-Schifferli, K. 2008. Selective Release of Multiple DNA Oligonucleotides from Gold Nanorods. *ACS Nano*, **3**(1), 80–86.

[181] Wu, A. Q., Kubota, Y., Klemmer, T., Rausch, T., Peng, C., Peng, Y., Karns, D., Zhu, X., Ding, Y., Chang, E. K. C., Zhao, Y., Zhou, H., Gao, K., Thiele, J. U., Seigler,

M., Ju, G., and Gage, E. 2013. HAMR Areal Density Demonstration of 1+ Tbpsi Spinstand. *IEEE Trans. Magn.*, **49**, 779–782.

[182] Wu, X., Ming, T., Wang, X., Wang, P., Wang, J., and Chen, J. 2010. High-Photoluminescence-Yield Gold Nanocubes: For Cell Imaging and Photothermal Therapy. *ACS Nano*, **4**(1), 113–120.

[183] Yang, K., Zhang, S., Zhang, G., Sun, X., Lee, S. T., and Liu, Z. 2010. Graphene in Mice: Ultrahigh In Vivo Tumor Uptake and Efficient Photothermal Therapy. *Nano Lett.*, **10**, 3318–3323.

[184] Yang, X., Skrabalak, S. E., Li, Z. Y., Xia, Y., and Wang, V. W. 2007. Photoacoustic Tomography of a Rat Cerebral Cortex in Vivo with Au Nanocages as an Optical Contrast Agent. *Nano Lett.*, **7**(12), 3798–3802.

[185] Yang, X., Stein, E. W., Ashkenazi, S., and Wang, L. V. 2009. Nanoparticles for Photoacoustic Imaging. *Adv. Rev.*, **1**, 360.

[186] Yen, C. W., and El-Sayed, M. A. 2009. Plasmonic Field Effect on the Hexacyanoferrate (III)-Thiosulfate Electron Transfer Catalytic Reaction on Gold Nanoparticles: Electromagnetic or Thermal? *J. Phys. Chem. C*, **113**, 19585–19590.

[187] Zhang, W., Guo, Z., Huang, D., Liu, Z., and Zhong, H. 2011. Synergistic Effect of Chemo-Photothermal Therapy Using PEGylated Graphene Oxide. *Biomaterials*, **32**(33), 8555–8561.

[188] Zharov, V. P., and Galitovsky. 2003. Photothermal Detection of Local Thermal Effects During Selectrive Nanophotothermolysis. *Appl. Phys. Lett.*, **83**(24), 4897–4899.

[189] Zhou, N., Xu, X., Hammack, A. T., Stipe, B. C., Gao, K., Scholz, W., and Gage, E. C. 2014. Plasmonic Near-Field Transducer for Heat-Assisted Magnetic Recording. *Nanophotonics*, **3**(3), 141–155.

[190] Zhu, M., Baffou, G., Meyerbröker, N., and Polleux, J. 2012. Micropatterning Thermoplasmonics Gold Nanoarrays to Manipulate Cell Adhesion. *ACS Nano*, **6**(8), 7227–7233.

A   Appendix: Dimensional Analysis

Dimensional analysis is a widely used and powerful method to determine actual orders of magnitudes of the physical quantities involved in a problem just from the expression of the governing differential equations, but without solving them.

In most cases, it consists in replacing the partial time and space derivatives respectively with $1/T$ and $1/L$ where τ and L are the typical time and space scales of the problem.

Example 1: Heat Diffusion Equation

As an example, let us consider the heat diffusion equation see Equation (2.16):

$$\partial_t T - D\nabla^2 T = 0 \tag{A.1}$$

If the evolution of the system is characterized by a typical time τ (the duration or the period of the evolution), the derivative $\partial_t T$ is supposed to be naturally on the order of T_0/τ, where T_0 is the typical temperature variation occurring during τ. Identically, $\nabla^2 T$ can be replaced with T_0/L^2, where L is the typical extension of the temperature increase T_0. Since the two terms of Equation (A.1) are equal, their orders of magnitude can be assimilated:

$$\frac{T_0}{\tau} \sim D\frac{T_0}{L^2} \tag{A.2}$$

which yields interesting relations such as

$$\tau \sim \frac{L^2}{D}. \tag{A.3}$$

This expression means that a system of typical size L, featuring a thermal diffusivity D, is supposed to thermalize within a typical duration τ, a valuable piece of information that we derived without conducting any numerical simulation.

Example 2: Kinetic Energy

Another approach of dimensional analysis consists in determining an order of magnitude of a physical quantity from the other known physical quantities of the problem. For instance, let us say that nobody ever told you what the kinetic energy is of a mass m animated with a velocity v; you can guess that an order of magnitude of this energy has to be mv^2 as this is the only means of making a quantity that has the dimension of an energy out of these two numbers. You will know that this is an order of magnitude, but you won't be able to determine the actual prefactor $1/2$.

Example 3: 1945 Atomic Bomb Test

Let us consider a final example, illustrated by an actual historical event. In 1945, in New Mexico, the first atomic bomb test explosion took place: the so-called Trinity test. Although a series of photographs of the event has been made public, the exact power of the bomb was a closely guarded secret. However, a physicist, G. I. Taylor, was able to estimate that the actual power delivered by the bomb was about 20 kt. First considered as a possible spy by the CIA, it turned out that Taylor just applied dimensional analysis from the series of photographs. Here is how he did it. From the photographs, he was able to estimate the typical size L of the explosion and the related timescale T. Then, he realized that the magnitude of the phenomenon was also dependent on the mass density ρ of the surrounding medium (air), as the atomic energy is ultimately changed into kinetic energy in the surrounding medium. From these three quantities, the only way to make a quantity that has the dimension of an energy E is

$$E = \frac{L^5 \rho}{T^2}. \tag{A.4}$$

Example 4: Casimir Effect

The power of dimensional analysis stems from the fact that valuable information can be obtained without solving the equation. The price to pay is that we can only derive orders of magnitude, or expressions up to a constant factor. The question is: *How do we know that the constant factor is not much different from unity?* It turns out that in physics, constant factors are rarely much different from unity. A famous exception is the Casimir force. The force per unit area acting between two plates separated by a distance a, and originating from quantum field theory, reads

$$f = \frac{\pi}{480} \frac{h\,c}{L^4}. \tag{A.5}$$

In this case, a constant factor as big as $480/\pi \approx 150$ comes out of the theory.

Appendix: Thermodynamical Constants

B.1 Thermodynamical Constants of Common Materials

When working in thermoplasmonics, and especially when writing code for numerical simulations, it is very common to look for thermodynamical constants of water, glass and gold. In the following table, I gather all the useful information related to thermodynamical constants of common materials in plasmonics, namely glass, water, glycerol, gold, silver and aluminum. I leave some lines free so that the reader can append their favorite materials.

Table B.1 Thermodynamical constants of common materials used in plasmonics, at $T = 20°C$.

	thermal conductivity (κ) $\mathrm{W \cdot m^{-1} \cdot K^{-1}}$	specific heat capacity (c_p) $\mathrm{J \cdot K^{-1} \cdot kg^{-1}}$	mass density (ρ) $\times 10^3 \mathrm{kg \cdot m^{-3}}$	thermal diffusivity (D) $\times 10^{-6} \mathrm{m^2 \cdot s^{-1}}$	dynamic viscosity (η) $\mathrm{Pa \cdot s}$	melting temperature (T_m) $°C$
glass[a]	1.2	840	2.2	0.62		
water	0.591	4180	1.000	0.146	1.00×10^{-3}	
glycerol	0.285	2430	1.261	93	1.20	
gold	318	129	19.32	127		1064
silver	429	240	10.50	166		962
aluminum	237	900	2.700	97		660

[a] borosilicate glass

Note that the thermal diffusivity D and the kinematic viscosity ν can be expressed using other thermodynamical constants according to these expressions:

$$D = \frac{\kappa}{\rho \, c_\mathrm{p}}, \qquad \nu = \frac{\eta}{\rho}$$

B.2 Temperature Dependence of Thermodynamical Constants of Water

The temperature dependence of the physical parameters of water is a common concern in thermoplasmonics. This section is intended to give all the information needed to handle this problem. Figure B.1 plots the temperature dependence of the mass density, the dynamic

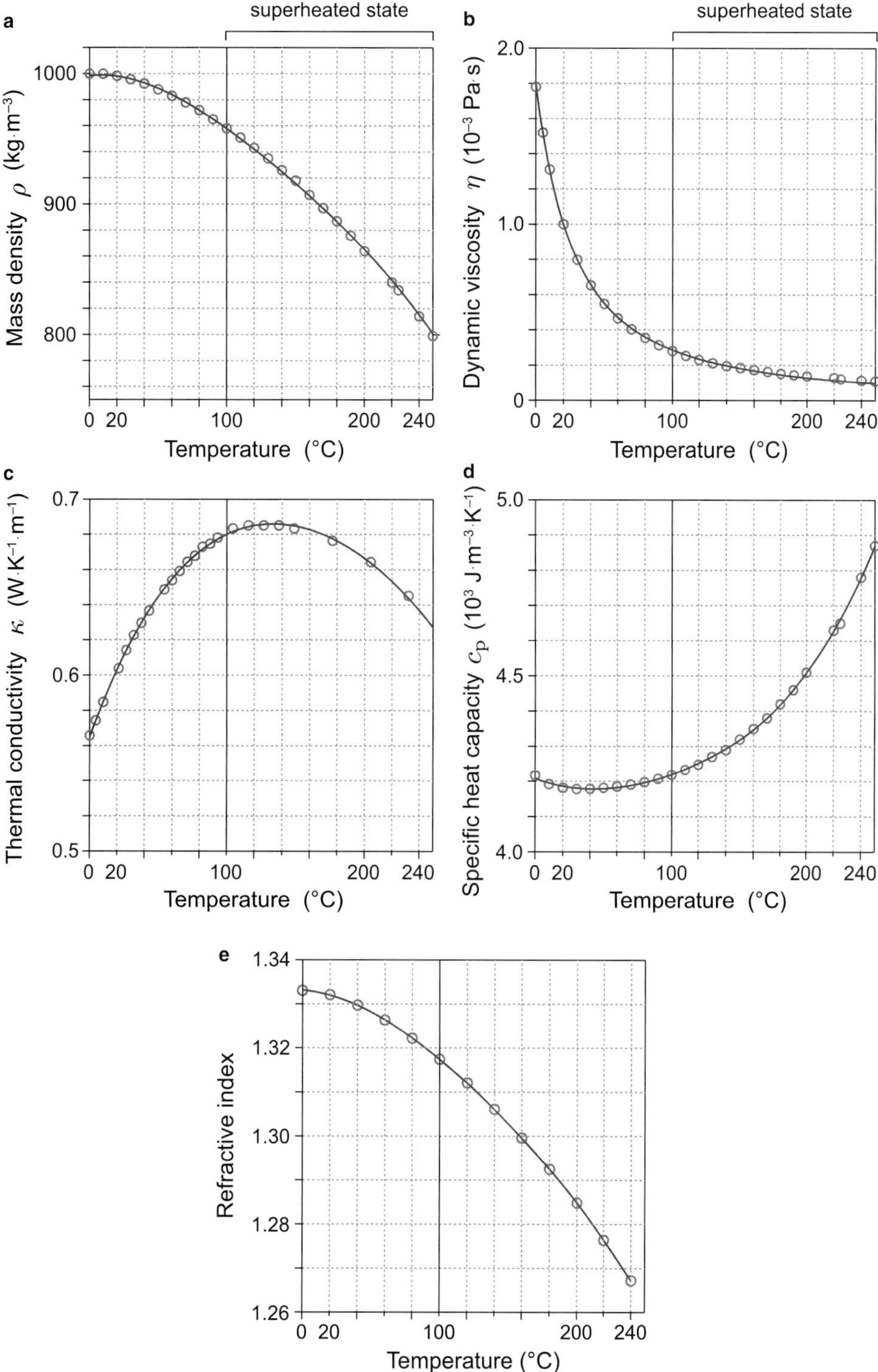

Fig. B.1   Temperature dependence of thermodynamical constants of water. Solid lines represent the fitting functions listed hereinafter. (a) Mass density. (b) Dynamic viscosity. (c) Thermal conductivity. (d) Specific heat capacity. (e) Refractive index.

viscosity, the thermal conductivity and the specific heat capacity of liquid water up to 240°C (source: www.engineeringtoolbox.com).

Here are some fitting functions corresponding to these data:

$$\rho = 999.8 - 0.0059988T^2 + 2.1496 \times 10^{-5}T^3 - 4.1009 \times 10^{-8}T^4$$

$$\eta = \frac{1.792 \times 10^{-3}}{(0.02392T + 1)^{1.5}}$$

$$\kappa = 0.56479 + 0.0021627T - 1.2719 \times 10^{-5}T^2 + 3.0073 \times 10^{-8}T^3$$
$$-3.9208 \times 10^{-11}T^4$$

$$c_{\mathrm{p}} = 4.2089 - 0.0015838T + 2.4335 \times 10^{-5}T^2 - 1.0385 \times 10^{-7}T^3$$
$$+2.9634 \times 10^{-10}T^4$$

$$n = 1.3345 - 5.3864e \times 10^{-6}T - 2.0985 \times 10^{-6}T^2 + 6.8405 \times 10^{-9}T^3$$
$$-1.250 \times 10^{-11}T^4 \tag{B.1}$$

where T is degrees Celsius. $n(T)$ was considered here for a light wavelength of $\lambda = 598$ nm. For more information regarding the dependence of n as a function of the temperature, pressure and wavelength, see Section 5.9 on page 205.

Appendix: Thermal Green's Function for a Three-Layer System

In this section, we present a Green's function related to the Laplace equation:

$$\nabla \cdot \kappa(\mathbf{r})\nabla T(\mathbf{r}) = -q(\mathbf{r}).$$ (C.1)

Physically, this equation governs the temperature distribution created by a heat source distribution $q(\mathbf{r})$ in a medium of thermal conductivity $\kappa(\mathbf{r})$, or the electric potential created by a charge distribution $q(\mathbf{r})$ in a medium of electric permittivity $\kappa(\mathbf{r})$.

The Green's function related to this equation, *i.e.*, the solution of the equation for $q(\mathbf{r}) = \delta(\mathbf{r})$, depends on the distribution of $\kappa(\mathbf{r})$. In this appendix, we shall address three different cases related to a system composed of three uniform layers of different conductivities separated by flat interfaces [Eng. Anal. Bound. Elem. 23, 777–786 (1999)].

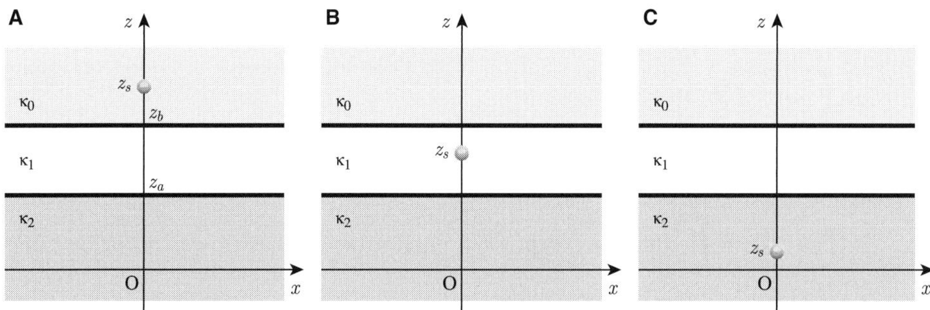

Fig. C.1 Three possible cases: (A) The source of heat lies in the upper layer. (B) The source of heat lies in the middle layer. (C) The source of heat lies in the bottom layer.

Case A: z_s is in the Top Layer

- $\infty < z \leq z_\mathrm{a}$

$$G = \int_0^\infty \kappa_0 \exp{-h(z_\mathrm{s} - z)}\frac{J_0(hr)}{\pi \det A(h)}\,\mathrm{d}h$$

where

$$\det A(h) = (\kappa_0 + \kappa_1)(\kappa_0 + \kappa_2) + (\kappa_0 - \kappa_2)(\kappa_1 - \kappa_0) \exp -2h\Delta$$

$$\Delta = z_b - z_a$$

$$r = \sqrt{(x - x_s)^2 + (z - z_s)^2}$$

- $z_a < z \leq z_b$

$$G = \int_0^\infty [(\kappa_0 - \kappa_2)\exp(-h(-2z_a + z_s + z)) + (\kappa_0 + \kappa_2)\exp(-h(z_s - z))]\frac{J_0(hr)}{2\pi \det A(h)}\,dh$$

- $z_b < z \leq \infty$

$$G = \frac{1}{4\pi \kappa_1 R} + \int_0^\infty [(\kappa_0 + \kappa_2)(\kappa_1 - \kappa_0)\exp(-h(-2z_a + z_s + z))\cdots$$

$$+ (\kappa_0 - \kappa_2)(\kappa_0 + \kappa_1)\exp(-h(-2z_a + z_s + z))]\frac{J_0(hr)}{4\pi \kappa_1 \det A(h)}\,dh$$

where

$$R = \sqrt{r^2 + (z - z_s)^2}$$

Case B: z_s is in the Middle Layer

- $-\infty < z \leq z_a$

$$G = \int_0^\infty [(\kappa_0 + \kappa_1)\exp(-h(z_s - z)) + (\kappa_0 - \kappa_1)\exp(-h(2z_b - z_s - z))]\frac{J_0(hr)}{2\pi \det A(h)}\,dh$$

- $z_a < z \leq z_b$

$$G = \frac{1}{4\pi \kappa_0 R} + \int_0^\infty \{(\kappa_0 - \kappa_1)(\kappa_0 - \kappa_2)\exp[-h(2\Delta + z_s - z)]\ldots$$

$$+ (\kappa_0 - \kappa_1)(\kappa_0 + \kappa_2)\exp[-h(2z_b - z_s - z)]\}\frac{J_0(hr)}{2\pi \kappa_0 \det A(h)}\,dh\ldots$$

$$+ \int_0^\infty \{(\kappa_0 - \kappa_2)(\kappa_0 + \kappa_1)\exp[-h(z_s - 2z_a + z)]\ldots$$

$$+ (\kappa_0 - \kappa_1)(\kappa_0 - \kappa_2)\exp[-h(2\Delta + z - z_s)]\}\frac{J_0(hr)}{2\pi \kappa_0 \det A(h)}\,dh$$

- $z_b < z \leq \infty$

$$G = \int_0^\infty [(\kappa_0 - \kappa_2)\exp(-h(z_s + z - 2z_a)) + (\kappa_0 + \kappa_2)\exp(-h(z - z_s))]\frac{J_0(hr)}{2\pi \det A(h)}\,dh$$

Case C: $z_{\rm s}$ is in the Bottom Layer

- $\infty < z \leq z_{\rm a}$

$$G = \frac{1}{4\pi\,\kappa_2\,R} + \int_0^\infty [(\kappa_0 + \kappa_1)(\kappa_2 - \kappa_0)\exp(-h(2z_{\rm a} - z_{\rm s} - z))$$
$$+ (\kappa_0 - \kappa_1)(\kappa_0 + \kappa_2)\exp(-h(2z_{\rm b} - z_{\rm s} - z))]\frac{J_0(hr)}{4\pi\kappa_2\det A(h)}{\rm d}h$$

- $z_{\rm a} < z \leq z_{\rm b}$

$$G = \int_0^\infty [(\kappa_0 - \kappa_1)\exp(-h(2z_{\rm b} - z_{\rm s} - z)) + (\kappa_0 + \kappa_1)\exp(-h(z - z_{\rm s}))]\frac{J_0(hr)}{2\pi\det A(h)}{\rm d}h$$

- $z_{\rm b} < z \leq \infty$

$$G = \int_0^\infty \kappa_0 \exp -h(z - z_{\rm s})\frac{J_0(hr)}{\pi\det A(h)}{\rm d}h$$

In all these expression J_0 is the Bessel function of the 0th order.

Index